CAMBRIDGE LIBRARY COLLECTION

Books of enduring scholarly value

Mathematical Sciences

From its pre-historic roots in simple counting to the algorithms powering modern desktop computers, from the genius of Archimedes to the genius of Einstein, advances in mathematical understanding and numerical techniques have been directly responsible for creating the modern world as we know it. This series will provide a library of the most influential publications and writers on mathematics in its broadest sense. As such, it will show not only the deep roots from which modern science and technology have grown, but also the astonishing breadth of application of mathematical techniques in the humanities and social sciences, and in everyday life.

Electricity and Magnetism

A.S. Ramsey (1867-1954) was a distinguished Cambridge mathematician and President of Magdalene College. He wrote several textbooks 'for the use of higher divisions in schools and for first-year students at university'. This book on electricity and magnetism, first published in 1937, and based upon his lectures over many years, was 'adapted more particularly to the needs of candidates for Part I of the Mathematical Tripos'. It covers electrostatics, conductors and condensers, dielectrics, electrical images, currents, magnetism and electromagnetism, and magnetic induction. The book is interspersed with examples for solution, for some of which answers are provided.

Cambridge University Press has long been a pioneer in the reissuing of out-of-print titles from its own backlist, producing digital reprints of books that are still sought after by scholars and students but could not be reprinted economically using traditional technology. The Cambridge Library Collection extends this activity to a wider range of books which are still of importance to researchers and professionals, either for the source material they contain, or as landmarks in the history of their academic discipline.

Drawing from the world-renowned collections in the Cambridge University Library, and guided by the advice of experts in each subject area, Cambridge University Press is using state-of-the-art scanning machines in its own Printing House to capture the content of each book selected for inclusion. The files are processed to give a consistently clear, crisp image, and the books finished to the high quality standard for which the Press is recognised around the world. The latest print-on-demand technology ensures that the books will remain available indefinitely, and that orders for single or multiple copies can quickly be supplied.

The Cambridge Library Collection will bring back to life books of enduring scholarly value across a wide range of disciplines in the humanities and social sciences and in science and technology.

Electricity and Magnetism

An Introduction to the Mathematical Theory

Arthur Stanley Ramsey

CAMBRIDGE UNIVERSITY PRESS

Cambridge New York Melbourne Madrid Cape Town Singapore São Paolo Delhi

Published in the United States of America by Cambridge University Press, New York

www.cambridge.org
Information on this title: www.cambridge.org/9781108002592

This edition first published 1937
This digitally printed version 2009

ISBN 978-1-108-00259-2

ELECTRICITY AND MAGNETISM

LONDON
Cambridge University Press
FETTER LANE

NEW YORK · TORONTO
BOMBAY · CALCUTTA · MADRAS
Macmillan

TOKYO
Maruzen Company Ltd

ELECTRICITY AND MAGNETISM

An Introduction to the Mathematical Theory

by

A. S. RAMSEY, M.A.

President of Magdalene College, Cambridge
formerly University Lecturer in Mathematics

CAMBRIDGE
AT THE UNIVERSITY PRESS
1937

PRINTED IN GREAT BRITAIN

PREFACE

This book has been written in response to suggestions from friends who have asked for a text-book on the subject adapted more particularly to the needs of candidates for Part I of the Mathematical Tripos.

A complete study of the theory of electricity and magnetism, as a logical mathematical development from experimental data, requires a knowledge of the methods of mathematical analysis far beyond what can reasonably be expected from most readers of an elementary text-book. The knowledge of pure mathematics assumed in the present volume amounts to little more than some elementary calculus and a few properties of vectors. The ground is restricted by this limitation. It covers the schedule for Part I of the Tripos, including the fundamental principles of electrostatics, Gauss's theorem, Laplace's equation, systems of conductors, homogeneous dielectrics and the theory of images; steady currents in wires; elementary theory of the magnetic field and the elementary facts about the magnetic fields of steady currents. There are also short chapters on induced magnetism and induction of currents.

From one standpoint it would be preferable that a book on a branch of Natural Philosophy should consist of a continuous logical development uninterrupted by 'examples'. But experience seems to indicate that mathematical principles are best understood by making attempts to apply them; and, as the purpose of this book is didactic, I have had no hesitation in interspersing examples through the chapters and giving the solutions of some of them. The text is based upon lectures given at intervals over a period of many years, and the examples are part of a collection which I began to make for the use of my pupils about forty years ago, drawn from Tripos and College Examination papers.

As regards notation, I felt much hesitation about abandoning the use of V for the potential of an electrostatic field; but the custom of using a Greek letter to denote the scalar potential of a vector field has become general, and the matter was decided for me when I found '$\boldsymbol{E} = -\textbf{\textit{grad}}\,\phi$' in the Cambridge syllabus.

I am greatly indebted to Mr E. Cunningham of St John's College for reading a large part of the text and making many appropriate criticisms and useful suggestions; and also to Dr S. Verblunsky of the University of Manchester for reading and correcting the proofs, and to the printers and readers of the University Press for careful composition and correction.

A. S. R.

Cambridge
November 1936

CONTENTS

Chapter I: PRELIMINARY MATHEMATICS

Chapter II: INTRODUCTION TO ELECTROSTATICS

Chapter III: CONDUCTORS AND CONDENSERS

Chapter VII: ELECTRIC CURRENTS

Chapter VIII: MAGNETISM

Chapter IX: ELECTROMAGNETISM

Chapter X: MAGNETIC INDUCTION AND INDUCED MAGNETISM

Chapter XI: ELECTROMAGNETIC INDUCTION

Table of Units

C.G.S. absolute unit of force = 1 dyne
C.G.S. absolute unit of work or energy = 1 erg

ELECTRICAL UNITS

Practical units		Equivalent absolute C.G.S. units	
		Electrostatic	Electro-magnetic
Charge	1 coulomb	3×10^9	10^{-1}
Potential or electromotive force	1 volt	$3^{-1} \times 10^{-2}$	10^8
Current	1 ampère	3×10^9	10^{-1}
Resistance	1 ohm	$3^{-2} \times 10^{-11}$	10^9
Capacity	1 farad	$3^2 \times 10^{11}$	10^{-9}
Inductance	1 henry	$3^{-2} \times 10^{-11}$	10^9
Rate of working	1 watt	10^7	10^7

One microfarad is one-millionth of a farad.

An electromotive force of 1 volt drives a current of 1 ampère through a resistance of 1 ohm and work is then being done at the rate of 1 watt or 10^7 ergs per second.

Chapter I

PRELIMINARY MATHEMATICS

1·1. We propose in this chapter to give a brief account of some mathematical ideas with which the reader must be familiar in order to be able to understand what follows in this volume.

1·2. Surface and volume integrals. Though the process of evaluating surface and volume integrals in general involves double or triple integration and must be learnt from books on Analysis, yet in theoretical work in Applied Mathematics considerable use is made of surface and volume integrals without evaluation, and we propose here merely to explain what is implied when such symbols as

$$\int f(x,y,z)\,dS \quad \text{and} \quad \int f(x,y,z)\,dv$$

are used to denote integration over a surface and throughout a volume.

A definite integral of a function of one variable, say $\int_a^b f(x)\,dx$, may be defined thus: let the interval from a to b on the x-axis be divided into any number of sub-intervals $\delta_1, \delta_2, \ldots \delta_n$, and let f_r denote the value of $f(x)$ at some point on δ_r; let the sum $\sum_{r=1}^{n} f_r\delta_r$ be formed and let the number n be increased without limit. Then, provided that the limit as $n\to\infty$ of $\sum_{r=1}^{n} f_r\delta_r$ exists and is independent of the method of division into sub-intervals and of the choice of the point on δ_r at which the value of $f(x)$ is taken, this limit is the definite integral of $f(x)$ from a to b.

In the same way we may define $\int f(x,y,z)\,dS$ over a given surface; let the given surface be divided into any number of small parts $\delta_1, \delta_2, \ldots \delta_n$ and let f_r denote the value of $f(x,y,z)$

at some point on δ_r, then the limit as $n \to \infty$ of $\sum_{r=1}^{n} f_r \delta_r$, provided the limit exists under the same conditions as aforesaid, is defined to be the integral $\int f(x, y, z)\, dS$ over the given surface.

Any difficulty as to the precise meaning to be attributed to 'area of a curved surface' may be avoided thus: after choosing the point on each sub-division δ_r of the surface at which the value of $f(x, y, z)$ is taken, project this element of surface on to the tangent plane at the chosen point, and take the plane projection of the element as the measure of δ_r in forming the sum.

The integral $\int f(x, y, z)\, dv$ through a given volume may be defined in the same way.

1·3. Solid angles. The solid angle of a cone *is measured by the area intercepted by the cone on the surface of a sphere of unit radius having its centre at the vertex of the cone.*

The solid angle subtended at a point by a surface of any form is measured by the solid angle of the cone whose vertex is at the given point and whose base is the given surface.

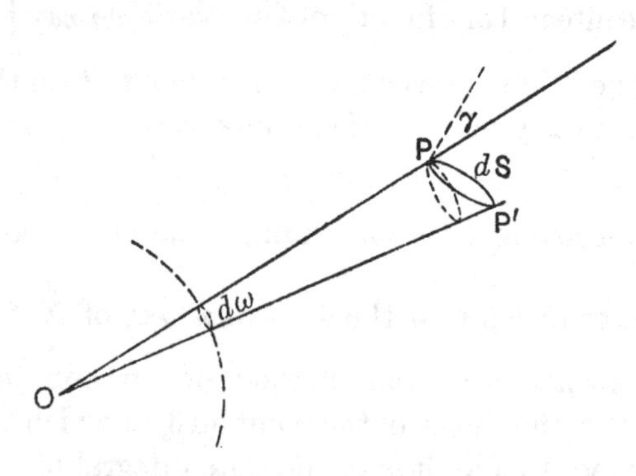

Let PP' be a small element of area dS which subtends a solid angle $d\omega$ at O.

Let the normal to dS make an acute angle γ with OP and let $OP = r$. Then the cross-section at P of the cone which PP' subtends at O is $dS \cos \gamma$, and this cross-section and the small area $d\omega$ intercepted on the unit sphere are similar figures, so that

$$dS\cos\gamma : d\omega = r^2 : 1.$$

Whence
$$\left.\begin{aligned} d\omega &= (dS\cos\gamma)/r^2 \\ dS &= r^2\sec\gamma\, d\omega \end{aligned}\right\} \qquad \text{.................(1).}$$
or

It follows that the area of a finite surface can be represented as an integral over a spherical surface, thus

$$S = \int r^2 \sec\gamma\, d\omega \qquad \text{.....................(2)}$$

with suitable limits of integration.

1·31. $d\omega$ in polar co-ordinates. $d\omega$ is an element of the surface of a unit sphere. Let the element be $PQRS$ bounded by meridians and small circles, where the angular co-ordinates of P are θ, ϕ. Then since the arc PS subtends an angle $d\phi$ at the centre of a circle of radius $\sin\theta$, therefore $PS = \sin\theta\, d\phi$; and $PQ = d\theta$, so that

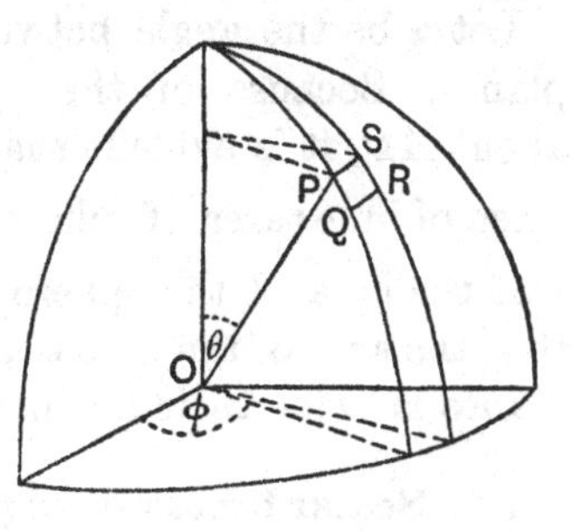

$$d\omega = PQ\,.\,PS = \sin\theta\, d\theta\, d\phi.$$

1·32. Solid angle of a right circular cone. A narrow zone of a sphere of radius a cut off between parallel planes may be regarded as a circular band of breadth $a\,d\theta$ and radius $a\sin\theta$, so that its area $= 2\pi a^2 \sin\theta\, d\theta$

$$= -2\pi a\, dx, \text{ where } x = a\cos\theta.$$

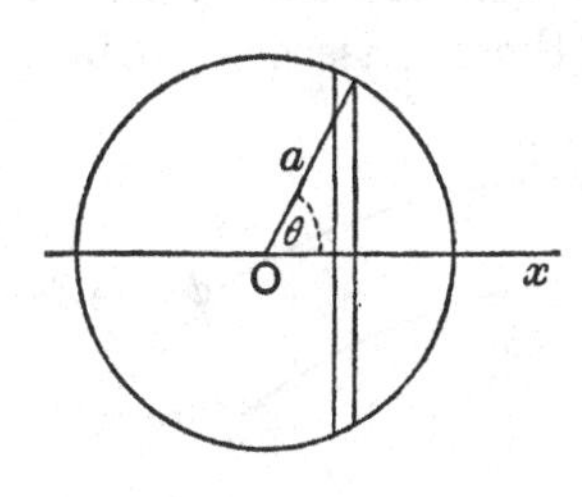

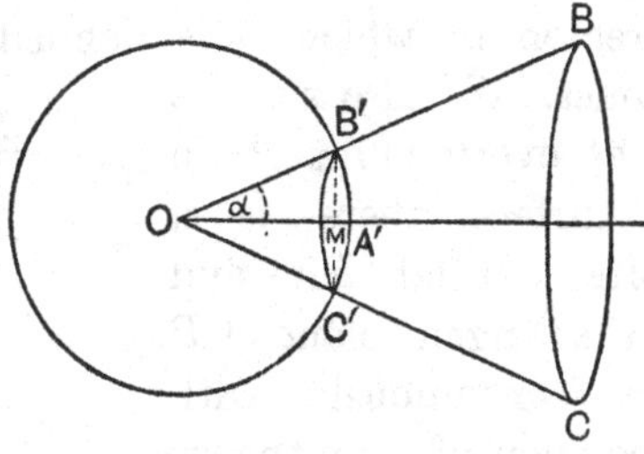

Hence the area of a zone of finite breadth

$$= 2\pi a\,(x_1 - x_2)$$
$$= \text{circumference of sphere} \times \text{axial breadth of zone.}$$

A right circular cone BOC of vertical angle 2α intercepts on a unit sphere of centre O a cap $B'A'C'$ of height $MA' = 1 - \cos\alpha$ and area

$$2\pi(1 - \cos\alpha) \quad \text{.....................(1)},$$

so that this is the measure of the solid angle of the cone.

1·33. The idea of the solid angle may easily be extended, if we observe that any bounded area on a unit sphere may be regarded as measuring a solid angle. Thus a lune bounded by semi-circles ABD, ACD may be taken as measuring the solid angle between the diametral planes ABD, ACD.

Let α be the angle between these planes. Because of the symmetry about AD, it is evident that

area of lune : area of sphere $= \alpha : 2\pi$.

But the area of the sphere is 4π, so that the area of the lune is 2α; or the solid angle between two planes is twice their inclination to one another.

1·4. Scalar functions of position and their gradients. Let $\phi(x, y, z)$ be a continuous single-valued function of the position of a point in some region of space. Suppose that the function ϕ is not constant throughout any region, so that the equation

$$\phi(x, y, z) = \text{const.}$$

represents a surface. We assume that through each point of the region in which ϕ is defined, there passes a surface $\phi = \text{const.}$ We also assume that at every point P on this surface there is a definite normal PN and that the tangent plane at P varies continuously with the position of P on the surface.

From the definition of ϕ two surfaces

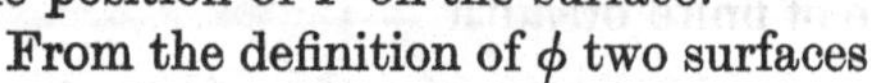

$$\phi(x, y, z) = a \quad \text{and} \quad \phi(x, y, z) = b$$

cannot intersect; for if they had a common point it would be a point at which ϕ had more than one value, in contradiction to the hypothesis that ϕ is a single-valued function.

Consider two neighbouring surfaces

$$\phi = a \quad \text{and} \quad \phi = a + \delta a.$$

Let P, P' be points on each and let the normal at P to $\phi = a$ meet $\phi = a + \delta a$ in N. For small values of δa PN will also be normal to $\phi = a + \delta a$.

Then using ϕ_P to denote the value of ϕ at P, we have

$$\frac{\phi_{P'} - \phi_P}{PP'} = \frac{\delta a}{PP'} = \frac{\phi_N - \phi_P}{PP'} = \frac{\phi_N - \phi_P}{PN} \cdot \frac{PN}{PP'}$$

$$= \frac{\phi_N - \phi_P}{PN} \cos\theta,$$

where θ is the angle NPP'.

Now if $PP' = \delta s$ and $PN = \delta n$, and we make δa and therefore δs and δn tend to zero, the limit of $(\phi_{P'} - \phi_P)/PP'$ is the rate of increase of ϕ in the direction δs and is denoted by $\frac{\partial \phi}{\partial s}$; and similarly the limit of $(\phi_N - \phi_P)/PN$ is the rate of increase of ϕ in the normal direction δn and is denoted by $\frac{\partial \phi}{\partial n}$, and we have

$$\frac{\partial \phi}{\partial s} = \frac{\partial \phi}{\partial n} \cos\theta \quad \text{.......................(1).}$$

Thus we have proved that the space rate of increase of ϕ in any direction δs is the component in that direction of its space rate of increase in the direction normal to the surface $\phi = \text{const.}$; or that if we construct a vector of magnitude $\partial\phi/\partial n$ in direction PN, then the component of this vector in any direction is the space rate of increase of ϕ in that direction.

The vector $\partial\phi/\partial n$ with its proper direction is called the **gradient of ϕ** and written **grad ϕ**.

To recapitulate: ϕ is a continuous *scalar* function of position having a definite single value at each point of a certain region of space, and the *gradient of ϕ* is defined in this way: through any point P in the region there passes a surface $\phi = \text{const.}$, then *a vector* normal to this surface at P whose magnitude is the space rate of increase of ϕ in this normal direction is defined to be the *gradient of* ϕ at P, and it has the property that its component in any direction gives the space rate of increase of ϕ at P in that direction. It is clear that the gradient measures the greatest rate of increase of ϕ at a point.

1·5. A vector field. If to every point of a given region there corresponds a definite vector **A**, generally varying its magnitude and direction from point to point, then the region is called a **vector field**, or the field of the vector **A**; e.g. electric field, magnetic field.

1·51. Flux of a vector. If a surface S be drawn in the field of a vector **A** and A_n denotes the component of **A** normal to an element dS of the surface, then the integral $\int A_n dS$ is called the *flux of A through S*. Since a surface has two sides the sense of the normal must be taken into account, and the sign of the flux is changed when the sense of the normal is changed. The flux of a vector through a surface is clearly a scalar magnitude.

1·52. Divergence of a vector field. Let **A** denote a vector field which has no discontinuities throughout a given region of space. Let δv denote any small element of volume containing a point P in the region and let $\int A_n dS$ denote the outward flux of **A** through the boundary of δv, then the limit as $\delta v \to 0$ of $\dfrac{\int A_n dS}{\delta v}$ is defined to be the **divergence** of **A** at the point P and denoted by div **A**.

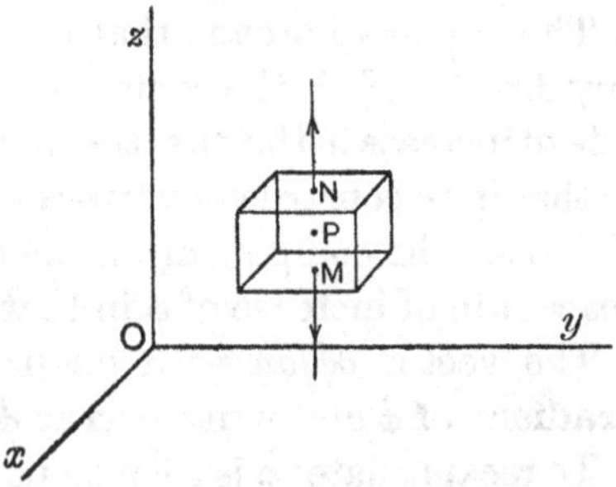

It can be shewn that, subject to certain conditions, this limit is independent of the shape of the element of volume δv, but for our present purpose, which is to obtain a Cartesian form for div **A**, it will suffice to calculate the limit for a rectangular element of volume. Using rectangular axes let P be the centre (x, y, z)

of a small rectangular parallelepiped with edges parallel to the axes of lengths δx, δy, δz.

Let the vector **A** have components A_x, A_y, A_z parallel to the axes at P.

Consider the contributions of the faces of the parallelepiped to the flux of the vector out of the element of volume. The two faces parallel to the xy plane are of area $\delta x \, \delta y$, the component of **A** normal to these at the centre (x, y, z) of the parallelepiped is A_z. The co-ordinates of the centres M, N of these faces are $x, y, z-\frac{1}{2}\delta z$ and $x, y, z+\frac{1}{2}\delta z$; so that if the magnitude of A_z at P is $f(x, y, z)$, its magnitude at M is $f(x, y, z-\frac{1}{2}\delta z)$ or $f(x, y, z)-\frac{1}{2}\frac{\partial f}{\partial z}\delta z$, to the first power of δz, i.e. $A_z-\frac{1}{2}\frac{\partial A_z}{\partial z}\delta z$, and similarly the magnitude at N is $A_z+\frac{1}{2}\frac{\partial A_z}{\partial z}\delta z$, and both these components are in the direction Oz. Then assuming what can easily be proved, that, subject to certain conditions, the average value of the component over each small rectangle is the value at its centre, the contributions of these two faces to the total *outward* flux are

$$-\left(A_z-\frac{1}{2}\frac{\partial A_z}{\partial z}\delta z\right)\delta x\,\delta y \quad \text{and} \quad \left(A_z+\frac{1}{2}\frac{\partial A_z}{\partial z}\delta z\right)\delta x\,\delta y,$$

giving a sum
$$\frac{\partial A_z}{\partial z}\delta x\,\delta y\,\delta z.$$

Finding similarly the contributions of the other two pairs of faces, we have for the total outward flux

$$\left(\frac{\partial A_x}{\partial x}+\frac{\partial A_y}{\partial y}+\frac{\partial A_z}{\partial z}\right)\delta x\,\delta y\,\delta z$$

to this order of small quantities.

But the volume δv of the small element is $\delta x\,\delta y\,\delta z$, so that in accordance with our definition, dividing the flux by the volume and proceeding to the limit in which the terms of higher order in the numerator disappear, we have

$$\operatorname{div}\mathbf{A}=\frac{\partial A_x}{\partial x}+\frac{\partial A_y}{\partial y}+\frac{\partial A_z}{\partial z} \quad \text{..............(1).}$$

1·53. Divergence in polar co-ordinates. Let P be the point (r, θ, ω) and suppose it to be the centre of an element of volume $ABCDA'B'C'D'$, whose faces $ABCD$, $A'B'C'D'$ are portions of spheres of radii $r \mp \frac{1}{2}\delta r$, $ADD'A'$, $BCC'B'$ are portions of cones of angles $\theta \mp \frac{1}{2}\delta\theta$, and $ABB'A'$, $DCC'D'$ are planes $\omega \mp \frac{1}{2}\delta\omega$. The lengths of the edges of the element of volume are δr, $r\,\delta\theta$ and $r \sin\theta\,\delta\omega$ and its volume is $r^2 \sin\theta\,\delta r\,\delta\theta\,\delta\omega$.

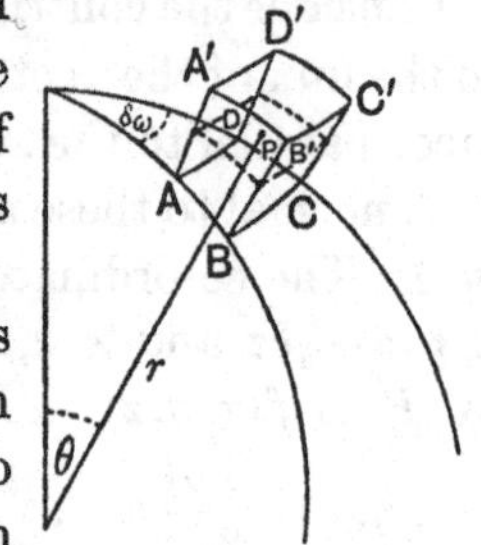

Let A_r, A_θ, A_ω denote the components of the vector $\mathbf{A}$ at P in the directions in which r, θ, ω increase, i.e. perpendicular to the faces of the element. The cross-section of the element through P at right angles to A_r is of area $r^2 \sin\theta\,\delta\theta\,\delta\omega$, so that the flux of $\mathbf{A}$ through this cross-section is $A_r r^2 \sin\theta\,\delta\theta\,\delta\omega$. Hence the outward flux across the parallel section $ABCD$ which only differs from that through P by having $r - \frac{1}{2}\delta r$ instead of r is

$$-\left\{A_r r^2 \sin\theta\,\delta\theta\,\delta\omega - \tfrac{1}{2}\delta r \,.\, \frac{\partial}{\partial r}(A_r r^2 \sin\theta\,\delta\theta\,\delta\omega)\right\} + \epsilon_1,$$

and the flux across $A'B'C'D'$ is in like manner

$$+\left\{A_r r^2 \sin\theta\,\delta\theta\,\delta\omega + \tfrac{1}{2}\delta r \,.\, \frac{\partial}{\partial r}(A_r r^2 \sin\theta\,\delta\theta\,\delta\omega)\right\} + \epsilon_2,$$

where ϵ_1, ϵ_2 are small quantities of the fourth order in δr, $\delta\theta$, $\delta\omega$.

Hence this pair of opposite faces contribute an amount

$$\frac{\partial}{\partial r}(r^2 A_r) \sin\theta\,\delta r\,\delta\theta\,\delta\omega + \epsilon_1 + \epsilon_2$$

to the total outward flux.

It may be shewn in the same way that the faces $ADD'A'$ and $BCC'B'$ contribute

$$-\left\{A_\theta r \sin\theta\,\delta r\,\delta\omega - \tfrac{1}{2}\delta\theta \,.\, \frac{\partial}{\partial\theta}(A_\theta r \sin\theta\,\delta r\,\delta\omega)\right\} + \epsilon_3$$

and $$+\left\{A_\theta r \sin\theta\,\delta r\,\delta\omega + \tfrac{1}{2}\delta\theta \,.\, \frac{\partial}{\partial\theta}(A_\theta r \sin\theta\,\delta r\,\delta\omega)\right\} + \epsilon_4;$$

and that the faces $ABB'A'$, $DCC'D'$ contribute

$$-\left\{A_\omega r\,\delta\theta\,\delta r - \tfrac{1}{2}\delta\omega\,.\frac{\partial}{\partial\omega}(A_\omega r\,\delta\theta\,\delta r)\right\} + \epsilon_5$$

and
$$+\left\{A_\omega r\,\delta\theta\,\delta r + \tfrac{1}{2}\delta\omega\,.\frac{\partial}{\partial\omega}(A_\omega r\,\delta\theta\,\delta r)\right\} + \epsilon_6,$$

where ϵ_3, ϵ_4, ϵ_5, ϵ_6 have like meanings.

Hence the total outward flux from the six faces is

$$\left\{\frac{1}{r^2}\frac{\partial}{\partial r}(r^2 A_r) + \frac{1}{r\sin\theta}\frac{\partial}{\partial\theta}(\sin\theta A_\theta) + \frac{1}{r\sin\theta}\frac{\partial A_\omega}{\partial\omega}\right\} r^2\sin\theta\,\delta r\,\delta\theta\,\delta\omega + \epsilon,$$

where ϵ is of the fourth order in δr, $\delta\theta$, $\delta\omega$.

Therefore if we divide the flux by the volume and then make δr, $\delta\theta$, $\delta\omega$ tend to zero, we get for the divergence at P

$$\operatorname{div}\mathbf{A} = \frac{1}{r^2}\frac{\partial}{\partial r}(r^2 A_r) + \frac{1}{r\sin\theta}\frac{\partial}{\partial\theta}(\sin\theta A_\theta) + \frac{1}{r\sin\theta}\frac{\partial A_\omega}{\partial\omega} \quad \text{...(1).}$$

1·54. Divergence in cylindrical co-ordinates. Using cylindrical co-ordinates r, θ, z and taking an element of volume of edges δr, $r\,\delta\theta$, δz with its centre at (r, θ, z), it can be shewn in the same way that

$$\operatorname{div}\mathbf{A} = \frac{1}{r}\frac{\partial}{\partial r}(rA_r) + \frac{1}{r}\frac{\partial A_\theta}{\partial\theta} + \frac{\partial A_z}{\partial z},$$

where A_r, A_θ, A_z are the components of $\mathbf{A}$ in the directions of the increments in r, θ and z.

Chapter II

INTRODUCTION TO ELECTROSTATICS

2·1. The electric field. It was known to the Greeks and Romans that when pieces of amber are rubbed they acquire the power of attracting to themselves light bodies. There are other substances which possess the same property; thus, if a stick of sealing-wax is rubbed on a piece of dry cloth it will attract bran or small scraps of paper sufficiently near to it. The same result is obtained if a glass rod is rubbed with a dry piece of silk. It is also found that the cloth and the silk acquire the same property as the sealing-wax and the glass rod. Further, if the sealing-wax is suspended so that it is free to move, it is found that the cloth attracts the sealing-wax, but two pieces of sealing-wax similarly treated repel one another. In the same way the glass rod and the silk attract one another, but two such glass rods repel one another.

We describe bodies in such a state as *electrified* or *charged with electricity*. The word was derived by William Gilbert* from ἤλεκτρον or *amber*, the first substance upon which such experiments were performed.

If we experiment further we find that an electrified stick of sealing-wax is attracted by an electrified glass rod, but repelled by the piece of silk with which the rod has been rubbed.

These and kindred phenomena are explained by the statement that electricity is of two kinds, and that charges of the same kind repel while charges of opposite kinds attract one another.

It is convenient to describe the two kinds of electricity as *positive* and *negative*, that produced as above on the sealing-wax is called negative and that on the glass rod positive; but it must be pointed out that this is merely a convenient arbitrary nomenclature and that the opposite would have answered all purposes equally well.

* William Gilbert (1540–1603), a native of Colchester, Fellow of St John's College, Cambridge.

If we try the same experiment using a metal rod held in the hand instead of the glass rod or stick of sealing-wax, no result is obtained.

If a small pith ball coated with gold leaf is suspended by a silk fibre and allowed to touch an electrified rod, it appears to acquire a charge of the same kind as the rod, for after contact it is repelled from the rod.

If such a charged pith ball is removed to a distance from all other bodies, it does not appear to be acted upon by any forces save its weight and the tension of the supporting fibre; but when brought into the neighbourhood of other charged bodies it appears to be subject to an additional force, which at every point has a definite direction.

It follows that the properties of space in the neighbourhood of charged bodies appear to differ from those of the rest of space. The space in the neighbourhood of charged bodies is called an **electric field**. The field in general extends through all space, but its intensity (2·2) diminishes as distance from charged bodies increases. There are natural electric fields of great intensity, such as those due to the presence of highly charged regions in the atmosphere; e.g. thunder-clouds.

It is our object to formulate a working hypothesis and build upon it a mathematical theory which will describe correctly the phenomena which take place in an electric field. But before we attempt to do this we must make a further appeal to experiment.

2·11. The nature of electricity. The modern answer to the question 'what is electricity?' is that it is 'a fundamental entity of nature', an answer which gives the minimum of information. For a long time electricity was considered to be of continuous fluid form, though it only resembles a fluid in that it moves in 'currents'. It is now known to be of 'atomic structure'. The existence of the **electron** or ultimate indivisible negative charge was demonstrated by J. J. Thomson in 1897. The corresponding positive charge is called a **proton**. In accordance with the electron theory of matter, an atom of an element consists of a central *nucleus* composed of protons and electrons with a total positive charge, round which electrons move in orbits like planets round the sun, the number of electrons in an uncharged atom being just sufficient to balance the positive charge of the nucleus. The nucleus is small in size compared to the dimensions of the orbits of the revolving electrons; but the mass of the atom is almost entirely concentrated in the nucleus. Recent experimental work in the Cavendish Laboratory has demonstrated the existence of the proton and also of an uncharged particle of the same mass called a **neutron**.

A *charged body* means one which contains either protons or electrons, or both protons and electrons, apart from those which compose the atoms of the body.

2·12. Conductors and non-conductors. An electrified metal-coated pith ball looses its charge if touched by a metal rod held in the hand, but not if touched by a dry glass or ebonite rod.

The explanation of the disappearance of the charge is that a metal rod acts as a 'conductor' and a passage of electricity takes place along it. Thus a positive charge disappears either because it escapes to earth along the rod, or because an equal quantity of negative electricity passes from the earth along the rod to neutralize the charge on the metal-coated ball, or partly from the one cause and partly from the other. Dry glass and ebonite do not permit of a like passage of electricity.

This and kindred experiments enable us to classify substances either as '*conductors*' of electricity, or as '*non-conductors*', '*insulators*', or '*dielectrics*'. Among conductors are the earth, metals, water, carbon and the human body; and among insulators are dry air, glass, sulphur and ebonite. Some substances are better conductors or better insulators than others, and there are some substances such as wood which may be classed either among bad conductors or among bad insulators.

2·13. Electricity produced by induction. Uncharged conducting bodies contain both positive and negative electricity, i.e. protons and electrons, so distributed as to neutralize one another's effects. But when such a body is brought into an electric field the distribution of the electricity is altered and becomes positive and negative charges on different parts of the surface of the body.

Thus let A represent a metal body supported on an insulating stand. When a positively charged body B is brought into the neighbourhood of A, negative electricity in A is attracted towards B and positive is repelled from B to the remoter parts of A. The presence of the latter can be demonstrated by allowing a metal-coated pith ball C to come into contact with A; part of the positive charge on A then passes to C and it is repelled from A.

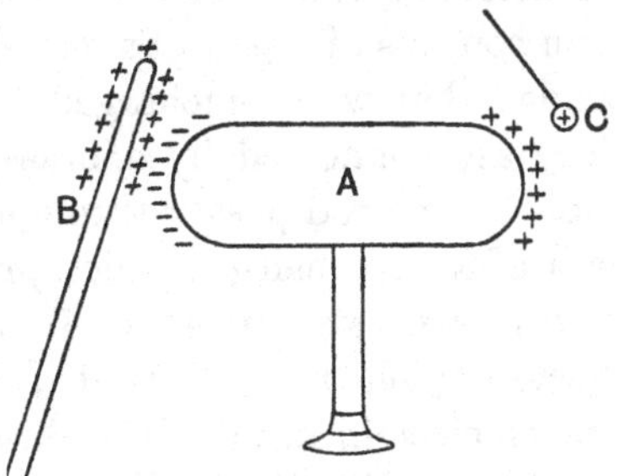

If, further, the conductor A is touched by the finger, the positive charge on A is repelled by the charge on B through the human body to the earth, and when contact is broken the conductor A is left with a negative charge which it retains after A is moved out of the range of the influence of the charged body B. The body A is then said to have been charged by electrostatic induction.

2·14. The electrophorus of Volta.* This is the simplest machine for producing a succession of electric charges. It consists of a circular metal plate A to one side of which an insulating handle B is attached, and a slightly larger circular slab of shellac or resin C. The diagram shews the apparatus in section. The upper surface of the slab of shellac is electrified by rubbing with cat's skin, and the metal plate is then placed upon it. Owing to the roughness of the surface, actual contact only exists at a few points. The electrified shellac plays the part of the body B of **2·13** and the metal plate the part of A. The negative electricity on the shellac attracts positive

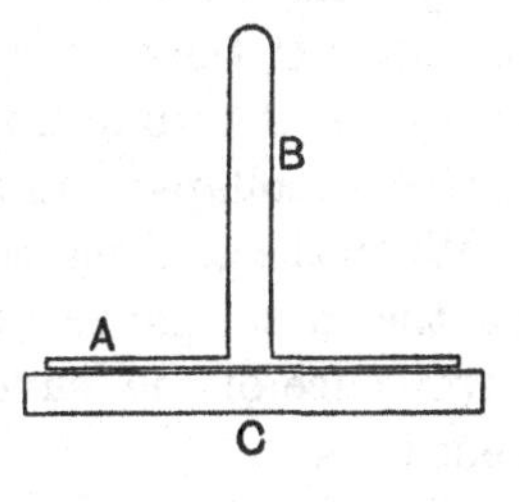

* Alessandro Volta (1745–1827), Italian physicist.

electricity to the under side of the metal plate A and repels negative electricity to the upper side. The plate is touched by the finger so that the negative electricity on its upper side escapes to earth. The plate can then be lifted off the shellac by the insulating handle and it now possesses a positive charge.

2·15. The electroscope. An electroscope is an apparatus for detecting the presence of a charge of electricity. A simple form consists of a pair of small sheets of gold leaf a, a, so suspended that when uncharged they hang in a vertical plane. They are connected by a brass rod b to a brass plate or a knob C. The rod passes through a cork in the neck of a glass bell-jar which protects the gold leaves from air currents. To avoid irregular effects which might result from a possible uneven electrification of the glass, a cylinder of metal gauze slightly smaller than the cylindrical part of the bell-jar is placed inside it and connected to earth. This effectively screens the leaves from any electrification of the glass. When a charged body is brought near to the plate C, electricity is repelled into the gold leaves and causes them to repel one another and separate.

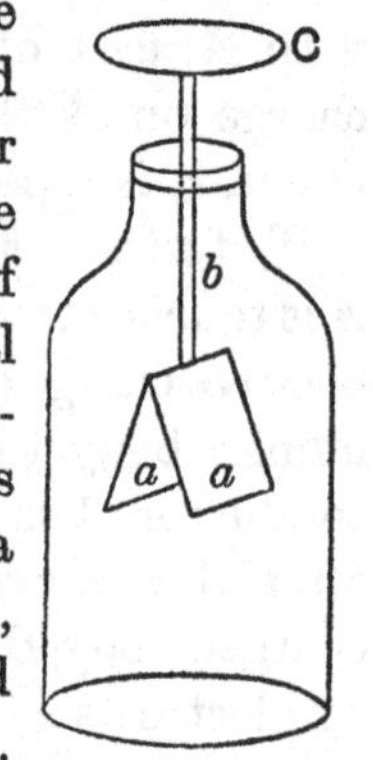

In fact the parts a, b, C of the apparatus can be charged by induction in the manner described in **2·13**. If when the gold leaves are positively charged a negatively charged body such as a stick of sealing-wax is made to approach the plate C, the divergence of the leaves will diminish, and the approach of a sufficiently strong negative charge will cause the leaves to collapse and then re-separate charged negatively.

When the gold leaves become charged, electricity is induced on the metal gauze cylinder and this tends to increase the divergence of the leaves and so renders the instrument more sensitive.

2·16. Comparison of electric charges. Experiments may be performed by standing on the plate of an electroscope a metal vessel which may be closed by a metal lid in which are a few small holes through which silk threads can be passed.

By this means a charged body can be suspended inside the closed vessel and raised or lowered at will without contact with the vessel.

The effect of suspending a body charged with electricity inside the vessel is to attract electricity of the opposite kind to the inner surface of the vessel, and repel electricity of the same kind as the given charge to the outer surface of the vessel and so to the gold leaves. And it is found that the amount of deflection of the gold leaves does not depend on the position of the charged body inside the vessel so long as the body and the vessel are not in contact. It follows that the external electrification of the vessel only depends on the total charge of the body inside it and not on the position of this charge. If two charged bodies suspended in turn inside the vessel produce the same divergence of the gold leaves, we say that their charges are equal; and if two charged bodies when suspended simultaneously inside the vessel produce no divergence of the gold leaves, we say that their charges are equal and opposite.

By experiments of this kind it can be shewn that when electricity is produced by friction the amounts of the two kinds of electricity produced are equal and opposite.

We can also shew that the total charge on two bodies is unaltered by allowing them to touch or by connecting them with one another by a conductor.

Again, if the outer surface of the vessel is connected to the earth (e.g. by touching it with a finger) its external charge disappears and at the same time the gold leaves collapse, so that the total charge inside the vessel is now zero, which implies that the charge on the inner surface of the vessel is equal and opposite to the charge on the body suspended within it.

Such experiments therefore give us a standard of comparison, in that, if a charged body C suspended inside the vessel produces the same divergence of the gold leaves as is produced by the joint effect of two bodies A and B similarly suspended, we say that the charge on C is equal to the sum of the charges on A and B, and hence that *electricity is measurable as regards quantity*. We shall have occasion later to define a unit of electricity; for the present it is sufficient to recognize the fact that charges are measurable in terms of some unit of charge.

2·2. The electric vector. When an electric field is explored by placing a small charged body at different points in turn, it is found that the body is acted upon by a force having a definite magnitude and direction at every point of the field, and that the magnitude F of this force is the product of two quantities, one of which e denotes the charge on the small body and the other E depends upon the field; so that, written as a vector equation,

$$\mathbf{F}=e\mathbf{E}.$$

E therefore denotes the force per unit charge at any point of the field; it is called the **electric intensity**, or the **electric vector**, or simply **the field**.

The foregoing exploration of the field involves the carrying out of a number of experiments, in some of which the charge on the small body is constant while its position in the field is altered; and in others the charge on the body is altered but its position remains fixed.

There may be points or regions of equilibrium where **E** is zero, but in general **E** is variable from point to point and has a definite magnitude and direction at each point of the field.

It must be understood that when we speak of exploring a given field as above with a small charged body, we assume that the charge carried by the body does not disturb the field by its presence. We also assume the existence of a mechanical support for the small charged body capable of balancing the force F due to the electric field. If there were no such mechanical support, the small charged body would move under the influence of the electric field.

We may now define a **conductor** explicitly as a substance in which, if there is an electric field, then there is a flow of electricity; or alternatively, as a substance in which, if there is no flow, then there is no field. A **non-conductor** is a substance in which there is no flow, though there may be a field.

Later we shall have to consider the motion of electricity, but our present object is to establish a theory of electrostatics; i.e. of electric fields in which there is no motion of electricity.

2·21. Experimental basis. We shall assume that the following are demonstrable facts, though in some cases the strongest evidence for the truth of such statements lies not in direct experiment but in the accord of the general theory deduced therefrom with experiments of a more general kind:

(*a*) When electricity is produced by friction or by induction the quantities of positive and negative electricity produced are equal.

In a state of equilibrium

(*b*) there is no electric field inside a hollow conductor which contains no charge;

(*c*) there is no free electric charge within the substance of a conductor;

(*d*) at the surface of a conductor the electric intensity is normal to the surface.

It will appear later that (*d*) is contained in the definition of a conductor. For it will be shewn that, in crossing a boundary surface, the tangential component of the electric intensity is continuous, and, as there is no tangential component inside the conductor, neither can there be a tangential component on the outside of the surface.

It follows that in equilibrium electric charges reside on the surfaces of conductors. Also that there are abrupt changes in the electric field at the surfaces of conductors, there being a field outside a charged conductor and no field in its substance, and these abrupt changes are due to the presence of charges on the surfaces of the conductors. We may therefore say that an electric field is bounded by electric charges.

2·22. Lines and tubes of force. A line in an electric field such that its direction at every point is the direction of the electric vector at that point is called a **line of force**. It follows that lines of force cannot intersect one another, for there is only one direction for the electric vector at any point in the field. If lines of force are drawn through every point of a small closed curve, they lie on a tubular surface called a **tube of force**.

In order to be explicit we shall assume that the small charged body which we use for exploring an electric field has a *positive* charge, so that lines of force are considered as directed in the sense in which a small positive charge would move if it were free to do so.

It appears from 2·21 (d) that if a conducting surface is charged positively lines of force start away from the surface along the normals, and since these lines cannot intersect and are always in the sense in which a positive charge would move they must continue through space until they arrive at a negatively charged surface.

Consider the field of a conductor with a charge $+e$, and suppose that we divide its surface into areas each of which carries a unit of charge and draw the tubes of force through the boundary curves of these areas; we may call these unit tubes since each starts from a unit of charge. These tubes will continue through space until they fall on some other conducting surface or surfaces.

2·23. Coulomb's Law of Force. It is possible and has been customary to build up a theory of electrostatics starting from the fact that *the force between two small charges is proportional to the product of the charges and inversely proportional to the square of the distance between them*, without inquiry at the outset as to how this *action at a distance* is produced. This law is generally known as **Coulomb's Law of Force**, since Coulomb* was the first to publish an experimental verification.

The graphic representation of an electric field obtained by drawing lines of force was introduced by Faraday,† who was the first to endeavour to explain electrical effects as the result of interactions in the medium in which charges are placed rather than as the result of charges acting upon one another at a distance.

* Charles Augustin Coulomb (1736–1806), French physicist.

† Michael Faraday (1791–1867), Director of the Laboratory of the Royal Institution. Author of *Experimental Researches in Electricity and Magnetism*.

2·231. The ether. According to modern views actions in an electric field require a medium for their transmission, and charges at a distance can only affect one another by means of stresses in this intervening medium. It is usual to speak of this medium as the *ether*. The ether is regarded as filling all space including the space occupied by material bodies. It *transmits* radiation (heat, light, electromagnetic waves), but does not emit or absorb it. On this hypothesis it is necessary to consider electric effects as effects *localized in a medium* and express the mathematical formulae of the subject in a localized form.

We have seen how by means of Faraday's lines of force we can make a picture of an electric field which shews the *direction* of the field at every point, and, in order to proceed further on this basis, we must make a hypothesis concerning the *magnitude* of the electric intensity, and it must be a hypothesis which will accord with the facts of observation so far as we can state them. In the following article we shall state this fundamental hypothesis.

2·3. Gauss's Theorem. *The outward flux of electric intensity through any closed surface is proportional to the total charge within the surface.*

Thus if **E** denotes the electric intensity, the outward flux of **E** through a closed surface S means the surface integral over S of the normal component of **E**, i.e. $\int E_n dS$, where E_n is the component of **E** normal to dS (**1·51**); and this fundamental theorem, known as Gauss's Theorem,* states that this integral is proportional to the sum of the charges inside S, or

$$\int E_n dS = h\Sigma e,$$

where h is a constant and Σe denotes the sum of the charges. In electrostatic units, which we shall presently define, the constant h is 4π, and the theorem is

$$\int E_n dS = 4\pi\Sigma e \quad \text{.....................(1).}$$

* Karl Friedrich Gauss (1777–1855), German mathematician and physicist.

If the electric charge be considered to exist as a volume distribution of density ρ inside S, including the possibility of $\rho = 0$ through the whole or any part of the region, the theorem may be written

$$\int E_n dS = 4\pi \int \rho \, dv \quad \text{.................(2)},$$

where the integration on the right extends to all parts of the region inside S at which there is electric charge.

2·31. Consequences of Gauss's Theorem. The reader will notice that we do not offer any proof of Gauss's Theorem but state it as a fundamental hypothesis. The justification for doing so is that the theory which we are able to build up on this hypothesis accords with experiment. It will appear later that Coulomb's Law of Force can be deduced from Gauss's Theorem, or *vice versa*, and on the whole it is more satisfactory to take the latter rather than the former as our fundamental hypothesis and so avoid basing our theory on the idea of action at a distance.

We shall now indicate how Gauss's Theorem accords with some results already obtained.

(i) Let the surface S of the theorem be drawn in the substance of a conductor not containing a cavity; then, by the definition of a conductor, **E** is zero at every point of S; hence by Gauss's Theorem the total charge inside S is zero, and this is true however small the space enclosed by S. It follows that there can be no charge at any point in the substance of a conductor in equilibrium, thus confirming 2·21 (c).

(ii) If the region bounded by S contains no charge, then the total flux of **E** through S is zero, or the flux out of the region is balanced by an equal flux into the region.

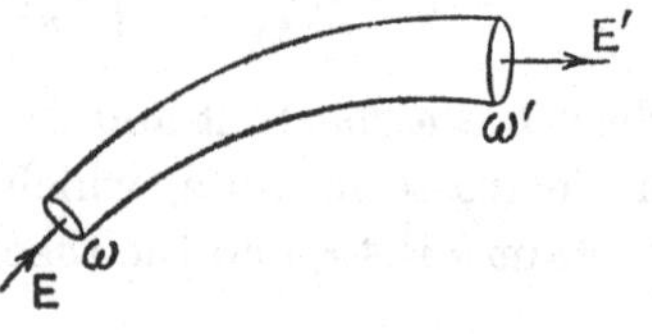

In particular if the region be a portion of a narrow tube of force between two cross-sections of areas ω, ω' and E, E' denote the magnitudes of the electric intensity at these cross-sections, then since **E** has no component at right angles to the sides of the tube and the tube contains no charge, Gauss's Theorem gives

$$-E\omega + E'\omega' = 0,$$

or

$$E\omega = E'\omega';$$

so that the electric intensity along a tube of force varies inversely as the cross-section of the tube.

It follows that the mapping of an electric field by unit tubes of force gives an indication not only of the direction but also of the magnitude of the intensity at any point.

Again consider the whole length of a tube of force starting from one charged conductor and ending on another, and suppose it to be prolonged at each end a short distance into each conductor and then closed. The electric intensity **E** has no component at right angles to the sides of the tube in air and **E** is zero inside the conductors, hence for the complete boundary of this prolonged tube $\int E_n dS = 0$, and therefore the total charge contained is zero. But the only charges are those on the elements of the conductors which form the ends of the tube, so that these charges must be equal and opposite.

(iii) Let a body A with a charge $+e$ be surrounded by an uncharged closed conductor whose inner and outer surfaces are S_1 and S_2. Let the surface S of Gauss's Theorem be drawn in the conductor between the surfaces S_1 and S_2 so as to surround S_1. Then at every point of S there is no electric intensity, so that $\int E_n dS = 0$, and therefore S contains no total charge. Hence there must be on S_1 a charge equal and opposite to the given charge e on the inner conductor A. This is also a consequence of (ii) above, since every tube which starts from A must end on S_1, and the charges on the ends of a tube are equal and opposite.

Since the closed conductor has no total charge, its outer surface S_2 must have a charge $+e$, and the tubes of force which proceed from this charge must continue until they reach some other conductor, possibly the walls of a room.

In this case there are two electric fields, one inside the conductor and the other outside, and though they both owe their origin to the charge on A, they have an independent existence; for the surface S_2 can be discharged by touching it with the finger. The field outside S_2 will then disappear, but the field

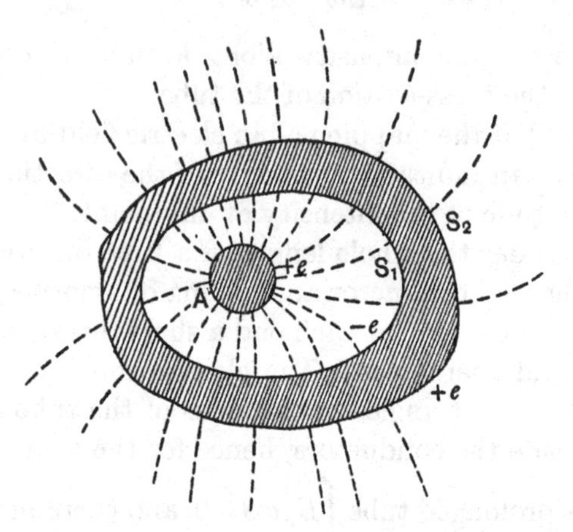

inside S_1 will remain unaltered; or, A may be allowed to touch S_1, when the charges on A and S_1 will neutralize one another and the field inside S_1 will disappear, but the charge on S_2 and the external field will remain unaltered.

2·32. A uniformly charged sphere. Let a charge e be uniformly distributed over the surface of a sphere of radius a. To find the electric intensity **E** at a point P at a distance r from the centre of the given sphere, draw a concentric sphere of radius r. By symmetry the electric intensity has the same value at all points of this sphere and is directed radially, so that the total flux of intensity out of this sphere of radius r is $4\pi r^2 E$, and, by Gauss's Theorem, if $r > a$,

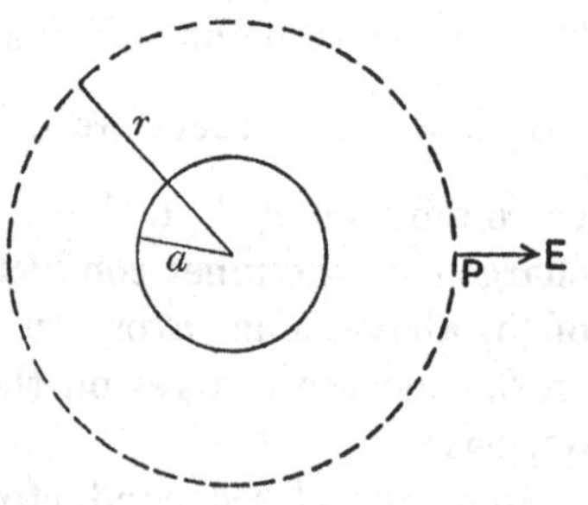

$$4\pi r^2 E = 4\pi e \qquad (1)$$

or

$$E = e/r^2 \qquad (2).$$

But, if $r < a$, the sphere of radius r contains no charge so that

$$4\pi r^2 E = 0$$

or

$$E = 0 \quad \ldots\ldots\ldots\ldots\ldots\ldots\ldots\ldots\ldots(3);$$

a result in accordance with **2·21** (b).

There is therefore no field inside the charged sphere and the external field does not depend upon the radius a of the sphere, and would have the same intensity e/r^2 no matter how small the radius of the given sphere. We infer that 'a point charge' e produces a radial field of intensity e/r^2 and that it would repel a like charge e' at a distance r with a force ee'/r^2; so that by this reasoning Coulomb's Law of Force is a consequence of Gauss's Theorem.

We shall shortly prove the converse and thereafter for special problems in electrostatics we shall use whichever is more convenient for the purpose in hand, but for the logical development of the general theory we shall regard Gauss's Theorem as the basis.

2·321. The electrostatic unit of charge may now be defined in the language of action at a distance, as one which will produce a field of unit intensity at unit distance, i.e. the charge which will repel an equal charge at unit distance with unit force.

It is now apparent why in Gauss's Theorem (**2·3**) the constant h was taken to be 4π. For, if we retained h in **2·32**, then (1) would read

$$4\pi r^2 E = he,$$

and, if units are so chosen that $E = 1$ when $e = 1$ and $r = 1$, we must have $h = 4\pi$.

2·33. The local form of Gauss's Theorem. Let us apply Gauss's Theorem to a small element of volume δv enclosing a point P. If $\mathbf{E}$ denotes the electric vector at P, we have from **1·52**

$$\operatorname{div} \mathbf{E} = \lim_{\delta v \to 0} \left(\int E_n \, dS \right) \Big/ \delta v,$$

where the integral is the outward flux of E through the boundary of δv. Therefore

$$\int E_n \, dS = (\operatorname{div} \mathbf{E} + \epsilon) \, \delta v \quad \ldots\ldots\ldots\ldots\ldots\ldots(1),$$

where ϵ is a small quantity which vanishes with δv. But if ρ represents the average density of electricity inside δv, the left-hand side of (1) is, by Gauss's Theorem, equal to $4\pi\rho\,\delta v$.

Therefore $$(\text{div}\,\mathbf{E}+\epsilon)\,\delta v = 4\pi\rho\,\delta v.$$

Whence, by dividing by δv and then making $\delta v \to 0$, we get

$$\text{div}\,\mathbf{E} = 4\pi\rho \quad\text{......................(2)},$$

where the left side is the divergence of the intensity and the right side is the density at any point P in the field at which $\mathbf{E}$ is continuous.

This is a fundamental equation of electrostatics. It includes the fact that at all points at which there is no charge, i.e. at which $\rho = 0$,

$$\text{div}\,\mathbf{E} = 0 \quad\text{......................(3)}.$$

We must comment on the fact that equation (2) contains a *volume density* ρ of electricity; and we have stated that charges reside upon surfaces and we should therefore expect that they would enter into calculations as *surface density* or charge per unit area. But a charge per unit area is really a charge occupying a volume though concentrated by the fact that one dimension of the volume is indefinitely diminished, and it is often convenient in theoretical work to assume the existence of volume densities and treat surface densities as limiting cases.

We must also comment on the fact that when we use such integral forms as $\int \rho\, dv$ we imply that ρ represents something mathematically continuous and having a definite value at each point of a region of space. But electricity exists in the form of electrons and protons and is not continuous physically or mathematically, so that we are compelled to put a special interpretation upon such an integral as $\int \rho\, dv$; and we take it to mean what the value of such an integral would be if we imagined a continuous ρ to exist and to have at every point of the region of integration a value which is the average density of the actual charges in a small *but finite* element of volume surrounding the point.*

2·34. Gauss's Theorem for a surface distribution. We have seen that discontinuities in the electric field, i.e. in the vector $\mathbf{E}$, arise at charged surfaces. We assume in **2·33** that

* On this subject see J. G. Leathem's Tract, *Volume and Surface Integrals used in Physics*.

there is no surface of discontinuity of **E** within the region considered. We shall now consider the field in the neighbourhood of a surface of discontinuity.

Let a charge of electricity be spread over a surface. To make the argument as general as possible we shall not limit the surface to be that of a conductor but include the case of the surface of a dielectric body such as a piece of glass, so that there may be an electric field on both sides of the surface.

About a point P on the surface draw a short narrow cylinder at right angles to the surface with its ends parallel to the surface and its length small compared to the linear dimensions of its cross-section.

Let ω be the cross-section of the cylinder and σ the surface density of the charge at P, i.e. the charge per unit area, so that $\omega\sigma$ is the total charge within the cylinder.

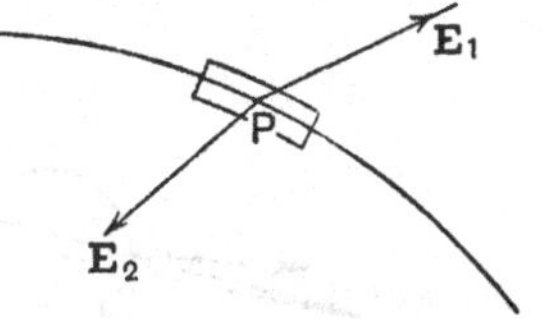

The charge on the surface causes a discontinuity in the electric field. Let $\mathbf{E}_1$, $\mathbf{E}_2$ denote the electric vector at points close to P on opposite sides of the surface. Now apply Gauss's Theorem to the region bounded by the cylinder and take the cylinder so short that the flux of **E** through its sides is negligible. Hence we get

$$(E_1)_n \omega + (E_2)_n \omega = 4\pi\sigma\omega,$$

where the suffix n indicates a normal component. Therefore

$$(E_1)_n + (E_2)_n = 4\pi\sigma \quad \text{..................(1).}$$

This is the form which Gauss's Theorem takes for a surface distribution.

It will be noticed that the normal components are directed away from the surface, and that if we change the sign of $(E_2)_n$ so that both are directed in the same sense, the theorem shews a jump in the value of the normal component of the electric vector of amount $4\pi\sigma$ in crossing the surface.

In the special case in which the surface is that of a conductor, if we take $\mathbf{E}_1$ and $\mathbf{E}_2$ as referring to the regions outside and inside the conductor, we have $\mathbf{E}_2 = 0$, since there is no field in the substance of a conductor, and in this case suppressing the suffix, the result may be written

$$E_n = 4\pi\sigma \quad \text{.......................(2).}$$

This is *Coulomb's expression for the normal electric intensity just outside the surface of a charged conductor.*

2·35. Examples. (i) *Deduce Gauss's Theorem from Coulomb's Law of Force; i.e. to prove that for any closed surface S*

$$\int E_n dS = 4\pi\Sigma e.$$

Consider the field due to a single point charge e at a point O. Draw a cone of small solid angle $d\omega$ and vertex O and let it cut S in small elements of area dS_1, dS_2, dS_3, ... at the points P_1, P_2, P_3, ...; and

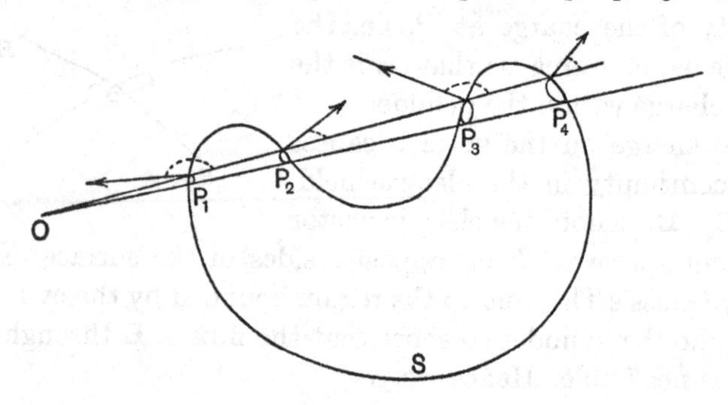

let the outward drawn normals to these elements make angles θ_1, θ_2, θ_3, ... with the line OP_1P_2.... The θ's are obtuse or acute angles according as the line OP_1P_2... is entering or leaving the region bounded by S.

The point charge e at O produces at a point P a field of intensity e/OP^2 directed along OP (Coulomb's Law), so that the contributions to the outward flux of $\mathbf{E}$ which arise from the elements of area in which the cone cuts the surface amount to

$$\frac{e\cos\theta_1}{OP_1^2}dS_1 + \frac{e\cos\theta_2}{OP_2^2}dS_2 + \frac{e\cos\theta_3}{OP_3^2}dS_3 + \ldots \quad \text{.........(1).}$$

But from **1·3** (1), for each of these elements

$$\frac{\cos\theta}{OP^2}dS = \pm d\omega,$$

$+$ or $-$ according as θ is acute or obtuse, i.e. according as the cone drawn from O is leaving or entering the closed region. Hence expression (1) is of the form

$$-e d\omega + e d\omega - e d\omega + \ldots;$$

and if O is outside S the number of entrances of the cone is equal to the number of its exits from the region and the sum is zero; but if O is inside S the cone makes a final exit from the region and the sum is $ed\omega$.

Then taking cones in all directions round O, since $\int d\omega = 4\pi$, it follows that, for a point charge e,

$$\int E_n dS = 0 \quad \text{or} \quad 4\pi e \quad \ldots\ldots\ldots\ldots\ldots\ldots\ldots (2),$$

according as the point charge is outside or inside S.

Proceeding in this way and adding together the effects of all the point charges in the field, we get

$$\int E_n dS = 4\pi \Sigma e \quad \ldots\ldots\ldots\ldots\ldots\ldots\ldots\ldots (3),$$

where Σe means the sum of the charges inside S.

(ii) *Prove that the electric intensity at all points on either side of a uniformly charged plane is $2\pi\sigma$, where σ is the charge per unit area.*

Shew that, of the total intensity $2\pi\sigma$ at a point A at a distance of half an inch from the plane, one-half is due to the charge at points within an inch of A. [M. T. 1918]

Here it is assumed that the medium on both sides of the plane sheet of electricity is air and that the plane is of infinite extent. By symmetry the tubes of force are at right angles to the plane everywhere and of constant cross-section, so that by **2·31** (ii) the electric intensity has the same value everywhere, and, taking σ to be positive, it is directed away from the plane on both sides. Then if as in **2·34** we apply Gauss's Theorem to a short cylinder of cross-section ω at right angles to the plane with plane ends parallel to the plane on opposite sides of it, denoting by E the constant value of the electric intensity, we get

$$E\omega + E\omega = 4\pi\sigma\omega,$$

or

$$E = 2\pi\sigma.$$

This is in fact a special case of **2·34** (1), in which $(E_1)_n = (E_2)_n = E$; and taking note of the sense of the intensity on opposite sides of the plane it accords with the general rule that there is a jump of $4\pi\sigma$ in the value in crossing the plane.

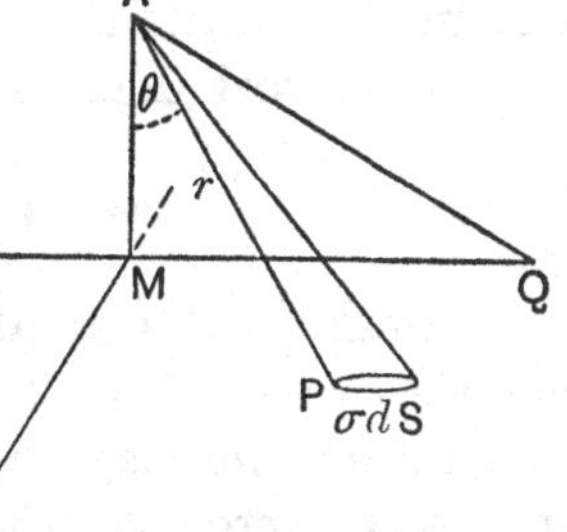

For the second part of the question, the charge on an element of area dS at P is σdS, and if r is its distance from A it produces at A an intensity $\sigma dS/r^2$ along PA, and resolving this along the normal MA gives $\sigma dS \cos\theta/r^2$, where θ is the angle MAP, or, from **1·3** (1), $\sigma d\omega$, where $d\omega$ is the solid angle subtended at A by the element dS.

But the points of the plane which lie within an inch of A, lie within a circle of centre M and radius MQ, where $AQ = 1$ inch. The charge on this circular area therefore contributes to the total intensity at A an amount $\sigma\omega$, where ω is the solid angle which the circle subtends at A. But $AM=\frac{1}{2}$ inch, so that the solid angle is that of a right circular cone of semi-vertical angle 60°, and therefore

$$\omega=2\pi(1-\cos 60°) \quad (1\cdot22)$$
$$=\pi.$$

Therefore an intensity $\pi\sigma$ at A, or half the total intensity, is due to the charges which lie within an inch of A.

2·4. The potential function. In order to complete the basis of the theory of electrostatic fields in air we must now introduce another function—a scalar function of position like that discussed in **1·4**—called the **potential function.** We assert as our **second fundamental hypothesis** for the building up of the theory of the electrostatic field that *there is, for every electrostatic field, a single-valued potential function such that the electric intensity is its negative gradient;* i.e. we assume that there exists a single-valued function ϕ such that, at every point of the field,

$$\mathbf{E}=-\mathbf{grad}\,\phi.$$

We proceed to shew that this function ϕ bears an interpretation in terms of work.

Suppose that a unit charge could be displaced in an electrostatic field without causing a disturbance of the other charges; then the work done by the field on the unit charge as it moves from P to Q by any path would be given by $\int_P^Q E_s ds$, where δs denotes an element of the path and E_s the component of $\mathbf{E}$ in the direction δs. But, by hypothesis, $E_s=-\partial\phi/\partial s$, so that the work done

$$=-\int_P^Q \frac{\partial\phi}{\partial s}ds=\phi_P-\phi_Q;$$

or the work done is equal to the excess of the potential at P over the potential at Q. If we assume that the field is of finite extent, we may suppose the potential at a great distance to be zero, and the potential of the field at a point P is then the work that would be done by the forces of the field on a unit charge as it moved from P to an infinite distance, always supposing the field to be undisturbed by the presence of the unit charge.

Physically therefore potential is of dimensions 'work per unit charge', or 'energy per unit charge'.

Since **E** is zero in the substance of a conductor in electrostatic equilibrium, therefore the potential has a constant value throughout such a conductor.

For laboratory purposes the potential of the earth may be regarded as invariable, for it is obvious that any charge which might in the course of an experiment be added to that of the earth could not affect its potential appreciably.

Further, it is clear that the addition of a constant to the function ϕ will not affect **E**, neither will it affect the difference of potential between any two points. It is therefore permissible and convenient to take the potential of the earth to be zero.

2·41. Equations for the potential. The relation

$$\mathbf{E} = -\operatorname{grad} \phi \quad \text{.....................(1)}$$

implies that for rectangular components

$$E_x, E_y, E_z = -\frac{\partial \phi}{\partial x}, -\frac{\partial \phi}{\partial y}, -\frac{\partial \phi}{\partial z} \quad \text{...........(2).}$$

Hence by substituting in the fundamental relation **2·33** (2), viz.

$$\operatorname{div} \mathbf{E} = 4\pi\rho,$$

or (see **1·52**)

$$\frac{\partial E_x}{\partial x} + \frac{\partial E_y}{\partial y} + \frac{\partial E_z}{\partial z} = 4\pi\rho,$$

we get

$$\frac{\partial^2 \phi}{\partial x^2} + \frac{\partial^2 \phi}{\partial y^2} + \frac{\partial^2 \phi}{\partial z^2} = -4\pi\rho \quad \text{..............(3).}$$

This is known as *Poisson's Equation.* It is satisfied by the potential at every point of the field at which there is a volume density of electricity. At points at which there is no electricity the equation becomes

$$\frac{\partial^2 \phi}{\partial x^2} + \frac{\partial^2 \phi}{\partial y^2} + \frac{\partial^2 \phi}{\partial z^2} = 0 \quad \text{..................(4),}$$

known as *Laplace's Equation.*

The operator $\frac{\partial^2}{\partial x^2}+\frac{\partial^2}{\partial y^2}+\frac{\partial^2}{\partial z^2}$, known as Laplace's operator, is usually abbreviated into the form ∇^2, so that (3) and (4) are written

$$\nabla^2\phi = -4\pi\rho \quad \text{and} \quad \nabla^2\phi = 0.$$

To obtain Laplace's equation *in polar co-ordinates* r, θ, ω, we take polar components of the electric vector, viz.

$$E_r,\ E_\theta,\ E_\omega = -\frac{\partial\phi}{\partial r},\ -\frac{\partial\phi}{r\partial\theta},\ -\frac{\partial\phi}{r\sin\theta\,\partial\omega}$$

and substitute in the equation

$$\operatorname{div}\mathbf{E} = 0,$$

using the expression for divergence given in **1·53** (1); this gives

$$\frac{1}{r^2}\frac{\partial}{\partial r}\left(r^2\frac{\partial\phi}{\partial r}\right)+\frac{1}{r^2\sin\theta}\frac{\partial}{\partial\theta}\left(\sin\theta\frac{\partial\phi}{\partial\theta}\right)+\frac{1}{r^2\sin^2\theta}\frac{\partial^2\phi}{\partial\omega^2}=0 \quad \text{...(5).}$$

Similarly, in *cylindrical co-ordinates* r, θ, z, we take

$$E_r,\ E_\theta,\ E_z = -\frac{\partial\phi}{\partial r},\ -\frac{\partial\phi}{r\partial\theta},\ -\frac{\partial\phi}{\partial z}$$

and substitute in the equation

$$\operatorname{div}\mathbf{E} = 0,$$

using the expression for divergence given in **1·54**; this gives

$$\frac{1}{r}\frac{\partial}{\partial r}\left(r\frac{\partial\phi}{\partial r}\right)+\frac{1}{r^2}\frac{\partial^2\phi}{\partial\theta^2}+\frac{\partial^2\phi}{\partial z^2}=0 \quad \text{..............(6).}$$

2·411. For a surface distribution. As explained in **2·34**, there are different fields on opposite sides of an electrified surface, and the potential function has therefore different forms ϕ_1, ϕ_2 on opposite sides of the surface, so that

$$\mathbf{E}_1 = -\mathbf{grad}\,\phi_1 \quad \text{and} \quad \mathbf{E}_2 = -\mathbf{grad}\,\phi_2.$$

Therefore **2·34** (1) may be written

$$\frac{\partial\phi_1}{\partial n_1}+\frac{\partial\phi_2}{\partial n_2} = -4\pi\sigma \quad \text{.................(1),}$$

where ∂n_1 and ∂n_2 are elements of the normal directed away from the surface on each side.

As we have already seen therefore, at every electrified surface there is a discontinuity in the electric vector or, what is the same thing, a discontinuity in the gradient of the potential function, and the potential functions on opposite sides of the surface must satisfy (1).

At the surface of a charged conductor we have **2·34** (2), which may now be written

$$\frac{\partial \phi}{\partial n} = -4\pi\sigma \quad \text{.......................(2),}$$

where the differentiation is along the outward normal and ϕ is the potential outside the conductor.

Further, the potential function is assumed to be physically continuous, i.e. between any two assigned values it assumes all intermediate values, save, as we shall see later, possibly at a common boundary of two different substances where there is a constant difference. At all points of such a surface we have

$$\phi_1 - \phi_2 = C \quad \text{.......................(3),}$$

where C is a constant, which is in most cases negligible; and ϕ_1, ϕ_2 denote the potential functions on opposite sides of the common interface.

Let δs be a small arc drawn in *any* direction on the surface of separation, then by differentiating (3) we get

$$\frac{\partial \phi_1}{\partial s} - \frac{\partial \phi_2}{\partial s} = 0 \quad \text{.......................(4)}$$

or

$$E_{1s} = E_{2s} \quad \text{.......................(5);}$$

i.e. the tangential component of the electric vector is continuous in crossing the common surface of two media, whether this surface be electrified or not.

To summarize: the potential function must satisfy **2·41** (3) at all points where there is a volume density and (4) at all points in empty space. At a charged surface the normal components of its gradient must satisfy (1) above but the tangential components of its gradient are continuous. The function itself must be continuous everywhere save in exceptional cases where there may be a constant difference at an interface between two media.

COROLLARIES. (i) If the potential is constant through any region, e.g. in the substance of a conductor in equilibrium, then from **2·41** (3), since $\phi = \text{const.}$, therefore $\nabla^2\phi = 0$ and $\rho = 0$; so that there can be no charge in the substance of a conductor in equilibrium.

(ii) If the normal component of the gradient of the potential is continuous across any surface, then from (1) there can be no charge on the surface.

2·42. Equipotential surfaces. A surface at every point of which the potential ϕ has the same constant value is an *equipotential surface*.

In any case in which the potential ϕ has been determined as a function of x, y, z, the equation

$$\phi(x, y, z) = \alpha \quad \text{......................(1)},$$

for different values of the constant α, represents the family of equipotential surfaces. It is clear that two such surfaces of different potentials cannot intersect.

The electric vector at every point is normal to the equipotential surface through the point. This follows from the fact that the vector is the negative gradient of the potential, or simply from the fact that in any direction along the surface ϕ is constant, so that $d\phi = 0$, and there is therefore no component of the vector in that direction.

Conversely a surface to which the electric vector is everywhere normal must be an equipotential surface, for $\partial\phi/\partial s = 0$ in every direction at right angles to the vector, and therefore $\phi = \text{const.}$ over the surface.

It follows that if space is mapped out by tubes of force and equipotential surfaces, the former cut the latter at right angles wherever they intersect.

2·43. On lines and tubes of force. Since the electric intensity is the negative gradient of the potential, therefore a line of force, which is at every point in the direction of the electric intensity, may be regarded as having a positive sense in the direction in which potential decreases. Hence we regard lines and tubes of force as drawn from places of higher to places of lower potential; and as in **2·22** lines and tubes of force start from positive charges and end on negative charges.

(i) Since the potential of a conductor is constant it follows that no line of force can begin and end on the same conductor, nor can it begin and end on conductors at the same potential.

(ii) In a field containing several conductors the conductor of highest potential must be positively charged; for a negative charge would imply tubes of force arriving there having started from a place of still higher potential.

(iii) The conductor of lowest potential must be negatively charged; for a positive charge would imply tubes of force starting thence and proceeding to a place of still lower potential.

(iv) Apart from the earth there must be at least one conductor whose charge is wholly positive or wholly negative, for if there is an electric field its potential must have either a highest or lowest value or both a highest and a lowest.

(v) On a conductor insulated and uncharged the distribution of electricity, if any, is partly positive and partly negative and as many unit tubes of force fall on the conductor as proceed from it. These are separated by a line of no electrification on the surface, and along this line, since $\sigma=0$, therefore by **2·411** (2) $\partial\phi/\partial n=0$, or ϕ is constant in the direction normal to the conductor, and therefore at all points of this line of no electrification the conductor cuts an equipotential surface at right angles.

(vi) When all the conductors but one are connected to earth (i.e. at zero potential) and that one has a positive charge, the charges induced on the others are all negative and their numerical sum cannot exceed the positive charge on the insulated charged conductor. This follows because the unit tubes of force which start from the positively charged conductor have at their further ends to account for all the negative charges in the field.

2·44. Theorems on the potential. (i) *If a closed equipotential surface contains no charge, the potential is constant throughout the region bounded by the surface.*

If there are no charges inside a closed surface S, no lines of force can begin or end inside S; so that, if there is any field inside S, the lines of force must be continuous lines traversing the region bounded by S. But this is impossible since all points on S have the same potential, and no two points on a line of force can have the same potential. Therefore there is no field inside S, or the potential is constant.

(ii) *The potential cannot have a maximum or minimum value at a point at which there is no charge.*

Let the potential have a maximum at a point P; then the potential decreases in every direction round P, so that if we draw a small sphere round P, the electric vector is directed outwards at every point of the sphere, so the outward flux of the electric vector is a positive number, and by Gauss's Theorem there must be a positive charge inside the sphere. Hence the potential cannot have a maximum at P unless there is a positive charge there. Similarly at a point where the potential has a minimum there must be a negative charge.

2·5. Special fields. Point charges. It is possible to give a formal proof that if the potential function in any field satisfies the relations **2·41** (3), (4) and **2·411** (1), (3) at different points of the field, then the potential at any point is equal to the sum of each element of charge divided by its distance from this point. But the proof requires too much analysis to be introduced here, and we shall content ourselves with shewing that the definition of potential in terms of work in **2·4** leads to the result stated above in the case of a field due to point charges repelling in accordance with Coulomb's Law.

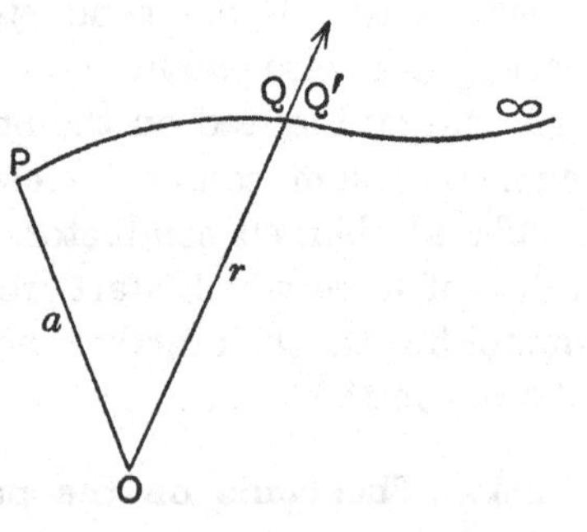

Thus let there be a charge e at O. To find its contribution to the potential at a point P, let QQ' be an element ds of a path from P to an infinite distance in any direction. Let $OP=a$ and $OQ=r$. Then taking the electric intensity at Q to be e/r^2 along OQ, the work done in repelling a unit charge from P to infinity along this path is represented by

$$\int_P^\infty \frac{e}{r^2}\frac{dr}{ds}\,ds = \int_a^\infty \frac{e}{r^2}\,dr = \frac{e}{a}.$$

Similarly if there are other point charges e', e'', ... at distances a', a'', ... from P, they give like contributions to the potential at P, so that the potential is

$$\frac{e}{a}+\frac{e'}{a'}+\frac{e''}{a''}+\dots \quad \text{or} \quad \Sigma\frac{e}{a}.$$

2·51. The field due to a set of collinear point charges. Let a number of charges $e_1, e_2, e_3, \dots$ be situated at collinear points $A_1, A_2, A_3, \dots$; let P be a point at distances $r_1, r_2, r_3, \dots$ from $A_1, A_2, A_3, \dots$, and let A_1P, A_2P, A_3P, ... make angles $\theta_1, \theta_2, \theta_3, \dots$ with the line $A_1A_2 \dots$.

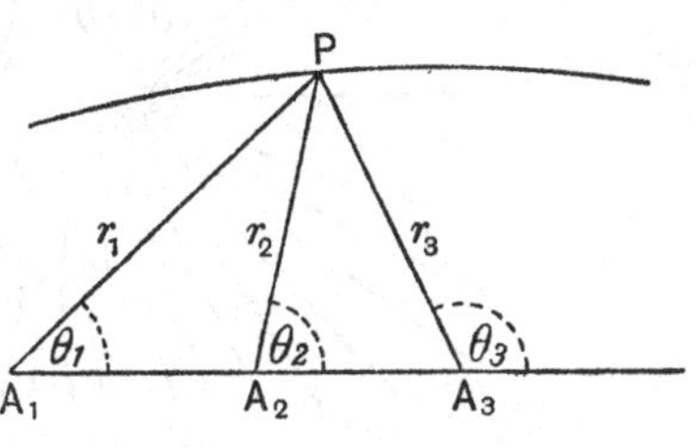

The point charges make radial contributions e_1/r_1^2, e_2/r_2^2, e_3/r_3^2, ... to the intensity at P, and if we resolve these along the normal to the line of force through P, the sum of the components is zero, since the resultant intensity is tangential to the line of force. Hence

$$\frac{e_1}{r_1^2}.r_1\frac{d\theta_1}{ds}+\frac{e_2}{r_2^2}.r_2\frac{d\theta_2}{ds}+\frac{e_3}{r_3^2}.r_3\frac{d\theta_3}{ds}+\dots=0,$$

where ds is an element of the line of force.

Therefore $$\frac{e_1}{r_1}d\theta_1+\frac{e_2}{r_2}d\theta_2+\frac{e_3}{r_3}d\theta_3+\dots=0.$$

But $$r_1\sin\theta_1=r_2\sin\theta_2=r_3\sin\theta_3=\dots,$$

so that

$$e_1\sin\theta_1 d\theta_1+e_2\sin\theta_2 d\theta_2+e_3\sin\theta_3 d\theta_3+\dots=0,$$

and by integration

$$e_1\cos\theta_1+e_2\cos\theta_2+e_3\cos\theta_3+\dots=\text{const.} \quad \dots\dots(1),$$

and for different values of the constant this equation represents all the lines of force.

2·52. Examples. (i) *Two point charges e_1, e_2 of the same sign.*

Since the charges have the same sign, no line of force can pass from one to the other and all the lines of force must go to infinity.

Let the charges e_1, e_2 be at the points A, B and let P be a point on the line of force which starts from A at an inclination α to BA produced.

The equation of the lines of force is

$$e_1 \cos \theta_1 + e_2 \cos \theta_2 = \text{const.},$$

where θ_1, θ_2 are the inclinations to the line BA of the radii from A, B to any point on such a line. But as $P \to A$ on the line considered we have $\theta_1 \to \alpha$ and $\theta_2 \to 0$; therefore its equation is

$$e_1 \cos \theta_1 + e_2 \cos \theta_2 = e_1 \cos \alpha + e_2.$$

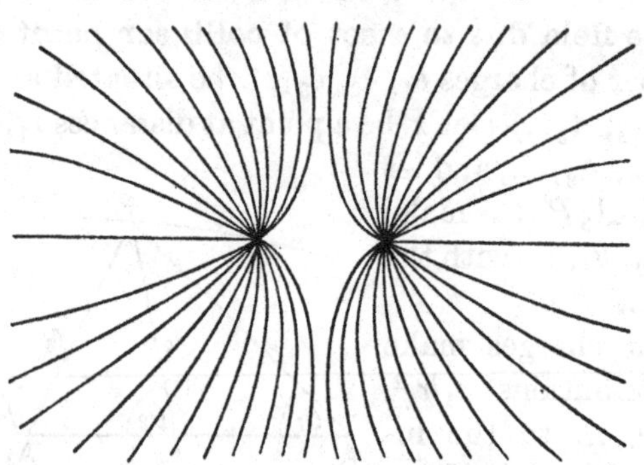

The direction of the line at infinity is got by putting $\theta_1 = \theta_2 = \theta$, so that

$$(e_1 + e_2) \cos \theta = e_1 \cos \alpha + e_2$$

gives the slope of the asymptote.

Again if the tangent at P cuts AB in T, since there is no electric intensity at P at right angles to TP, therefore

$$\frac{e_1}{AP^2} \sin APT = \frac{e_2}{BP^2} \sin BPT.$$

But $$\frac{\sin APT}{\sin BPT} = \frac{\sin APT}{\sin ATP} \cdot \frac{\sin BTP}{\sin BPT} = \frac{AT}{AP} \cdot \frac{BP}{BT},$$

so that $$\frac{e_1}{AP^3} \cdot AT = \frac{e_2}{BP^3} \cdot BT.$$

Now let P move along the curve to infinity, then $AP/BP \to 1$ and T moves up to a point C on AB such that

$$AC : CB = e_2 : e_1.$$

And as this is independent of α, it follows that the asymptotes to all the lines of force pass through this fixed point C.

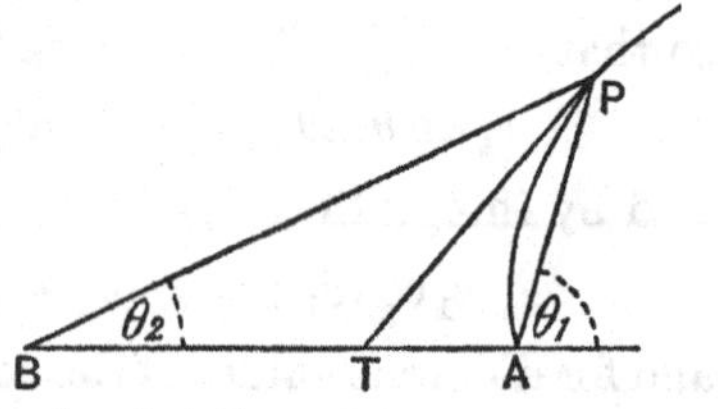

Further there is a point of equilibrium O on the line AB, where the repulsions e_1/AO^2 and e_2/BO^2 are equal and opposite.

The equipotential surfaces are surfaces of revolution about the line AB given by the equation

$$\phi \equiv \frac{e_1}{r_1} + \frac{e_2}{r_2} = \text{const.},$$

where r_1, r_2 denote distances from A and B.

For small values of r_1 and of r_2, i.e. for large values of the constant, these are disconnected surfaces surrounding the points A, B respectively. The potential decreases as we proceed outwards from either

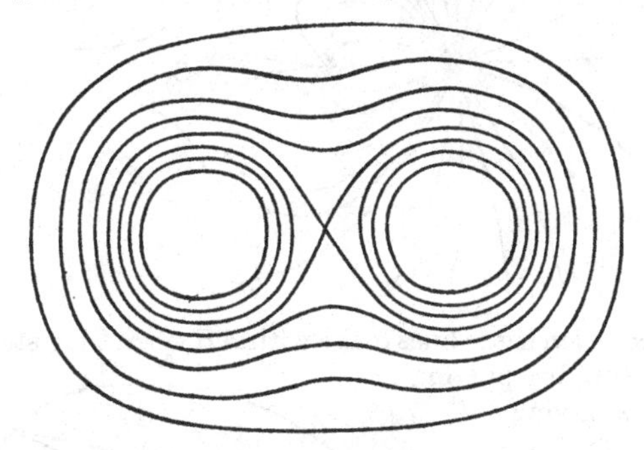

charge. For a particular value of the constant the two surfaces become a single surface passing through O, the point of equilibrium; the vanishing of the components of intensity $\frac{\partial\phi}{\partial x}, \frac{\partial\phi}{\partial y}, \frac{\partial\phi}{\partial z}$ at this point being the conditions that the surface $\phi(x, y, z) = \text{const.}$ should have there a conical or nodal point and therefore no definite normal. For smaller values of the potential the equipotential surfaces are single sheets surrounding both charges.

(ii) *Charges* $4e$ *and* $-e$.

Let the charges $4e$ and $-e$ be at A and B. There is a point of equilibrium C on AB produced so that $BC = AB$, for this makes

$$4e/AC^2 = e/BC^2.$$

Of the unit tubes of force which start from A only one-quarter will end on the negative charge at B and the rest will go to infinity, and the equilibrium point C will separate those which go to B from those which go to infinity.

If P be any point on a line of force and $PAC = \theta$ and $PBC = \theta'$, the equation of the lines of force is given by

$$4e\cos\theta - e\cos\theta' = \text{const.} \quad \text{.....................(1).}$$

Consider the possibility of a line of force passing through C. At C we have $\theta = \theta' = 0$, so that its equation would be

$$4\cos\theta - \cos\theta' = 3 \quad \text{........................(2);}$$

and as the point P approaches C, θ, θ' are small and (2) may be written

$$4(1-\tfrac{1}{2}\theta^2)-(1-\tfrac{1}{2}\theta'^2)=3,$$

which gives $\theta'=2\theta$. From this we deduce that $BP=AB=BC$; and therefore as P approaches C the curve cuts BC at right angles. In the plane of the paper a similar line of force would approach BC from below.

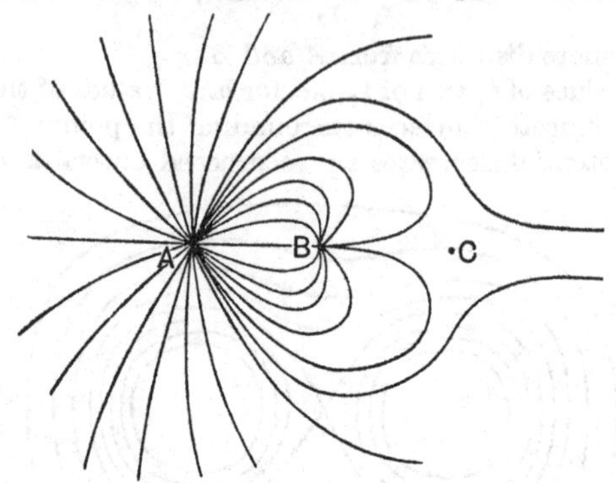

The angle α which these lines make with AB when they start from A is got from (2), by putting $\theta=\alpha$ and $\theta'=\pi$, giving

$$4\cos\alpha+1=3,$$

so that $\alpha=60°$.

The equipotential surfaces are surfaces of revolution about the line AB, given by the equation $\frac{4}{r}-\frac{1}{r'}=\text{const.}$, where r, r' denote distances from A and B.

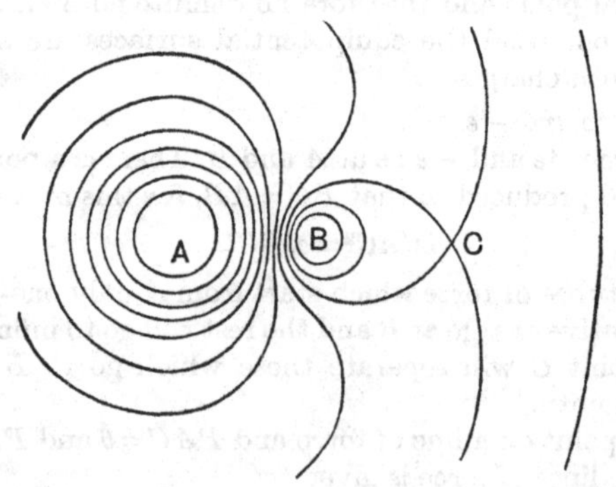

For small values of r' the potential is negative and the surfaces surround the point B. The section of the surfaces by a plane through AB consists of loops round the point B increasing in size, until we reach the value of the potential at C, viz. $\frac{4e}{AC}-\frac{e}{BC}$ or $\frac{e}{AB}$, for which the equipotential surface is a surface which crosses itself with a conical point at C; the section consisting of two loops the larger of which encloses loops which surround the point A with potentials which continually increase as the loops shrink round the point A.

Outside the surface through C there are equipotential surfaces which surround both A and B with potentials which decrease steadily to zero as the surfaces expand to infinity.

2·53. A two-dimensional field. When a field is produced by a number of long uniformly charged cylinders, with parallel axes and ends so far distant from the region of space considered that the field may be deemed to be the same in all planes at right angles to the cylinders, we may regard the field as two-dimensional.

Consider a single circular cylinder of radius a carrying a charge e per unit length. To find the electric intensity at a point P at a distance r from the axis of the cylinder, describe a coaxial cylinder of radius r and consider the flux out of a unit length of this cylinder. By symmetry the field E is radial so that the total flux is $2\pi rE$, and if $r > a$ the charge enclosed in unit length of the cylinder is e, therefore by Gauss's Theorem

$$2\pi rE = 4\pi e$$

or

$$E = 2e/r \qquad \text{.......................(1).}$$

But if $r < a$, no charge is contained in the cylinder of radius r and therefore

$$E = 0 \qquad \text{..........................(2).}$$

If ϕ denotes the potential of which E is the negative gradient, we have

$$-\frac{\partial \phi}{\partial r} = E = \frac{2e}{r} \quad (r > a)$$

or

$$\phi = \text{const.} - 2e \log r \qquad \text{.................(3)}$$

and, for $r < a$,

$$\phi = \text{const.} \qquad \text{........................(4).}$$

We notice that the formulae (1) and (3) are independent of the radius a of the given cylinder.

2·531. Field produced by a set of long charged parallel fine wires. Let long parallel fine wires carrying charges e_1, e_2, e_3, ... per unit length cut the plane of the paper at right angles at the points A_1, A_2, A_3, ...

We shall speak of the flux across a line on the paper, meaning thereby the flux across a unit length of a cylindrical surface of which the line is a cross-section.

With this understanding the total flux of intensity across a curve which surrounds the point A_1 is, by Gauss's Theorem, $4\pi e_1$; and the amount of flux which lies between two radii inclined at an angle $d\theta_1$ is $\frac{d\theta_1}{2\pi} . 4\pi e_1$ or $2e_1 d\theta_1$.

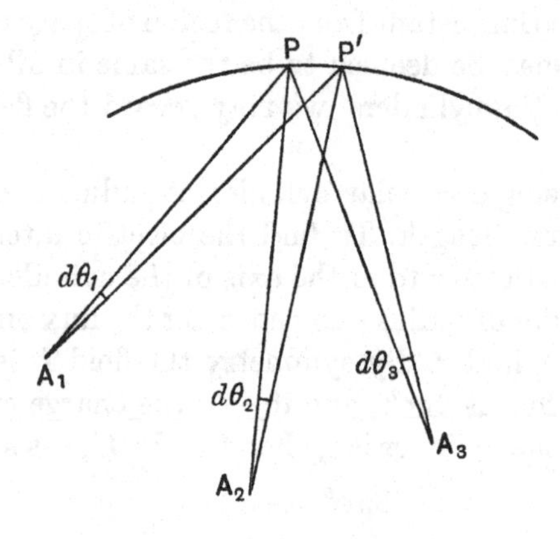

Let an element PP' of a line of force subtend angles $d\theta_1$, $d\theta_2$, $d\theta_3$, ... at A_1, A_2, A_3, ..., then since the total flux of intensity across PP' is zero, therefore

$$2e_1 d\theta_1 + 2e_2 d\theta_2 + 2e_3 d\theta_3 + \dots = 0,$$

or, by integration,

$$e_1 \theta_1 + e_2 \theta_2 + e_3 \theta_3 + \dots = \text{const.} \quad \dots\dots\dots\dots(1),$$

where the θ's are the angles made by the lines $A_1 P$, $A_2 P$, $A_3 P$, ... with any fixed direction.

For different values of the constant, equation (1) represents the lines of force in any cross-section of the field.

It should be noticed that if any point A lies on the opposite side of the line of force to the other points the sign of the corresponding term in the equation must be changed.

In like manner by using **2·53** (3) and assuming that the potential due to a number of charged wires is got by adding the potentials which they produce separately, we get for the potential of the field

$$\phi = \text{const.} - 2e_1 \log r_1 - 2e_2 \log r_2 - 2e_3 \log r_3 - \dots \quad \dots(2),$$

where the r's are the distances of a point from $A_1, A_2, A_3, \dots$.

2·532. Examples. (i) *Two wires with equal and opposite charges.*

Let e, $-e$ denote the charges per unit length on wires which cut the paper at right angles at A, B. Let radii from A, B to a point P make angles θ, θ' with BA produced. Then the lines of force are given by

$$e\theta - e\theta' = \text{const.},$$

or the angle APB is constant. So the lines of force are coaxial circles passing through A and B.

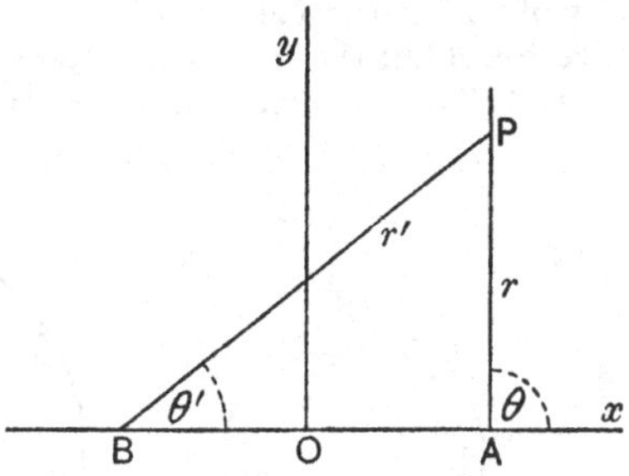

The potential is given by

$$\phi = \text{const.} - 2e \log AP + 2e \log BP,$$

so that the equipotential curves are given by $AP/BP = \text{const.}$, i.e. the family of coaxial circles which have A, B as limiting points and cut the former family at right angles.

(ii) *Two wires with equal charges of the same sign.*

Here the lines of force are seen to be

$$\theta + \theta' = \text{const.}$$

Taking rectangular axes so that A, B are the points $(\pm a, 0)$, this is

$$\tan^{-1} \frac{y}{x-a} + \tan^{-1} \frac{y}{x+a} = \tan^{-1} \frac{1}{k},$$

say, or
$$x^2 - 2kxy - y^2 = a^2,$$

a family of rectangular hyperbolas passing through A and B.

As above, the equipotential curves are the ovals

$$rr' = \text{const.}$$

and we thus have a physical proof that the orthogonal trajectories of the family of ovals is a family of rectangular hyperbolas.

2·6. Experimental verification of the Law of Inverse Squares. Cavendish's experiment. A metal globe rests on an ebonite ring inside another metal globe formed of two tightly fitting hemispheres supported on an insulating stand. There is a small hole in the outer globe which can be closed by a metal lid to which is affixed a short wire and this makes contact with the inner globe when the lid is closed. When the lid is lifted by a silk thread, a wire can be passed through the hole to connect the inner globe with an electrometer and so test whether it has a charge. The figure shows the essential part of the apparatus in section.

A charge is given to the outer sphere with the lid closed so that the inner and outer are in communication. The lid is then lifted, the outer sphere discharged and the inner sphere is tested and found to be without charge, thus demonstrating experimentally that there is no electric charge inside a uniformly charged spherical conductor.

Now let A be any point inside a sphere which is charged with electricity of uniform surface density σ. Let a cone of small solid angle $d\omega$ and vertex A cut the sphere in elements of area dS, dS' at P, P'. Let $AP=r$, $AP'=r'$, and let O be the centre of the sphere. The charges on

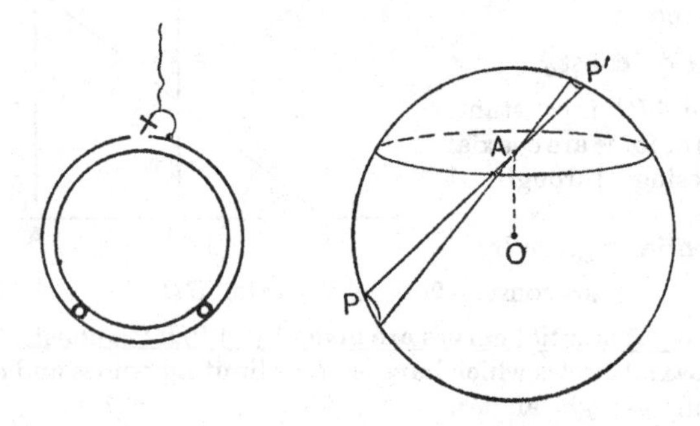

the elements are σdS and $\sigma dS'$. But $dS=r^2 d\omega \sec OPA$, and $dS'=r'^2 d\omega \sec OP'A$, and the angles OPA, $OP'A$ are equal. So if the law of force were the inverse square of the distance, the forces due to the elements, $\sigma dS/r^2$ and $\sigma dS'/r'^2$, would be equal and opposite, and by taking cones in all directions round A the whole surface of the sphere is divided into like pairs of elements exerting equal and opposite forces at A so that the resultant would be zero and there would be nothing to cause a transfer of electricity between the outer and inner spheres when they are in communication during the experiment.

Now consider the effect of taking a modified law of force r^{-2-p} instead of r^{-2}. A plane through A at right angles to AO divides the surface of the sphere in such a way that, for all such cones as we considered above, $AP'<AP$, where P' is above this plane and P is below it. The repulsions of the corresponding elements are $\dfrac{\sigma dS}{r^{2+p}}$ and $\dfrac{\sigma dS'}{r'^{2+p}}$, where as above $\dfrac{\sigma dS}{r^2}=\dfrac{\sigma dS'}{r'^2}$, but $r'<r$. Hence if p is positive, the force along PA is less than the force along $P'A$, and *vice versa* if p is negative: and therefore with this law of force the plane through A divides the surface into portions the charges on which exert unbalancing forces at A, and there would be a resultant force at any point A inside the sphere, other than the centre. Thus at all points on the inner

sphere in the experiment there would be a force in the same sense, inwards if p is positive and outwards if p is negative, which must result in the passage of electricity when the spheres are in communication and the presence of a charge on the inner sphere. Since experiment indicates that there is no such charge, we conclude that if the law of force is some power of the distance it must be the inverse square.

EXAMPLES

1. Prove that, if there be only one charged conductor in an electric field, the amount of electricity of either kind induced on any other conductor cannot exceed the charge on the first. [I. 1891]

2. An insulated conductor is under the influence of a point charge $-e$; if the total charge on the conductor is positive and greater than e, prove that the distribution of electrification on the conductor is everywhere positive.

3. Shew that, if a conductor A encloses a conductor B, and the two conductors are both charged, there will be no change in the field outside if B is connected to A by a wire. [St John's Coll. 1913]

4. The total flux of intensity through a closed surface of one sheet in an electric field is zero, also the surface encloses all points of the field at which there is any distribution of electricity and is not a surface of zero potential. Prove that it encloses points of positive and points of negative and points of zero potential. [M. T. 1909]

5. Prove that, if there are two positively charged conductors in the field, neither enclosed within the other, there is at least one point in the field where the intensity is zero. [I. 1912]

6. Two conductors carrying respectively a positive charge e_1 and a negative charge $-e_2$ are introduced into the interior of a larger conductor maintained at potential V. Shew that if $e_1 > e_2$ the potential of the first conductor is greater than V. [M. T. 1909]

7. A field of force is due to three quantities of electricity e_1, e_2, e_3 at A, B, C. Shew that the direction of the line of force at a point P is given by joining P to the centre of gravity of masses

$$\frac{e_1}{AP^3}, \quad \frac{e_2}{BP^3}, \quad \frac{e_3}{CP^3}$$

at A, B, C respectively, and find the magnitude of the force at P. [I. 1895]

8. Two particles each of mass m and charged with e units of electricity of the same sign are suspended from the same point by strings

each of length a. Prove that the inclination of the strings to the vertical is given by the equation

$$4mga^2 \sin^3 \theta = e^2 \cos \theta. \qquad \text{[I. 1894]}$$

9. It is required to hold four equal point charges e in equilibrium at the corners of a square. Find the point charge which will do this if placed at the centre of the square. [M. T. 1922]

10. Three equal small spheres each of mass m grams and each carrying a charge e electrostatic units are suspended by strings of length l cm. from the same point. If d cm. is the distance between a pair of the spheres when the system is in equilibrium, shew that

$$9e^4 l^2 = d^2 (m^2 g^2 d^4 + 3e^4).$$

Why, for a similar system consisting of charges λe and strings of length λl, do the masses have to be m (and not λm) for the charges to rest at distance λd apart? [M. T. 1930]

11. Assuming only that the potential due to a small quantity e of electricity is e/r, prove that the potential due to a uniform spherical distribution at any internal point is equal to the potential at the centre, and at any external point is the same as if all the electricity were at the centre. [I. 1894]

12. The potential is given at four points near each other and not all in one plane; obtain an approximate construction for the direction of the field in their neighbourhood. [M. T. 1894]

13. If any closed surface be drawn not enclosing a charged body or any part of one, shew that at every point of a certain closed line on the surface it intersects the equipotential surface through the point at right angles. [M. T. 1897]

14. Sketch the form of the equipotential surfaces and lines of force of two equal and opposite point charges. [M. T. 1927]

15. Positive and negative unit charges are situated at two points A, B. Find at what distance from AB the plane normal to and bisecting AB is cut by the lines of force which issue from A in a direction parallel to this plane. [M. T. 1919]

16. Draw a diagram of the lines of force and the equipotential surfaces in a field in which the only charges are $2e$ at a point A, and $-e$ at a point B. [I. 1905]

17. Sketch the lines of force and level surfaces of two positive point charges of 1 and 2 units respectively. [I. 1914]

18. Sketch roughly the field of force due to two point-charges $4e$ and $-e$ at A and B respectively. Sketch also the equipotential surfaces on a separate diagram.

What is the locus of the points at which the lines of force are parallel to AB? [I. 1914]

19. Two point charges e and $-e'$ $(e>e')$ are situated at A and B respectively. Prove that the extreme lines of force which pass from A to B make, on leaving A, an angle θ with AB, where $\cos\theta=(e-2e')/e$; and indicate by a figure the general form of the lines of force.

[M. T. 1907]

20. Draw a rough diagram to show the distribution of lines of force and equipotential surfaces due to two point charges, one of them being $4e$ and the other $-3e$. [I. 1909]

21. Charges $+1$, -4, $+1$ are placed at collinear points A, B and C, where $AB=BC$. Sketch the lines of force; and shew that any line of force leaving C will reach B at an inclination to BC less than $\frac{1}{3}\pi$.

[I. 1923]

22. Point-charges e, $-e'$ and $-e'$ are placed at O, A and B respectively, which are in a straight line, and $OA=OB$. If e is $>2e'$, shew that the greatest angle a line of force leaving O and entering A can make with OA is α, where

$$e\sin^2\frac{\alpha}{2}=e'.$$

Draw a diagram, giving the lines of force approximately, when $4e'=e$. [I. 1900]

23. Charges of electricity 1, -2, 8 units respectively are placed in a straight line at points A, B, C at unit distances apart. Prove that there is a point of equilibrium between A and B and make a rough drawing of the lines of force. [I. 1906]

24. Indicate by a sketch the forms of the equipotential surfaces and lines of force for three point charges $+e$ at $(a, 0)$, $+e$ at $(-a, 0)$ and $-2e$ at $(0, 0)$. [M. T. 1931]

25. Give careful sketches indicating the most significant features of the lines of force and equipotential lines in the following cases:

(i) Two equal point charges of the same sign.

(ii) Two equal and opposite point charges.

(iii) Two charges $4e$ and $-e$.

(iv) A point charge placed in a uniform field of force. [M. T. 1933]

26. Charges 1, 1, -2 are placed at three points A, B, C forming an equilateral triangle. Shew by a diagram the arrangement of the lines of force within the triangle and just outside it in its plane. [I. 1906]

27. If a total charge e were uniformly distributed throughout the volume of a sphere of radius a, what would be the electric intensity at a distance r from the centre of the sphere?

28. One side of a circular glass plate of radius a is uniformly charged with electricity, the charge per unit area being σ. Prove that the electric intensity at a point on the axis of the plate at a distance c from the centre is

$$2\pi\sigma\{1-c/\sqrt{(c^2+a^2)}\}.$$

29. Three infinite straight cylinders of negligible section, electrified to line densities e, e' and $-(e+e')$, cut a plane perpendicular to them in the points B, C, A respectively, and P is a variable point. Prove that the equations of the lines of force in the plane are

$$e.\widehat{BPA}\pm e'.\widehat{CPA}=\text{const.} \qquad \text{[I. 1892]}$$

30. If three infinitely long thin wires whose charges per unit length are 1, -2, 1 respectively cut a plane at right angles in collinear points A, B, C, where $AB=BC=a$, prove that the equation of the lines of force in this plane is

$$r^2=a^2\cos(2\theta+\alpha)\sec\alpha,$$

B being the origin, BC the axis from which θ is measured, and α a variable parameter. [I. 1904]

31. Three infinite parallel straight lines, with charges e, $-\frac{1}{2}e$, $-\frac{1}{2}e$ per unit length, meet a plane perpendicular to them in A, B, C respectively, where $AB=AC$. Prove that the lines of force reaching B, after having passed indefinitely close to the equilibrium point, make with BC an angle $=BAC$. Draw a rough sketch of the lines of force and the equipotential surfaces. [I. 1909]

32. Three infinitely long parallel straight wires carrying charges $2e$, $-e$, $-e$ per unit length cut a plane at right angles at the corners A, B, C of a triangle right-angled at A. Prove that no line of force which starts from A inside the triangle emerges from the triangle.

33. Three infinite parallel wires cut a plane perpendicular to them in the angular points A, B, C of an equilateral triangle and have charges e, e, $-e'$ per unit length respectively. Prove that the extreme lines of force which pass from A to C make at starting angles

$$\frac{2e-5e'}{6e}\pi \text{ and } \frac{2e+e'}{6e}\pi$$

with AC, provided that $e'\not>2e$. [M. T. 1904]

34. Three long thin straight wires, equally electrified, are placed parallel to each other so that they are cut by a plane perpendicular to them in the angular points of an equilateral triangle of side $\sqrt{3}c$; shew that the polar equation of an equipotential curve drawn on the plane is

$$r^6+c^6-2r^3c^3\cos 3\theta=\text{const.},$$

the pole being at the centre of the triangle and the initial line passing through one of the wires. [I. 1894]

35. The electric intensity in certain regions of a discharge tube under certain conditions is a linear function of x, the distance from one end of the tube. Shew that the volume density of electricity is constant. [M. T. 1927]

36. If the potential at a point outside a conducting sphere of radius a is

$$-F\cos\theta\,(r-a^3/r^2),$$

where r is the distance measured from the centre, θ is the angular distance measured from a fixed diameter and F is constant, find the surface density at any point of the sphere and shew that the total charge on the sphere is zero. [M. T. 1935]

37. Give a physical proof that the families of curves

$$e_1\theta_1+e_2\theta_2+e_3\theta_3+\ldots=\text{const.}$$

and
$$r_1^{e_1}r_2^{e_2}r_3^{e_3}\ldots=\text{const.}$$

cut one another at right angles at all their intersections.

ANSWERS

7. $\left\{\frac{e_1}{AP^3}+\frac{e_2}{BP^3}+\frac{e_3}{CP^3}\right\}$ GP, where G is the centre of gravity of the masses.

9. $-e(1+2\sqrt{2})/4$.

15. $AB\sqrt{3}/2$.

18. The sphere for which $AP=2^{\frac{2}{3}}BP$.

27. er/a^3, $r<a$; and e/r^2, $r>a$.

Chapter III

CONDUCTORS AND CONDENSERS

3·1. Mechanical force on a charged surface. Though electric charges can move freely through the substance and along the surface of a conductor, yet they cannot move normally outwards from the surface, although the field exerts on them a normal electric force. The mechanical effect in an electrostatic field is the same as if the conductor exerted a force on each element of charge on its surface preventing it from moving normally, and, since action and reaction are equal and opposite, the conductor is therefore subject to normal forces equivalent to the forces exerted by the field upon the charges on its surface.

Consider a surface distribution of density σ. The charge on an element of area dS is σdS, and we want to find the resultant force on this charge.

It is clear that the charge σdS will not contribute to the field which produces the force upon it; for if it existed alone it would have no tendency to move, hence the required force on dS is due to the charges in the field other than the charge on dS. Now let $\mathbf{E}_1$, $\mathbf{E}_2$ denote the electric vector at points P, Q close to dS near its centre and on opposite sides of it. The charge on dS may be regarded as an infinite plane sheet in relation to the points close to it near its centre, and therefore producing at these points a normal field of intensity $2\pi\sigma$ (**2·35** (ii)) away from itself on both sides. The rest of the charges in the field will make practically the same contribution $\mathbf{R}$ to the intensity at both P and Q, so that we have vector equations

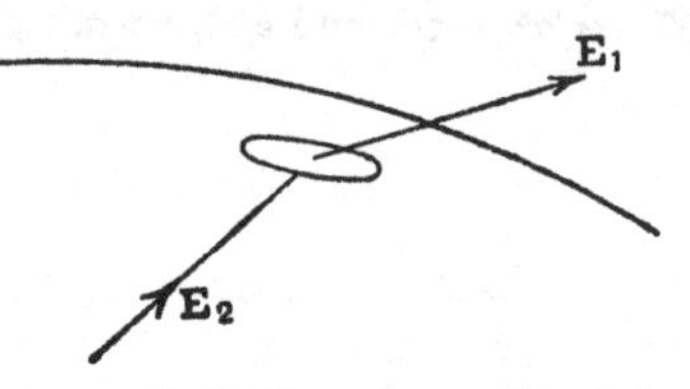

$$\mathbf{E}_1 = \mathbf{R} + 2\pi\sigma\mathbf{n}$$

and

$$\mathbf{E}_2 = \mathbf{R} - 2\pi\sigma\mathbf{n},$$

where $\mathbf{n}$ is a unit normal vector in the positive sense from 2 to 1.

Therefore $\mathbf{R} = \frac{1}{2}(\mathbf{E}_1 + \mathbf{E}_2)$, and the force on the charge σdS is

$$\mathbf{R}\sigma dS, \quad \text{or} \quad \tfrac{1}{2}(\mathbf{E}_1 + \mathbf{E}_2)\,\sigma dS.$$

Hence the mechanical force per unit area on the surface of the body is

$$\tfrac{1}{2}(\mathbf{E}_1 + \mathbf{E}_2)\,\sigma \quad \text{.......................(1).}$$

If the body is a conductor, and the sides 2, 1 are inside and outside, $\mathbf{E}_2 = 0$ and $\mathbf{E}_1$ is along the normal and equal to $4\pi\sigma$ (**2·34**). Hence there is an outward normal force $2\pi\sigma^2$ per unit of area over the surface of a charged conductor.

3·2. A uniformly charged conducting sphere. Reverting to **2·32**, if ϕ denotes the potential of the sphere at a distance r from its centre, then, for $r > a$,

$$-\frac{\partial \phi}{\partial r} = E = \frac{e}{r^2},$$

so that

$$\phi = \frac{e}{r} + \text{const.}$$

But, when the sphere is alone in the field, $\phi \to 0$ as $r \to \infty$, so that the constant is zero and, for $r > a$,

$$\phi = \frac{e}{r} \quad \text{.........................(1).}$$

Similarly, for $r < a$,

$$-\frac{\partial \phi}{\partial r} = E = 0,$$

so that

$$\phi = \text{const.}$$

inside the sphere, the constant value being the value at the centre, viz. e/a, or the value on the surface obtained by putting $r = a$ in (1), since the potential is continuous.

3·21. The **capacity** of a conductor is defined to be the charge necessary to raise it to unit potential when it is alone in the field, or when all other conductors are at zero potential. It follows that the capacity of a sphere is equal to its radius, and that the physical dimensions of *capacity* are one dimension in length.

3·3. Condensers. A *condenser* consists of a pair of conductors insulated from one another and generally so placed that a surface of one is parallel to a surface of the other; e.g.

parallel planes, concentric spheres, coaxial cylinders; but parallelism of the surfaces is not an essential condition.

When the two conductors are so arranged that they have equal and opposite charges, e.g. by earth-connecting one of them, *the capacity of the condenser* is defined to be the positive charge on one of the conductors which will make the difference of the potentials of the conductors equal to unity.

3·31. A parallel plate condenser. Let two parallel conducting plane plates be at a distance t apart and let us consider the field between them, neglecting any external field and the irregularities in the field caused by their edges and assume

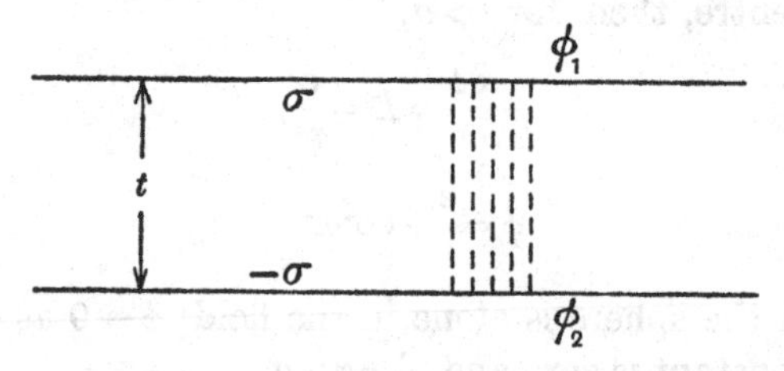

that the tubes of force are everywhere straight. Then if σ is the surface density of the charge on one plate, $-\sigma$ is that on the other, and the electric vector $E = 4\pi\sigma$ (**2·34** (2)) is at right angles to the plates and constant in value.

Then if ϕ_1, ϕ_2 denote the potentials of the plates $\phi_1 - \phi_2 = Et$ (the work done on a unit charge in moving from one plate to the other)

$$= 4\pi\sigma t.$$

Hence the capacity per unit area of the condenser, or the charge when $\phi_1 - \phi_2 = 1$, is $1/4\pi t$, and the capacity for area A is $A/4\pi t$, when the irregularities at the edges of the field are neglected. From **3·1**, there is a force of attraction between the plates which $= 2\pi\sigma^2 A$ or $\dfrac{A}{8\pi t^2}(\phi_1 - \phi_2)^2$.

3·32. The guard-ring. The effects of the edges may very largely be eliminated by the following device: a circular plate A is surrounded by a concentric annular plate B in the same plane, the inner radius of B being only slightly larger than the radius of A, a metallic connection between them is made by a piece of wire so that they will have the same potential, and they are placed parallel to a third conducting plate C. When the condenser is charged the distribution of the unit tubes of force which fall on C will be practically uniform save at its edges, so that so far as the plate A is concerned the 'edge effects' have been eliminated; but we must take into account the tubes which fall upon the part of C opposite to the gap between A and B and this we can do by assuming that half of them start from A and half from B, so that if A' denotes the area of A plus half the area of the gap between A and B the capacity of the condenser formed by A and C is $A'/4\pi t$.

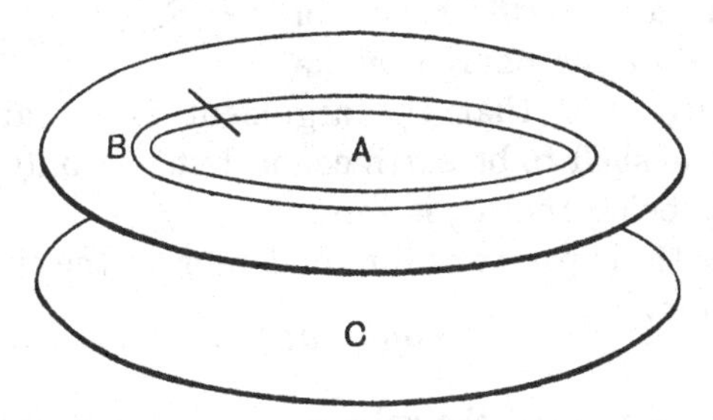

Again the force of attraction on the plate A towards C is $2\pi A\sigma^2$ or $\frac{A}{8\pi t^2}(\phi_1-\phi_2)^2$, where A denotes the area.

If the apparatus is so arranged that the plate A is suspended from one arm of a balance, this force can be measured, and we have thus a practical method of determining a potential difference. Such an apparatus is called the *Guard-Ring or Attracted Disc Electrometer*.

3·4. Spherical condenser. Let a conducting sphere of radius a be surrounded by a concentric conducting spherical shell of internal and external radii b, b'.

Let the sphere of radius a have a uniformly distributed charge e. The tubes of force which start from this charge all end on the sphere of radius b, so that its charge is $-e$. If the shell is insulated and has a total charge e' the charge on its outer surface must be $e+e'$, but by **2·32** the field between the conductors is independent of the charges on the surfaces of the shell and depends only on the charge on the inner sphere, having the value e/r^2 at a distance r from the centre. Hence, if ϕ_a, ϕ_b denote the potentials of the sphere and the shell, the difference $\phi_a - \phi_b$ is the work done on a unit charge as it moves from the sphere to the shell, i.e.

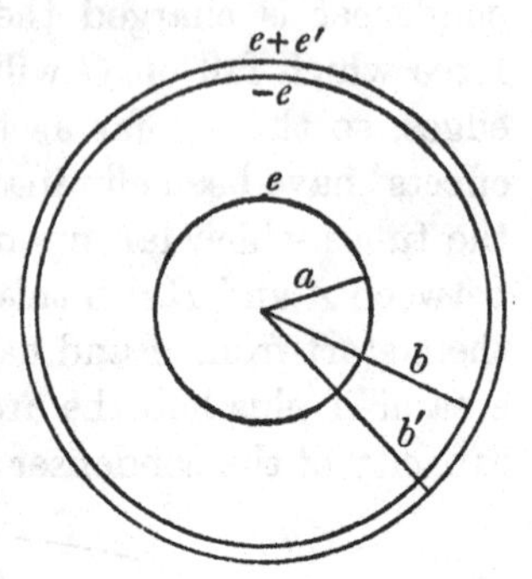

$$\phi_a - \phi_b = \int_a^b \frac{e}{r^2} dr = \frac{e}{a} - \frac{e}{b} \quad (1).$$

We notice that this is independent of the external radius of the shell and of the charge on its outer surface, and that the argument is not affected by supposing the shell to be earth connected. The only difference in that case being that ϕ_b is zero.

Therefore the capacity of the condenser, or the charge when $\phi_a - \phi_b$ is unity, is

$$ab/(b-a) \quad \text{.......................(2).}$$

If on the other hand the sphere of radius a is connected to earth by a fine wire passing through a small hole in the shell and insulated from it, so that the inner sphere is at zero potential, the capacity of the condenser will be the total charge on the shell when its potential is unity.

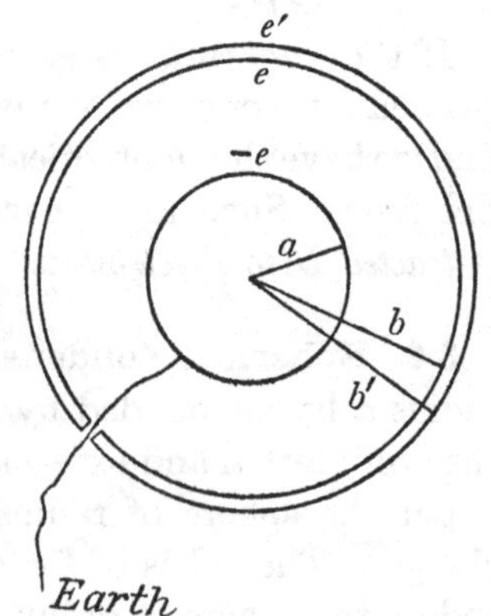

Let e, e' be the charges on the inner and outer surfaces of the shell. Since the tubes of force which start from the charge e all end on the sphere of radius a, its charge is $-e$.

Then, by **2·32**, the field in the space between the conductors is $-e/r^2$ in the direction of r increasing; and the field outside the shell, being due to the sum of all the charges, is e'/r^2.

Hence calculating the potential differences as the work done on a unit charge as it passes firstly from the sphere to the shell and secondly from the shell to an infinite distance, we get

$$0-\phi_b=\int_a^b -\frac{e}{r^2}dr=-\frac{e}{a}+\frac{e}{b}$$

and $$\phi_b-0=\int_{b'}^{\infty}\frac{e'}{r^2}dr=\frac{e'}{b'}.$$

Therefore $$e+e'=\phi_b\left(\frac{ab}{b-a}+b'\right);$$

and the charge, when the potential difference is unity, is

$$\frac{ab}{b-a}+b'.$$

3·41. We have seen in **3·21** that the capacity of a sphere alone in the field is equal to its radius, and in **3·4** that if a sphere of radius a is surrounded by a shell of internal radius b the capacity becomes $ab/(b-a)$ and this can be increased by diminishing the difference between b and a. To illustrate the utility of this increase of capacity, suppose that the wire in the second diagram in **3·4** is connected with the terminal of a voltaic cell whose potential ϕ is too low to cause a deflection of the leaves of an electroscope, and that the shell is connected to earth. The charge e on the sphere is then given by

$$\phi=\frac{e}{a}-\frac{e}{b}$$

or $$e=\frac{ab}{b-a}\phi \quad \text{..............................(1).}$$

Now let the wire be detached and removed, leaving this charge on the sphere; and let the shell, which might consist of two separable hemispheres, be also removed.

The potential of the sphere now alone in the field is e/a or $b\phi/(b-a)$; so that we have now got a conductor whose potential is the original potential ϕ multiplied by the factor $b/(b-a)$.

3·42. Concentric spheres. Consider a number of concentric conducting spheres. Let the radius of the innermost be a and let the inner and outer radii of the spheres next in order be b, b'; c, c'; etc.

If the charges on the spheres are given we can find the potentials, and conversely. Let the charges be $e, e', e'', \ldots$; then as in **3·4** there must be charges $-e$ and $e+e'$ on the spheres of radii b, b' respectively, and similarly $-e-e'$ and $e+e'+e''$ on the spheres of radii c and c' and so on if there are more spheres.

Again the spheres divide space into regions numbered 1, 2, 3, ... in the diagram, in which the potentials may be denoted by $\phi_1, \phi_2, \phi_3, \ldots$.

Then if r denotes distance from the centre, the field in region 2 is e/r^2, so that as in 3·4

$$\phi_1 - \phi_3 = \int_a^b \frac{e}{r^2}\, dr = e\left(\frac{1}{a} - \frac{1}{b}\right) \qquad \text{.................(1).}$$

Similarly, the field in region 4 is $(e+e')/r^2$, so that

$$\phi_3 - \phi_5 = \int_{b'}^{c} \frac{e+e'}{r^2}\, dr = (e+e')\left(\frac{1}{b'} - \frac{1}{c}\right) \qquad \text{...........(2);}$$

and the field in region 6 is $(e+e'+e'')/r^2$, so that

$$\phi_5 = \int_{c'}^{\infty} \frac{e+e'+e''}{r^2}\, dr = \frac{e+e'+e''}{c'} \qquad \text{..............(3).}$$

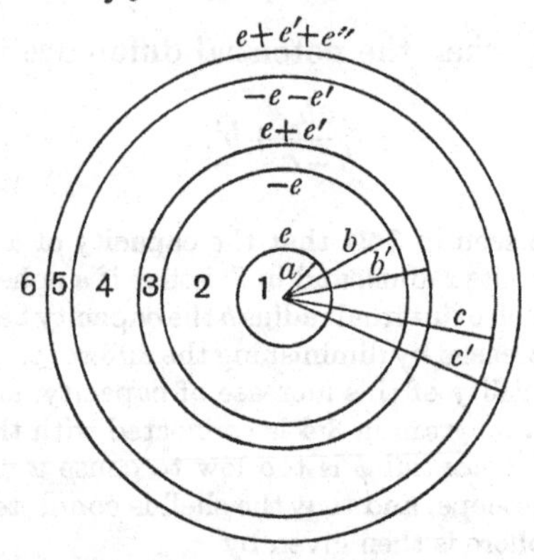

By adding (2) and (3) we find ϕ_3, and by adding (1), (2) and (3) we find ϕ_1, and these are the potentials of the conductors.

Alternatively, by using the fact that a uniformly distributed charge e on a sphere of radius a produces a field of potential e/a inside the sphere and e/r outside the sphere, we may take the sum of the effects of each spherical distribution in each of the regions in turn and write down

$$\phi_1 = e\left(\frac{1}{a} - \frac{1}{b}\right) + (e+e')\left(\frac{1}{b'} - \frac{1}{c}\right) + \frac{e+e'+e''}{c'},$$

$$\phi_2 = e\left(\frac{1}{r} - \frac{1}{b}\right) + (e+e')\left(\frac{1}{b'} - \frac{1}{c}\right) + \frac{e+e'+e''}{c'},$$

$$\phi_3 = (e+e')\left(\frac{1}{b'} - \frac{1}{c}\right) + \frac{e+e'+e''}{c},$$

$$\phi_4 = (e+e')\left(\frac{1}{r} - \frac{1}{c}\right) + \frac{e+e'+e''}{c},$$

$$\phi_5 = \frac{e+e'+e''}{c},$$

$$\phi_6 = \frac{e+e'+e''}{r}.$$

The formulae exhibit the continuity of the potential function, thus the variable function ϕ_2 becomes ϕ_1 or ϕ_3 according as we put $r=a$ or b

ϕ_4 becomes ϕ_3 or ϕ_5 according as we put $r=b'$ or c, and ϕ_6 becomes ϕ_5 or zero according as we put $r=c$ or ∞.

3·43. Condenser formed of coaxial cylinders. Let a be the radius of the inner cylinder and b the internal radius of the outer, then when a charge e per unit length is uniformly distributed on the inner and the cylinders are long enough for us to neglect the effects of their ends there is a field of intensity $2e/r$ between the cylinders (**2·53**), irrespective of whether the outer cylinder be charged or not. Hence for the potentials ϕ_a and ϕ_b we have a relation

$$\phi_a - \phi_b = \int_a^b \frac{2e}{r}\, dr = 2e \log \frac{b}{a},$$

and therefore the capacity of the condenser is $1\Big/\left(2\log\frac{b}{a}\right)$ per unit length.

3·5. Use of Laplace's equation. In **2·41** we saw that in empty space the potential function ϕ is a solution of Laplace's equation $\nabla^2\phi = 0$. There are many simple applications of this fact, especially where symmetry enables us to reduce the number of independent variables in the equation. We will take as illustrations the three types of condensers considered in **3·31**, **3·4** and **3·43**.

(i) *The parallel plate condenser, neglecting edge effects.* Let ϕ_1, ϕ_2 be the potentials of the plates P, Q ($\phi_1 > \phi_2$). Take an axis Ox at right angles to the plates, with the origin O on the plate P. Then if the plates are large enough for us to neglect the edge effects, the potential ϕ at a point between the plates will depend on its x co-ordinate alone, so that ϕ is independent of y and z, and Laplace's equation takes the simple form

$$\frac{\partial^2\phi}{\partial x^2} = 0 \quad \text{...........(1)}.$$

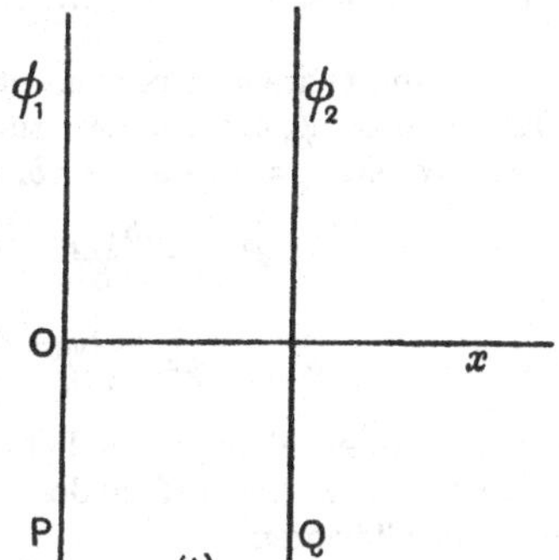

By integration we get

$$\frac{\partial\phi}{\partial x} = F \quad \text{...........(2)}$$

(i)

and
$$\phi = Fx + G \quad \text{..............................(3)},$$
where F and G are arbitrary constants.

But, if t is the distance between the plates, when $x=0$, $\phi=\phi_1$ and, when $x=t$, $\phi=\phi_2$. Substituting these values in turn in (3) we find that

$$G=\phi_1 \quad \text{and} \quad F=-(\phi_1-\phi_2)/t \quad \text{..............(4)},$$

so that

$$\phi=\phi_1-\frac{x}{t}(\phi_1-\phi_2) \quad \text{......................(5)}$$

gives the potential at any point between the plates.

Again the surface density σ on the positively charged plate P is given by Coulomb's Theorem, which is in this case

$$4\pi\sigma=-\left(\frac{\partial\phi}{\partial x}\right)_{x=0},$$

and, from (2), this is

$$4\pi\sigma=-F$$

or, from (4) or (5),

$$\sigma=\frac{\phi_1-\phi_2}{4\pi t} \quad \text{............................(6)};$$

in agreement with the result obtained in **3·31** from the consideration of tubes of force.

(ii) *Spherical condenser.* In the air space between two spheres of radii a, b uniformly charged, by symmetry the potential function ϕ depends upon r alone. Laplace's equation in polar coordinates (**2·41**) then takes the form

$$\frac{\partial}{\partial r}\left(r^2\frac{\partial\phi}{\partial r}\right)=0 \quad \text{...........(1)},$$

which gives, on integration,

$$\frac{\partial\phi}{\partial r}=\frac{F}{r^2} \quad \text{..............(2)}$$

and

$$\phi=-\frac{F}{r}+G \quad \text{...........(3)},$$

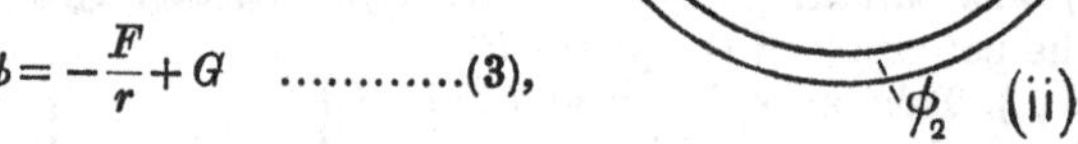

where F and G are arbitrary constants.

Hence, if ϕ_1, ϕ_2 are the potentials of the conductors, we have $\phi=\phi_1$ when $r=a$, and $\phi=\phi_2$ when $r=b$, so that

$$F=-\frac{ab(\phi_1-\phi_2)}{b-a}, \quad G=\frac{b\phi_2-a\phi_1}{b-a} \quad \text{..............(4)}$$

and

$$\phi=\frac{ab(\phi_1-\phi_2)}{r(b-a)}+\frac{b\phi_2-a\phi_1}{b-a} \quad \text{.................(5)}$$

gives the potential anywhere between the conductors.

Again, if σ be the surface density on the sphere $r=a$, we have, by Coulomb's Theorem,

$$4\pi\sigma=-\left(\frac{\partial\phi}{\partial r}\right)_{r=a}=-\frac{F}{a^2},$$

or

$$4\pi a^2\sigma=-F=\frac{ab(\phi_1-\phi_2)}{b-a}.$$

But $4\pi a^2\sigma$ is the total charge e on the inner sphere, so that

$$e=\frac{ab(\phi_1-\phi_2)}{b-a},$$

and the capacity
$$\frac{e}{\phi_1-\phi_2}=\frac{ab}{b-a},$$
as in **3·4**.

The reader will find it an instructive exercise to apply this method to a succession of concentric spheres as in **3·42**, obtaining the potentials in the successive regions as solutions of Laplace's equation.

(iii) *Coaxial cylinders.* Neglecting the effects of the ends of the cylinders, the problem is a two-dimensional one and there being symmetry about the axis the potential ϕ is a function of r alone, where we must now use Laplace's equation in cylindrical co-ordinates (**2·41**), i.e.

$$\frac{\partial}{\partial r}\left(r\frac{\partial\phi}{\partial r}\right)=0 \quad\text{.........................(1)}$$

giving
$$\frac{\partial\phi}{\partial r}=\frac{F}{r} \quad\text{..............................(2)}$$

and
$$\phi=F\log r+G \quad\text{..........................(3).}$$

Then, with the notation of **3·43**, $\phi=\phi_a$ when $r=a$ and $\phi=\phi_b$ when $r=b$, so that

$$F=-(\phi_a-\phi_b)\Big/\log\frac{b}{a} \quad\text{and}\quad G=(\phi_a\log b-\phi_b\log a)\Big/\log\frac{b}{a}\text{ ...(4),}$$

and
$$\phi=\left(\phi_a\log\frac{b}{r}-\phi_b\log\frac{a}{r}\right)\Big/\log\frac{b}{a} \quad\text{..............(5)}$$

gives the potential anywhere between the cylinders.

Then Coulomb's Theorem gives, for the surface density σ on the inner cylinder,

$$4\pi\sigma=-\left(\frac{\partial\phi}{\partial r}\right)_{r=a}=-\frac{F}{a};$$

so that the charge per unit length of cylinder is

$$e=2\pi a\sigma=-\tfrac{1}{2}F=(\phi_a-\phi_b)\Big/2\log\frac{b}{a}\text{ ....(6).}$$

This makes the capacity $\dfrac{e}{\phi_a-\phi_b}=\dfrac{1}{2\log(b/a)}$, as in **3·43**.

3·6. Sets of condensers. *A set of n condensers is said to be arranged in* **series** when the plate of lower potential of the first is connected with the plate of higher potential of the second, the plate of lower potential of the second to the plate of higher potential of the third and so on as in the figure.

Let $C_1, C_2, \ldots C_n$ be the capacities of the condensers, and let a charge e be given to the outer plate of the first. There will then be a charge $-e$ induced on the opposing plate of the first condenser and therefore a charge e on the plate of the second connected with it, and so on, so that the charges on all the plates are e and $-e$ alternately. Then since every pair of plates connected together have the same potential, we may denote the successive potentials by $\phi_1, \phi_2, \ldots \phi_{n+1}$ as in the figure; and then we have

$$\phi_1 - \phi_2 = \frac{e}{C_1}, \quad \phi_2 - \phi_3 = \frac{e}{C_2}, \quad \ldots \quad \phi_n - \phi_{n+1} = \frac{e}{C_n}$$

so that, by addition,

$$\phi_1 - \phi_{n+1} = e\left(\frac{1}{C_1} + \frac{1}{C_2} + \ldots + \frac{1}{C_n}\right).$$

But if C denotes the capacity of the compound condenser $\phi_1 - \phi_{n+1} = \frac{e}{C}$, therefore

$$\frac{1}{C} = \frac{1}{C_1} + \frac{1}{C_2} + \ldots + \frac{1}{C_n} \quad \ldots\ldots\ldots\ldots\ldots (1).$$

It follows that if the n condensers have the same capacity C, the capacity of the set arranged in series is C/n.

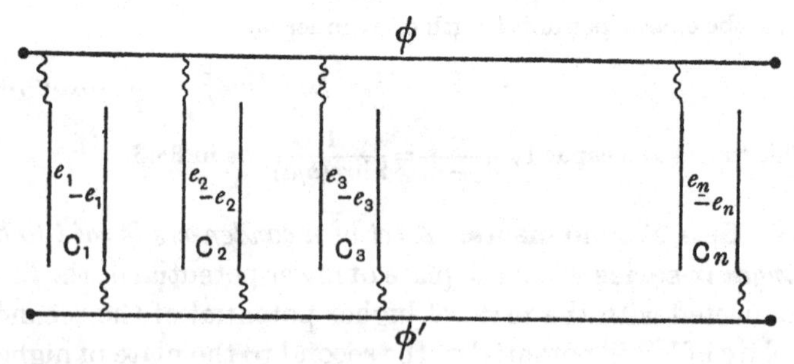

The n condensers may also be arranged in **parallel** by connecting all their plates of higher potential and also connecting all their plates of lower potential as in the figure.

Then if ϕ, ϕ' denote the higher and lower potentials, the charges on the successive condensers are given by

$$e_1 = C_1(\phi - \phi'),\ e_2 = C_2(\phi - \phi'),\ \dots\ e_n = C_n(\phi - \phi')$$

so that

$$e_1 + e_2 + \dots + e_n = (C_1 + C_2 + \dots + C_n)(\phi - \phi').$$

But if C denotes the capacity of the compound condenser

$$e_1 + e_2 + \dots + e_n = C(\phi - \phi'),$$

therefore $$C = C_1 + C_2 + \dots + C_n \quad\text{..................(2).}$$

It follows that if n condensers of the same capacity C are joined in parallel the capacity of the compound condenser is nC.

3·7. Energy of a charged conductor. The energy of a charged conductor is measured by the work done in charging it, and we assume that the work done in bringing a small charge δe to a place of potential ϕ in an electric field is $\phi\,\delta e$.

Consider a conductor alone in the field so that the potential is due only to the charge on the conductor. If C is the capacity, the charge e and potential ϕ are connected by the relation $e = C\phi$.

At any stage in charging the conductor let e' denote the charge and ϕ' the potential, so that $e' = C\phi'$. Then the work done in increasing the charge by a small amount de' is $\phi' de'$ or $\frac{1}{C}e'\,de'$. Hence the whole work done in placing a charge e on the conductor is

$$\frac{1}{C}\int_0^e e'de' = \frac{e^2}{2C},\ \text{or}\ \tfrac{1}{2}e\phi\ \text{or}\ \tfrac{1}{2}C\phi^2 \quad\text{.........(1),}$$

and this represents the energy.

3·71. Energy of a charged condenser. Let a condenser have one of its plates earth-connected. Let C denote the capacity of the condenser, and at any stage in charging the condenser let $\pm e'$ be the charges on the opposing surfaces and ϕ' their potential difference so that $e' = C\phi'$.

The work done in simultaneously increasing the positive charge by de' and the negative charge by $-de'$ is

$$\phi' de' \text{ or } \frac{1}{C} e' de'.$$

Hence the whole work done in placing charges $\pm e$ on the opposing surfaces is

$$\frac{1}{C}\int_0^e e' de' \text{ or } \frac{e^2}{2C} \text{ or } \tfrac{1}{2}e\phi \text{ or } \tfrac{1}{2}C\phi^2 \quad \text{.........(1)},$$

where ϕ is the final potential difference.

3·72. Application to the parallel plate condenser. Referring to the parallel plate condenser of **3·31**, taking A to be the area of either plate and neglecting edge effects, we see that the capacity $C = A/4\pi t$; also from **3·1** each plate is acted on by a normal force $2\pi A\sigma^2$ tending to draw the plates together.

Let us consider the mechanical work that would be required and the change in the energy that would result if the distance between the plates were increased from t to $t+\delta t$; according as the charges or the potentials are kept constant during the change.

Firstly, when the charges are constant, the work necessary to overcome the force of attraction is $2\pi A\sigma^2\,\delta t$; and, from **3·71** the initial energy is $e^2/2C$, where $e = A\sigma$ and $C = A/4\pi t$, so that the initial energy is $2\pi A\sigma^2 t$ and this undergoes an increment $2\pi A\sigma^2\delta t$, or the increase in energy is equal to the mechanical work done.

Secondly, when the potentials are kept constant; since, from **3·31**, $\sigma = (\phi_1 - \phi_2)/4\pi t$, the force of attraction $2\pi A\sigma^2$ is $A(\phi_1-\phi_2)^2/8\pi t^2$, and the mechanical work required is $A(\phi_1-\phi_2)\,\delta t/8\pi t^2$. But in this case the energy, expressed in terms of the potentials, is $\tfrac{1}{2}C(\phi_1-\phi_2)^2$ or $A(\phi_1-\phi_2)^2/8\pi t$, and the increment in this consequent on an increment δt in t is $-A(\phi_1-\phi_2)^2\,\delta t/8\pi t^2$, so that there is a loss of energy numerically equal to the mechanical work done. The explanation of this lies in the fact that to maintain the plates at a constant difference of potential they would have to be connected to the terminals of a battery, i.e. a source of electric energy, and in order, in this case, to get an *increase* in energy equal to the mechanical work done the battery would have to supply an amount of energy equal to twice the mechanical work done.

A general proof of the theorem of which this is a special case will be given in the next chapter.

3·8. Approximate expression for the capacity of a condenser. Let a condenser be formed of two nearly parallel conducting surfaces. Let σ be the surface density on an element of area dS of the positively charged surface, and t the distance between the surfaces at a point on dS. Then if t is small and ϕ_1, ϕ_2 are the potentials of the conductors, we have from **2·34**

$$4\pi\sigma = E = -\text{grad}\,\phi$$
$$= (\phi_1 - \phi_2)/t \text{ approximately.}$$

Hence the total positive charge

$$= \int \sigma\, dS = \frac{\phi_1 - \phi_2}{4\pi} \int \frac{dS}{t} \text{ approximately;}$$

and therefore an approximate expression for the capacity of the condenser is $\dfrac{1}{4\pi}\displaystyle\int \frac{dS}{t}$.

3·9. Electrostatic units for fields in air. The absolute electrostatic unit of charge in the C.G.S. system of units is, in accordance with **2·321**, the charge which will repel an equal charge at a distance of one centimetre with a force of one dyne.

If the fundamental units of mass, space and time are denoted by M, L, T, since Coulomb's Law may be expressed by a formula $ee'/r^2 =$ force, we have for the dimensions* $[e]$ of electric charge

$$[e]^2 = MLT^{-2} \times L^2, \text{ or } [e] = M^{\frac{1}{2}}L^{\frac{3}{2}}T^{-1}.$$

Then since *electric intensity* E is force per unit charge, we have for the dimensions of electric intensity

$$[E] = MLT^{-2} \div M^{\frac{1}{2}}L^{\frac{3}{2}}T^{-1} = M^{\frac{1}{2}}L^{-\frac{1}{2}}T^{-1};$$

and *potential difference*, or, as we shall call it later, *electromotive force*, being the line integral of E, we have for its dimensions

$$[\phi] = [E] \times L = M^{\frac{1}{2}}L^{\frac{1}{2}}T^{-1}.$$

Then *capacity* being the ratio of charge to potential, its dimensions are given by

$$[C] = [e] \div [\phi] = L,$$

in agreement with **3·21**.

The practical units are multiples or sub-multiples of the absolute units. For present purposes it is sufficient to state that

The practical unit of charge

1 Coulomb $= 3 \times 10^9$ absolute electrostatic C.G.S. units.

* See *Dynamics*, Pt. I, **4·8**.

The practical unit of potential difference or electromotive force

1 Volt = 1/300 absolute electrostatic C.G.S. unit.

The practical unit of capacity

1 Farad = 9×10^{11} absolute electrostatic C.G.S. units;

and the Microfarad is one-millionth of a Farad.

We remark that these units are consistent, in that

Coulomb = Volt × Farad,

or charge = potential difference × capacity.

The reader should notice that the arguments by which we deduce the dimensions of these physical quantities in terms of mass, space and time will need qualification when we consider fields in other media than air; also, that there is another system of units—the electromagnetic—which starts from a different basis and differs from the electrostatic system.

3·91. Examples. (i) *Two condensers of capacities* 0·1 *microfarad and* 0·01 *microfarad respectively are charged to voltages* 1 *and* 10 *respectively. Shew that if they are then connected in parallel there is a loss of energy amounting to* 3·7 *ergs.* [M. T. 1933]

Using the result of **3·71** for the energy of a charged condenser, and denoting the capacities of the condensers by C_1, C_2 and their potential differences by ϕ_1, ϕ_2, the initial energy is

$$\tfrac{1}{2}C_1\phi_1^2 + \tfrac{1}{2}C_2\phi_2^2,$$

where $C_1 = 0{\cdot}1$ microfarad $= 9 \times 10^4$ abs. units,

$C_2 = 0{\cdot}01$ microfarad $= 9 \times 10^3$ abs. units,

$\phi_1 = 1$ volt $= 1/300$ abs. unit, and $\phi_2 = 10$ volts $= 1/30$ abs. unit;

so that the energy = 5·5 ergs; and the charges on the upper plates are $C_1\phi_1 = 300$ and $C_2\phi_2 = 300$ abs. units.

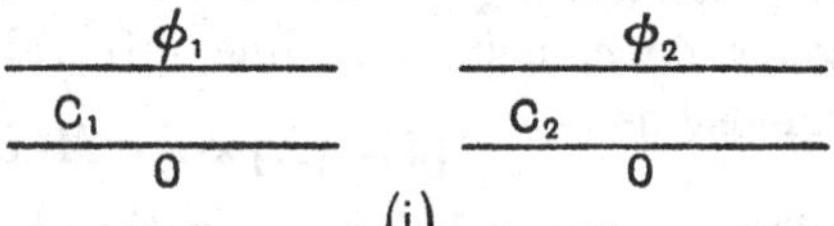

(i)

Now let the condensers be connected in parallel, i.e. the upper plates are connected and the lower plates remain connected through the earth. Let ϕ be the common potential of the upper plates, then their charges become $C_1\phi = 90000\phi$ and $C_2\phi = 9000\phi$.

But their total charge is 600 abs. units, so that $99000\phi = 600$ or $\phi = 1/165$.

And the energy of the compound condenser is half the charge multiplied by the potential difference, i.e.

$$\tfrac{1}{2} \times 600 \times \frac{1}{165} = \frac{20}{11} = 1{\cdot}8 \text{ ergs},$$

so that there is a loss of energy of 3·7 ergs.

The energy might also be obtained from the formula $\frac{1}{2}C\phi^2$, where C the capacity of the compound condenser $= C_1 + C_2$ (3·6 (2)).

(ii) *The two plates of a parallel plate condenser are at a distance d apart and are charged to a difference of potential of* 300 *volts* (1 *electrostatic unit*). *If one of the plates can turn about an axis through its centroid, prove that it will be in equilibrium when parallel to the other plate, but that the equilibrium will be unstable, the couple required to hold it at a small angle θ with the other plate being $I\theta/4\pi d^3$, where I is the moment of inertia of its area about the axis. It is assumed that to the approximation required the lines of force between the plates are nearly straight and the field outside is negligible.* [M. T. 1932]

When the plates are parallel the surface density is uniform and the mechanical force $2\pi\sigma^2$ per unit area is uniform and gives a resultant through the centroid of the plate, so that there is no force having a moment about the axis and the plate is in equilibrium. We shall prove that the equilibrium is unstable by shewing that when the movable plate is turned through a small angle θ it is acted on by a couple tending to increase the displacement.

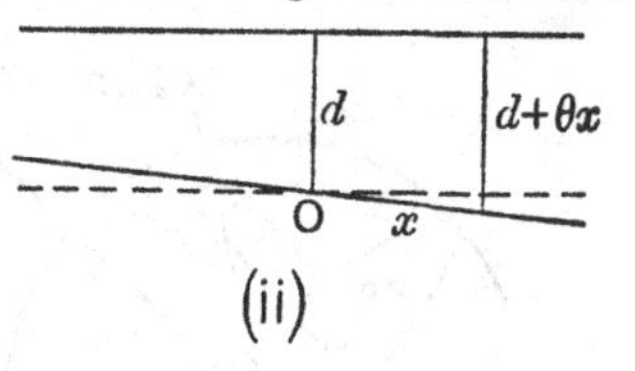

At a distance x from the axis the distance between the plates is $d + \theta x$, so that the surface density σ in this position is given by

$$4\pi\sigma = E = \frac{1}{d+\theta x}$$

and the force per unit area is

$$2\pi\sigma^2 = \frac{1}{8\pi(d+\theta x)^2}.$$

Let y be the breadth of the plate parallel to the axis at a distance x from the axis, and let the limiting values of x be $\pm a$; then the moment about the axis of the forces tending to *decrease* θ

$$= \int_{-a}^{a} 2\pi\sigma^2 yx\,dx = \frac{1}{8\pi}\int_{-a}^{a}\frac{yx\,dx}{(d+\theta x)^2} = \frac{1}{8\pi}\int_{-a}^{a}\frac{yx}{d^2}\left(1-\frac{2\theta x}{d}\right)dx,$$

to the first power of θ.

Since the centroid is on the axis, $\int_{-a}^{a} yx\,dx = 0$; also $\int_{-a}^{a} yx^2\,dx = I$. Therefore there is a couple of magnitude $I\theta/4\pi d^3$ tending to *increase* θ.

(iii) *A conducting sphere of radius a carrying a charge Q is divided into any two parts which remain in electrical contac with one another. Shew that the repulsion between them reduces to a single force through the centre of the sphere and that its component in any direction is equal to $Q^2S/(8\pi a^4)$, where S is the area of the curve formed by the projection on the plane normal to that direction of the curve dividing the two parts.*

What meaning must be attached to S if its boundary has a double point? [M. T. 1934]

The mechanical force per unit area on the surface of the sphere is $2\pi\sigma^2$ outwards (3·1), where in this case $\sigma = Q/4\pi a^2$.

Hence on an element of area $d\omega$ there is a force $2\pi\sigma^2 d\omega$ passing through the centre O of the sphere; and since the force on every element passes through O therefore the resultant force on any portion of the spherical surface passes through O. To find the component in an assigned direction Ox of the resultant force on a specified area of the spherical surface, we observe that the element of area $d\omega$ gives a contribution $2\pi\sigma^2 d\omega \cos\theta$, where θ is the angle between Ox and the normal to the element. But $d\omega \cos\theta$ is the projection dS of the element $d\omega$ on a plane perpendicular to Ox. Hence the element makes a contribution $2\pi\sigma^2 dS$ to the required force. And by summing for all such elements the required force is seen

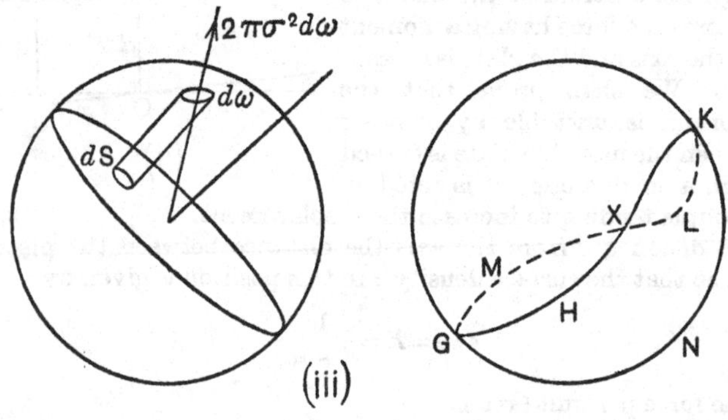

(iii)

to be $2\pi\sigma^2 S$, where S has the meaning assigned in the question, or $Q^2 S/(8\pi a^4)$.

Let the curve dividing the surface of the sphere be the curve $GHKLM$, as in the second figure, where GHK is on the front of the sphere and $KLMG$ is on the back. It is easy to see that if we take the direction of the required component force to be at right angles to the plane of the paper, the projection of the boundary will have a node. It will in fact be a curve roughly resembling the curve $GHKLM$ crossing itself at a point X. But the portion $GNKH$ of the sphere will contribute a force acting upwards through the paper, and the portion $KNGML$ will contribute a force acting downwards through the paper, so that the required component of the resultant force perpendicular to the plane of the paper will correspond to the difference of the projections of these areas, which will be the difference of the two loops into which S is divided through having a double point, and it is in this way that we must interpret S when its boundary has a double point.

(iv) *Two conducting concentric spherical surfaces, radii a, b $(a > b)$, insulated from each other and from earth, are given charges Q_1, Q_2. After the spheres have been charged, the inner sphere is connected to a* *distant uncharged gold leaf electroscope whose capacity is equal to that of a sphere of radius* c. *Find the charge* x *which the electroscope receives.*

The two spheres are now connected to each other without affecting the divergence of the leaves of the electroscope, and the final charge on the electroscope is thus either x *or* $-x$. *Shew that*

$$\frac{Q_1}{Q_2}=\frac{a}{c} \text{ or } \frac{Q_1}{Q_2}=-\frac{a(a+b+2c)}{bc+ca+2ab}. \qquad \text{[M. T. 1924]}$$

After the inner sphere has been connected to the electroscope let ϕ_2 be their common potential, and ϕ_1 that of the outer sphere. Since the inner sphere has given up a part x of its charge to the electroscope, it retains Q_2-x, and the charges on the outer sphere are therefore $-Q_2+x$ on its inner and Q_1+Q_2-x on its outer surface.

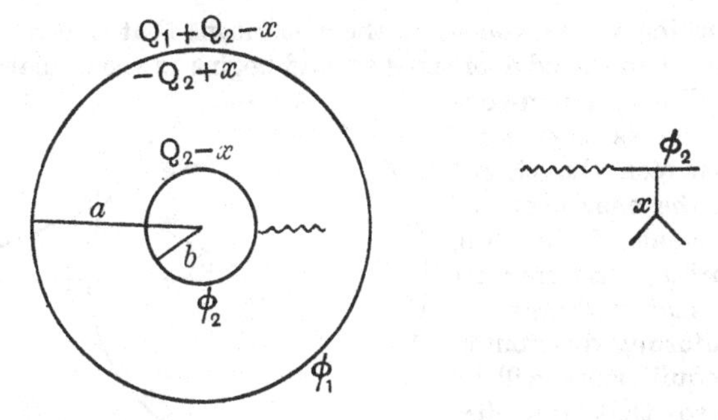

If we assume that the charge on the electroscope is too far distant to affect directly the potential at the centre of the spheres, we have

$$\phi_2=\frac{Q_2-x}{b}+\frac{Q_1}{a} \quad \text{.........................(1).}$$

But, since c is the capacity of the electroscope,

$$x=c\phi_2 \quad \text{...............................(2),}$$

so that by eliminating ϕ_2 from (1) and (2) we get

$$x=\frac{c(Q_1b+Q_2a)}{a(b+c)} \quad \text{......................(3).}$$

In the second case the two spheres are connected and the divergence of the leaves of the electroscope remains unaltered, so that the charge on the electroscope must be either x or $-x$.

The electricity on the inner sphere now passes to the outer sphere and its total charge becomes $Q_1+Q_2\mp x$ according as the charge on the electroscope is $\pm x$. By equating the potentials of the spheres and the electroscope, for they are all now in contact, we get

$$\frac{Q_1+Q_2\mp x}{a}=\frac{\pm x}{c}, \text{ or } \pm x=\frac{c(Q_1+Q_2)}{a+c} \quad \text{...........(4).}$$

Then by equating the values of x found in (3) and (4), we get either

$$\frac{Q_1}{Q_2}=\frac{a}{c} \text{ or } \frac{Q_1}{Q_2}=-\frac{a(a+b+2c)}{bc+ca+2ab}.$$

(v) *A circular disc of non-conducting material of radius a is free to rotate in its plane about its centre C. At a point in its plane at distance b (b > a) from C one terminal of an electrical machine is placed and this is charged up until sparks pass across to the edge of the disc. Explain why the disc commences to rotate.*

After a time the rate of rotation of the disc steadies down to a constant value, the moments of the electrical and frictional forces balancing. If the terminal at A be regarded as having a point charge E and if the charge on the disc be limited to its rim and at P is $E'e^{-\mu\theta}$ per unit length of rim, where $ACP = \theta$, shew that the moment of the electric forces round the axis of the disc is

$$\int_0^{2\pi} \frac{a^2 b E E' e^{-\mu\theta} \sin\theta \, d\theta}{(a^2+b^2-2ab\cos\theta)^{\frac{3}{2}}}.$$

[M. T. 1931]

The sparking across from A to the disc means that electric charge passes from A to the edge of the disc, and begins to accumulate there since the disc is a non-conductor. There is then a repulsion between the charge at A and the charge at the opposing point of the disc, and as sparking is an irregular process this latter charge will not be uniformly distributed and the equilibrium will be unstable so that the disc begins to rotate.

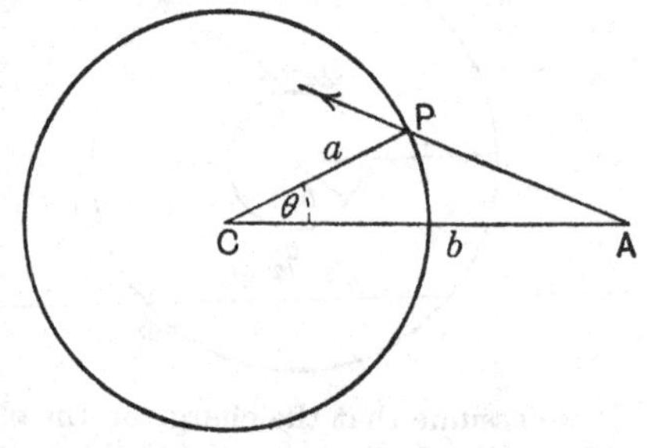

Taking the charge on an element $a\,d\theta$ of the rim at P to be $aE'e^{-\mu\theta}d\theta$, the repulsion between this charge and the charge E at A is

$$aEE'e^{-\mu\theta}d\theta/AP^2,$$

and its moment about C is got by multiplying by $a\sin CPA$ or $ab\sin\theta/AP$. But $AP^2 = a^2+b^2-2ab\cos\theta$, therefore the total moment is

$$\int_0^{2\pi} \frac{a^2 b E E' e^{-\mu\theta} \sin\theta \, d\theta}{(a^2+b^2-2ab\cos\theta)^{\frac{3}{2}}}.$$

EXAMPLES

1. Calculate the capacity of a condenser formed by two circular plates of tinfoil mounted on glass, taking the diameter of each plate to be 40 cm. and the thickness of the glass to be equivalent to an air thickness of 1·5 mm. Express the result in terms of a microfarad. [M. T. 1914]

2. One plate of a parallel plate condenser is a circle of radius 10 cm. and it is surrounded by a guard ring. The distance between the plates is 5 mm. Prove that when the force of attraction between the plates is 5 grams weight the difference of potential between the plates is approximately 3000 volts. [I. 1910]

3. The trap-door of an electrometer is a circle of 6 cm. diameter, and is at a distance of 2 mm. from the lower plate; it is found that a weight $\frac{1}{2}$ gram will balance the attraction between the plates. Calculate the potential difference in volts. [M. T. 1913]

4. The trap-door of a guard ring electrometer is circular, 6 cm. in diameter, and the difference of potential between the plates is 4 electrostatic C.G.S. units; if the force of attraction is 50 dynes, what is the distance between the plates? [St John's Coll. 1907]

5. What difference of potential (in electrostatic units) between the plates of a guard ring electrometer is required to produce an attraction on the trap-door of 5 dynes per square centimetre, when the distance between the plates is 0·3 centimetre? What is the potential difference in volts? [I. 1910]

6. Three parallel plates A, B, C, each 10 cm. square, are maintained at potentials 50, 100 and 10 volts respectively. B is between A and C, A and B being separated by 1 mm. and B and C by 3 mm. Find the magnitude and direction of the force required to prevent B moving. [M. T. 1925]

7. Shew that if the distance between the plates of a parallel plate condenser be halved the electric energy is halved or doubled according as the charges or the potentials are kept constant during the change. Explain this on physical grounds. [I. 1915]

8. Neglecting edge effects, find the capacity of a condenser in air consisting of two sets of parallel square plates of side a each at distance h from its neighbours, and connected together alternately as in the diagram.

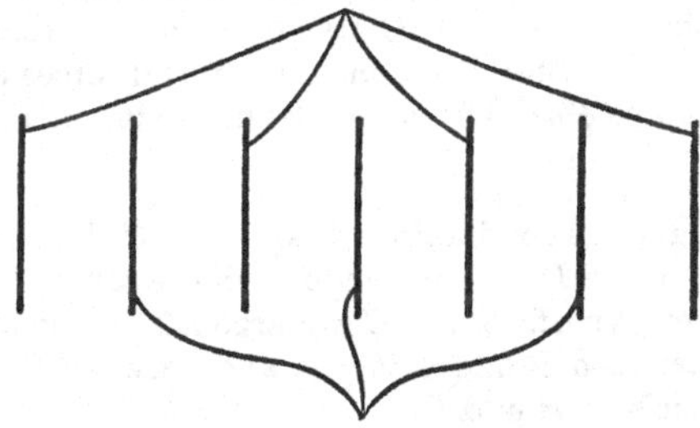

If the outer set are fixed in position and the inner set are free to move all together to the right or to the left, determine, from a consideration of the mechanical strains on the plates, whether the equilibrium of the central position is stable or unstable. [M. T. 1930]

9. A conductor is charged from an electrophorus by repeated contacts with a plate which after each contact is re-charged with a quantity E of electricity from the electrophorus. Prove that, if e is the charge of the conductor after the first operation, the ultimate charge is $Ee/(E-e)$. [M. T. 1893]

10. Two condensers AB, CD have capacities C_1, C_2 respectively. They are initially uncharged, and A is permanently connected to earth (potential zero). The two condensers are connected in series (Fig. a) and D is charged to a potential ϕ. CD is then disconnected, and joined in parallel with AB (Fig. b). Shew that there is a loss of energy equal to

$$\tfrac{1}{2}\phi^2 \frac{(C_1 - C_2)^2 C_1 C_2}{(C_1 + C_2)^3}.$$

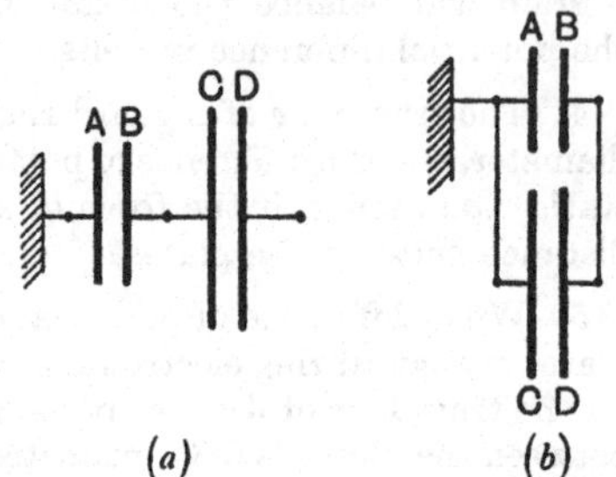

What becomes of this energy? Draw two diagrams similar to Figs. a and b, and mark against each plate (1) its potential, (2) its charge. [M. T. 1925]

11. Two condensers A and B both consist of a pair of parallel circular plates of radius r. Initially the plates of A are separated by a distance a and are charged to a potential difference ϕ, the plates of B are separated by b and are uncharged. The plates of A are now separated to a distance na and connected one to each of the plates of B. Shew that there is a gain of energy to the system of $\frac{r^2\phi^2}{8a} \cdot \frac{(n-1)b - na}{na + b}$ units.

[M. T. 1926]

12. A circular disc of radius a surrounded by a wide coplanar guard ring is placed with its plane inclined at a small angle ϵ to another plane. The centre of the disc is at a perpendicular distance d from the plane. The plane and the disc with its guard ring are maintained at a potential difference ϕ. Shew that the mechanical forces acting on the disc reduce to a force $\phi^2 a^2/(8d^2)$ perpendicular to its plane and a couple $\epsilon\phi^2 a^4 (16d^3)$. [M. T. 1920]

13. Four equal large conducting plates A, B, C, D are fixed parallel to one another. A and D are connected with the earth, B has a charge E per unit area of its surface, and C a charge E'. The distance between A and B is a, between B and C it is b, and between C and D it is c. Find the potentials of B and C. [I. 1905]

14. A spherical conductor of radius 10 cm. has initially a charge of 6284 units. Electrical connection is made, for an instant, with a distant insulated non-charged spherical conductor of radius 15 cm. What are the final charges of the two spheres? [M. T. 1917]

15. The radii of two concentric conducting spheres are 5·4 and 5·6 cm., the outer is earthed and the inner raised to a potential of 200 volts. Calculate the mechanical stress per sq. cm. on the inner sphere.

[I. 1914]

16. Two concentric spherical conductors of radii a and b $(a<b)$ carry charges e and e'; find the fields of force in the region between the spheres and in outside space. How would these fields be altered if the spheres were connected for an instant by means of a wire? What would be the loss of electrostatic energy? [St John's Coll. 1907]

17. On the outer of two concentric spherical conductors of radii a, b $(a<b)$ is a charge Q; what charge must be given to the inner to bring its potential to the value U? Obtain expressions for the potential function in all space. [St John's Coll. 1909]

18. Two insulated conducting spheres, of radii 10 and 20 cm. respectively, carry equal charges of 6 electrostatic units. They are so far apart that their influence on one another may be neglected. Shew that if they are joined by a wire the energy lost is equal to the work required to raise a weight of one milligram through a height of about 3·06 mm. [M. T. 1932]

19. A condenser is formed of a conducting sphere of radius r, and a conducting concentric spherical shell of internal radius R, insulated from it, the dielectric being air. Shew that, if there be placed between these two spherical conductors n conducting concentric spherical shells, the capacity becomes

$$1\Big/\left\{\frac{1}{r}-\Sigma\left(\frac{1}{a}\right)-\frac{1}{R}+\Sigma\left(\frac{1}{b}\right)\right\},$$

where $a_1, a_2, \ldots$ are the internal radii and $b_1, b_2, \ldots$ the external radii of the shells. [I. 1891]

20. A circular gold leaf of radius b is laid on the surface of a charged conducting sphere of radius a, a being large compared to b. Prove that the loss of electrical energy in removing the leaf from the conductor—assuming that it carries away its whole charge—is approximately $\frac{1}{4}b^2E^2/a^3$, where E is the charge of the conductor and the capacity of the leaf is comparable to b. [M. T. 1895]

21. Three concentric thin spherical shells are of radii a, b, c $(a<b<c)$; the first and third are connected by a fine wire through a small hole in the second, and the second is connected to earth through a small hole in the third. Shew that the capacity of the condenser so formed is

$$\frac{ab}{b-a}+\frac{c^2}{c-b}. \qquad \text{[M. T. 1909]}$$

22. Three insulated concentric spherical conductors, whose radii (in ascending order of magnitude) are a, b, c, have charges e_1, e_2, e_3 respectively. Find their potentials, and shew that, if the innermost sphere be connected to earth, the potential of the outermost is diminished by $\frac{a}{c}\left(\frac{e_1}{a}+\frac{e_2}{b}+\frac{e_3}{c}\right)$. [M. T. 1901]

23. A condenser is formed by a sphere B of radius b inside a concentric sphere A of radius a. A second condenser is formed by a sphere D of radius d inside a concentric sphere C of radius c. Connection can be made with B, D through small holes in A, C. Initially A, C are earthed and B, D are at potential Ψ. The sphere C is now insulated and afterwards B is joined to C by a fine wire and D is earthed. The spheres A and C are so far apart that the inductive effect of either on the other may be neglected. Shew that, if Φ is the final potential of B and C,

$$\Phi = \Psi \frac{ab(c-d) - cd(a-b)}{ab(c-d) + c^2(a-b)}.$$ [M. T. 1922]

24. If the middle conductor of a condenser, formed by a long cylinder (radius a) inside a coaxial shell (internal radius $2a$ and external $3a$) with another coaxial cylinder (internal radius d) outside, be charged and the other two put to earth, determine the minimum value of d in order that the parts of the middle shell may not separate, when it is cut into two parts by a plane through its axis. [I. 1912]

25. A condenser is formed of three concentric cylinders of which the inner and outer are connected together. Obtain a formula for the capacity, neglecting end effects, and shew that, if the middle plate is 10 cm. long and the radii of the cylinders are 3·9, 4·0, 4·1 cm., the capacity is equal approximately to that of a sphere of 4 metres radius. [M. T. 1920]

26. On a certain day the vertical electric force in the atmosphere at the earth's surface was 100 volts per metre and at a height of 1·5 kilometres it was 25 volts per metre. Prove that the total charge in the atmosphere per square kilometre of the earth's surface up to this height was $1{\cdot}99 \times 10^6$ electrostatic units, and that the effect of this charge is to increase the barometric pressure by about 4×10^{-7} dyne per sq. cm., assuming the charge uniformly distributed. [M. T. 1921]

27. Sketch the lines of force of the system of charges e at $(2a, 0, 0)$ and $-\frac{1}{2}e$ at $(\frac{1}{2}a, 0, 0)$. Prove that the sphere $r = a$ is the equipotential $\phi = 0$.

Prove that, if the sphere $r = a$ is a conductor at zero potential, the surface density at $(a, 0, 0)$ is $-3e/4\pi a^2$, and the mechanical stress on the conductor $9e^2/8\pi a^4$ dynes per (cm.)^2. [M. T. 1920]

28. Prove that the electric capacity of a conductor is less than that of any other conductor which can completely surround it. [M. T. 1904]

29. A condenser formed of two concentric spheres of radii a and b is divided into two halves by a diametral plane, the inner and outer surfaces being rigidly connected. Prove that the force required to keep the two halves together is $\frac{1}{8}e^2\left(\frac{1}{b^2} - \frac{1}{a^2}\right)$, where e and $-e$ are the total charges on the spheres. [M. T. 1922]

30. An electrified spherical soap bubble of radius a is surrounded by an insulated uncharged concentric spherical conducting shell whose internal and external radii are b and c; shew that when the bubble bursts the loss of electric energy is independent of c.

If the spherical shell consists of two hemispheres in contact, prove that the increase in the mechanical force between them is $\frac{1}{8}\frac{e^2}{b^2}$, where e is the original charge of the bubble. [I. 1899]

31. A conducting sphere of radius r is electrified to potential V. If the sphere consists of two separate hemispheres, shew that the force between them is $\frac{1}{8}V^2$; and if the whole be surrounded by an uninsulated concentric spherical conductor of internal radius R and the potential of the solid sphere is still V, prove that the force between the hemispheres is $\frac{1}{8}V^2R^2/(R-r)^2$. [I. 1895]

32. An insulated spherical conductor, formed of two hemispherical shells in contact, whose inner and outer radii are b and b', has within it a concentric spherical conductor of radius a and without it another concentric spherical conductor of which the internal radius is c. These two conductors are earth connected and the middle one receives a charge. Shew that the two shells will not separate if $2ac > bc + b'a$. [M. T. 1896]

33. Outside a spherical charged conductor there is a concentric insulated but uncharged conducting spherical shell, which consists of two segments; prove that the two segments will not separate if the distance of the separating plane from the centre is $< ab/(a^2+b^2)^{\frac{1}{2}}$, where a, b are the internal and external radii of the shell. [I. 1897]

34. An infinite circular conducting cylinder has charge q per unit length, and is surrounded by a thin coaxial cylindrical shell which is a conductor connected to earth. If the shell is divided along two generators subtending an angle 2θ at the axis of the cylinders, determine the resultant force between the portions. [Trinity Coll. 1898]

35. Calculate the capacity in fractions of a microfarad of a condenser formed by coating the inside and outside of a cylindrical jar with tinfoil. The radius of the jar is 12 cm., the thickness of the glass is equivalent to an air thickness of $\frac{1}{2}$ mm. and the tinfoil reaches to a height of 30 cm. as well as covering the base of the jar.

[The microfarad is equal to the capacity of a sphere whose radius is 9 kilometres.] [St John's Coll. 1915]

36. Shew that if a conductor is slightly deformed so that the element dS is displaced normally outwards a distance ζ the consequent change δC in the capacity is given by

$$\frac{1}{2}\frac{E^2\delta C}{C^2} = 2\pi\int \sigma^2\zeta dS,$$

E being the total charge. For example if a spherical conductor is slightly deformed the change in capacity is proportional to the change in volume. [St John's Coll. 1911]

ANSWERS

1. 0·00074. **3.** 1250 approx.

4. 6 mm. **5.** $3\cdot 36 = 1008$ volts.

6. 7 dynes towards C. **8.** $3a^2/2\pi h$. Unstable.

13. $\dfrac{8\pi a\{bE+c(E+E')\}}{a+b+c}$, $\dfrac{8\pi c\{bE'+a(E+E')\}}{a+b+c}$.

14. 2513·6 and 3770·4 units. **15.** 0·47 dyne per sq. cm.

16. $\frac{1}{2}e^2\left(\dfrac{1}{a}-\dfrac{1}{b}\right)$. **17.** $a\{U-(Q/b)\}$. $r<a$, $\phi=U$;

$$a<r<b,\ \phi=\frac{aU}{r}-\frac{aQ}{br}+\frac{Q}{b};\quad b<r,\ \phi=\frac{aU}{r}-\frac{aQ}{br}+\frac{Q}{r}.$$

24. $3a\times 2^{\sqrt{(2/3)}}$.

34. $(q^2/\pi b)\sin\theta$ per unit length, where b is the internal radius of the shell.

35. 0·0048.

Chapter IV

SYSTEMS OF CONDUCTORS

4·1. When an electric field is produced by charges on a number of conductors, it is clear that the distribution of the charge on each conductor is affected by inductive effects of the charges on the other conductors. The surface density is given at every point by Coulomb's Law (**2·411** (2))

$$4\pi\sigma = E_n = -\partial\phi/\partial n \quad \text{.................(1)},$$

so that a knowledge of the potential function ϕ would enable us to calculate the distribution of charge everywhere; and the total charge on any conductor may be represented by $-\frac{1}{4\pi}\int \frac{\partial\phi}{\partial n} dS$ integrated over the surface.

Generally either the total charges on the separate conductors are given, or else the constant values which the potential function takes on the conductors are given, and further, we know that the potential function ϕ must satisfy Laplace's equation $\nabla^2\phi = 0$ at all points at which there is no charge. While there is no general method for finding a value of ϕ which satisfies Laplace's equation and takes given values over a given set of surfaces, or provides for given total charges over a given set of conductors, yet it is possible to establish a good many general theorems about the kind of electric field which we are discussing, as will be seen in what follows; but, because the proof requires too much analysis, we shall assume that the potential at a point can always be represented by the sum of each element of charge divided by its distance from the point. (Cf. **2·5**.)

4·2. Principle of superposition. Let A_1, A_2, A_3, ... be a set of conductors and let a distribution of variable surface density σ give the conductors total charges e_1, e_2, e_3, ... and produce a field of potential ϕ which takes constant values ϕ_1, ϕ_2, ϕ_3, ... on the surfaces of the conductors. Let there be an alternative distribution of surface density σ' giving total charges e_1', e_2', e_3', ... and producing a field of potential ϕ' which takes constant values ϕ_1', ϕ_2', ϕ_3', ... on the conductors.

Then $$\phi = \int \frac{\sigma dS}{r} \text{ integrated over the conductors}$$

and $$\phi' = \int \frac{\sigma' dS}{r} \text{ integrated over the conductors.}$$

Hence a distribution in which the surface density is $\sigma + \sigma'$ will produce a field of potential $\int \frac{\sigma + \sigma'}{r} dS = \phi + \phi'$, and this will take the values $\phi_1 + \phi_1'$, $\phi_2 + \phi_2'$, $\phi_3 + \phi_3'$, ... on the conductors, and their total charges are now $e_1 + e_1'$, $e_2 + e_2'$, $e_3 + e_3'$, ...; and these charges are still in equilibrium because the surfaces are at constant potentials; so that we have proved that *if charges* e_1, e_2, e_3, ... *give rise to potentials* ϕ_1, ϕ_2, ϕ_3, ... *and charges* e_1', e_2', e_3', ... *give rise to potentials* ϕ_1', ϕ_2', ϕ_3', ..., *then charges* $e_1 + e_1'$, $e_2 + e_2'$, $e_3 + e_3'$, ... *will give rise to potentials* $\phi_1 + \phi_1'$, $\phi_2 + \phi_2'$, $\phi_3 + \phi_3'$, ... This is the principle of superposition of electric fields.

4·21. Uniqueness theorems. (i) *There is only one way in which given charges can be distributed in equilibrium over a given set of conductors.*

Let A_1, A_2, A_3, ... be the conductors and e_1, e_2, e_3, ... the charges placed on them. If possible let there be two different ways of distributing the charges on the conductors in equilibrium; and let σ, σ' denote the surface densities at the same point in the two distributions. Change the sign of the electrification at every point of the second distribution and then imagine the two to be superposed. This joint distribution would still be in equilibrium and it would be a distribution in which each conductor has a total charge zero. Hence, unless $\sigma' = \sigma$ at every point, each conductor has a charge partly positive and partly negative. Let ϕ_1, ϕ_2, ϕ_3, ... now denote the potentials of A_1, A_2, A_3, ... Then since there is no maximum of ϕ in empty space, one of these potentials, say ϕ_1, must be the highest in the field. Therefore tubes of force start from A_1 and none end on it; but this contradicts the hypothesis that it has both positive and negative electricity on its surface; therefore $\sigma' = \sigma$ at all points on A_1, and no tubes of force start from it or end on it. By similar reasoning we prove that $\sigma' = \sigma$ at all points on the rest of the conductors, or the equilibrium distribution of given charges is unique.

(ii) *There is only one distribution of electricity on a given set of conductors which will cause those conductors to have given potentials.*

If possible let σ, σ' denote surface densities at the same point in two different distributions of electricity which both cause the conductors $A_1, A_2, A_3, \ldots$ to have potentials $\phi_1, \phi_2, \phi_3, \ldots$. Then the potentials ϕ, ϕ' of the two distributions may be represented by

$$\phi = \int \frac{\sigma dS}{r} \quad \text{and} \quad \phi' = \int \frac{\sigma' dS}{r}$$

integrated over the surfaces of the conductors, where

$$\text{on } A_1,\ \phi = \phi' = \phi_1; \text{ on } A_2,\ \phi = \phi' = \phi_2; \text{ and so on.}$$

It follows that a distribution in which the surface density is $\sigma - \sigma'$ would produce a field in which the potential of each conductor would be zero. But the potential has no maximum or minimum in empty space, hence the potential of the charges of density $\sigma - \sigma'$ is zero everywhere. Therefore there can be no electricity in the field, i.e. $\sigma = \sigma'$ everywhere, or the distribution which produces the given potentials is unique.

4·3. Coefficients of potential, capacity and induction. Let an electric field be produced by charges residing on a set of n conductors $A_1, A_2, \ldots A_n$ of prescribed shapes occupying fixed relative positions in space.

In all calculations of potential made hitherto charge has entered linearly into the expression for potential, so we assume that if a charge e_1 is given to the conductor A_1, the other conductors being uncharged, the potentials of all the conductors will be proportional to e_1 and that they may be denoted by

$$p_{11}e_1, p_{12}e_1, p_{13}e_1, \ldots p_{1n}e_1.$$

Similarly when the conductor A_2 has a charge e_2 and all the other conductors are uncharged, the potentials may be denoted by

$$p_{21}e_2, p_{22}e_2, p_{23}e_2, \dots p_{2n}e_2,$$

and so on. Finally when the conductor A_n has a charge e_n and all the other conductors are uncharged, the potentials may be denoted by

$$p_{n1}e_n, p_{n2}e_n, p_{n3}e_n, \dots p_{nn}e_n.$$

Now imagine these n fields to be superposed and we have a field in which the charges on the conductors $A_1, A_2, \dots A_n$ are $e_1, e_2, \dots e_n$ and the potentials of the conductors may be denoted by $\phi_1, \phi_2, \dots \phi_n$, where, by the principle of superposition,

$$\left.\begin{array}{l} \phi_1 = p_{11}e_1 + p_{21}e_2 + \dots + p_{n1}e_n \\ \phi_2 = p_{12}e_1 + p_{22}e_2 + \dots + p_{n2}e_n \\ \dots\dots\dots\dots\dots\dots\dots\dots\dots\dots \\ \phi_n = p_{1n}e_1 + p_{2n}e_2 + \dots + p_{nn}e_n \end{array}\right\} \quad \dots\dots\dots\dots(1).$$

The p's are called the *coefficients of potential* of the system of conductors; they depend only on the shapes and sizes and relative positions of the conductors.

We see from the definition that p_{rs} denotes the potential of the conductor A_s when the conductor A_r has a unit charge and all the other conductors are uncharged. We shall prove later that $p_{rs} = p_{sr}$.

Since the equations (1) express the ϕ's as linear functions of the e's, they can be solved for the e's, so that the e's are expressible as linear functions of the ϕ's in the form

$$\left.\begin{array}{l} e_1 = q_{11}\phi_1 + q_{21}\phi_2 + \dots + q_{n1}\phi_n \\ e_2 = q_{12}\phi_1 + q_{22}\phi_2 + \dots + q_{n2}\phi_n \\ \dots\dots\dots\dots\dots\dots\dots\dots\dots\dots \\ e_n = q_{1n}\phi_1 + q_{2n}\phi_2 + \dots + q_{nn}\phi_n \end{array}\right\} \quad \dots\dots\dots\dots(2),$$

where the q's are functions of the p's.

The coefficients with double suffixes $q_{11}, q_{22}, \dots q_{nn}$ are called *coefficients of capacity*, and the other q's are called *coefficients of induction*. Thus q_{rr} is the charge on A_r when it is at unit potential and all the other conductors are at zero potential (earth connected). This agrees with the definition of the capacity of a conductor given in **3·21**. Also q_{rs} is the charge induced on the conductor A_s when the conductor A_r is raised to unit potential and all the conductors except A_r are earth connected.

4·31. *Coefficients of capacity are positive, those of induction are negative and the sum of the latter belonging to any given conductor is generally numerically less than the coefficient of capacity of that conductor.*

Let A_1 be insulated and raised to potential ϕ_1 and let all the other conductors be earth connected, so that

$$\phi_2 = \phi_3 = \ldots = \phi_n = 0.$$

Then from **4·3** (2)

$$e_1 = q_{11}\phi_1,\ e_2 = q_{12}\phi_1,\ \ldots\ e_n = q_{1n}\phi_1.$$

Now if e_1 is a positive charge, ϕ_1 is a positive potential, for the tubes of force all start from A_1 and proceed to places of lower potential (in this case zero). Therefore q_{11} is positive. But the charges induced on $A_2, A_3, \ldots$ are clearly all negative, being at the other ends of the tubes, therefore $q_{12}, q_{13}, \ldots q_{1n}$ are all negative. Also the sum of the charges $e_2, e_3, \ldots e_n$ cannot numerically exceed e_1, therefore

$$q_{12} + q_{13} + \ldots + q_{1n} \not> q_{11} \text{ numerically.}$$

4·32. *Coefficients of potential are all positive and p_{rs} cannot exceed p_{rr} or p_{ss}.*

Let A_r have a positive unit charge and let all the other conductors be uncharged, then, as in **2·43** (ii), p_{rr} is the highest potential in the field and is positive. If A_s lies inside A_r, then, since it has no charge, its potential is the constant potential inside A_r, i.e. in this case $p_{rs} = p_{rr}$. But when A_s lies outside A_r, then, since A_s has no total charge, as many unit tubes of force leave it as fall upon it, and those falling on it come from places of higher potential and those leaving go to places of lower potential, so that p_{rs} lies between the highest and lowest potential in the field, i.e. between p_{rr} and zero. Similarly p_{rs} cannot exceed p_{ss}.

4·4. The energy of a system of charged conductors. If it were possible to create a permanent electrostatic field in which there were no changes or movement of electricity owing to imperfect insulation or other cause, the maintenance of such a state would not require any expenditure of energy; but, in the creation of the field, work would have to be performed in the separation of the electric charges by some mechanical process. This work is to be regarded as conserved in the field in the form of electrical energy and is available for transformation into heat or mechanical work when the field ceases to be static.

The energy of the field of the n charged conductors of **4·3** may be calculated as the work required to place the charges on the conductors. In accordance with the definition of potential of **2·4** the work which an external agent would have to perform in order to bring a small charge de to a place of potential ϕ is $\phi\, de$. Now suppose the n conductors to be charged simultaneously in such a way that at any instant in the process each conductor has received the same fraction x of its final charge, so that the charges are

$$xe_1,\ xe_2,\ \dots\ xe_n.$$

Then, at this instant, the potentials being linear functions of the charges are

$$x\phi_1,\ x\phi_2,\ \dots\ x\phi_n,$$

where $\phi_1,\ \phi_2,\ \dots\ \phi_n$ denote the final potentials; and the work necessary to increase the charges by further small amounts

$$e_1 dx,\ e_2 dx,\ \dots\ e_n dx$$

is
$$e_1\phi_1 x\,dx + e_2\phi_2 x\,dx + \dots + e_n\phi_n x\,dx$$

or
$$\Sigma\,(e\phi)\, x\,dx.$$

And the whole process of charging is completed by summing for values of x between 0 and 1, so that the total work done or the energy of the field is

$$W = \Sigma\,(e\phi)\int_0^1 x\,dx$$

or
$$W = \tfrac{1}{2}\Sigma\,(e\phi) \qquad\qquad (1).$$

The theorems of **3·7** and **3·71** are particular cases of this result.

4·41. Green's Reciprocal Theorem.* *If the charges and potentials of a set of n fixed conductors are altered from*

$$e_1, \phi_1;\ e_2, \phi_2;\ \dots\ e_n, \phi_n$$

to
$$e_1', \phi_1';\ e_2', \phi_2';\ \dots\ e_n', \phi_n',$$

then $$e_1\phi_1' + e_2\phi_2' + \dots + e_n\phi_n' = e_1'\phi_1 + e_2'\phi_2 + \dots + e_n'\phi_n$$

or
$$\Sigma(e\phi') = \Sigma(e'\phi).$$

Let the alterations take place in such a way that at any instant each conductor has received the same fraction x of its final increment of charge, so that the charges are

$$e_1 + x(e_1' - e_1),\ e_2 + x(e_2' - e_2),\ \dots\ e_n + x(e_n' - e_n);$$

their potentials must then be

$$\phi_1 + x(\phi_1' - \phi_1),\ \phi_2 + x(\phi_2' - \phi_2),\ \dots\ \phi_n + x(\phi_n' - \phi_n),$$

and the work necessary to make further small increments of charge

$$(e_1' - e_1)\,dx,\ (e_2' - e_2)\,dx,\ \dots\ (e_n' - e_n)\,dx$$

is
$$\Sigma\{\phi + x(\phi' - \phi)\}(e' - e)\,dx,$$

and the total work necessary to alter the state of the conductors as prescribed is found by summing for values of x between 0 and 1, i.e.

$$\int_0^1 \Sigma\{\phi + x(\phi' - \phi)(e' - e)\}\,dx$$

or
$$\tfrac{1}{2}\Sigma(\phi + \phi')(e' - e) \quad \dots\dots\dots\dots(1),$$

i.e. *the sum of the increment in charge of each conductor multiplied by the mean of its initial and final potentials.*

But this work must measure the difference between the initial and final energy of the field, so that

$$\tfrac{1}{2}\Sigma(\phi + \phi')(e' - e) = \tfrac{1}{2}\Sigma e'\phi' - \tfrac{1}{2}\Sigma e\phi,$$

which reduces to

$$\Sigma(e\phi') = \Sigma(e'\phi) \quad \dots\dots\dots\dots(2).$$

* George Green (1793–1841), mathematician.

COR. By adding $\Sigma(e\phi') - \Sigma(e'\phi)$ to expression (1), it becomes $\frac{1}{2}\Sigma(e+e')(\phi'-\phi)$, expressing the work necessary to change the state as *the sum of the increment in potential of each conductor multiplied by the mean of its initial and final charges.*

4·42. To prove that $q_{rs} = q_{sr}$ and $p_{rs} = p_{sr}$. (i) To prove the first result, consider first the field in which A_r is insulated and given a charge which makes its potential unity while all the other conductors are earth connected, i.e. $\phi_r = 1$ and all other ϕ's are zero.

Then, from **4·3** (2),

$$e_1 = q_{r1},\ e_2 = q_{r2},\ \ldots\ e_r = q_{rr},\ \ldots\ e_s = q_{rs},\ \ldots\ e_n = q_{rn}.$$

Consider next the field in which A_s is insulated and given a charge which makes its potential unity while all the other conductors are earth connected, i.e. $\phi_s' = 1$ and all other ϕ''s are zero. Then, as above,

$$e_1' = q_{s1},\ e_2' = q_{s2},\ \ldots\ e_r' = q_{sr},\ \ldots\ e_s' = q_{ss},\ \ldots\ e_n' = q_{sn}.$$

But by Green's Reciprocal Theorem

$$\Sigma(e\phi') = \Sigma(e'\phi),$$

therefore $$q_{rs} = q_{sr}.$$

(ii) To prove the second result, consider first the field in which A_r has a unit charge and all the other conductors are insulated and uncharged, i.e. $e_r = 1$ and all other e's are zero.

Then, from **4·3** (1)

$$\phi_1 = p_{r1},\ \phi_2 = p_{r2},\ \ldots\ \phi_r = p_{rr},\ \ldots\ \phi_s = p_{rs},\ \ldots\ \phi_n = p_{rn}.$$

Consider next the field in which A_s has a unit charge and all the other conductors are insulated and uncharged, i.e. $e_s' = 1$ and all other e''s are zero. Then, as above,

$$\phi_1' = p_{s1},\ \phi_2' = p_{s2},\ \ldots\ \phi_r' = p_{sr},\ \ldots\ \phi_s' = p_{ss},\ \ldots\ \phi_n' = p_{sn}.$$

But $$\Sigma(e\phi') = \Sigma(e'\phi),$$

so that $$p_{sr} = p_{rs}.$$

It would be sufficient to prove one of these theorems because then the other would follow from the relations which express the p's in terms of the q's or *vice versa.*

4·43. Further applications of the Reciprocal Theorem. (i) *Shew that the locus of the positions in which a unit charge will induce a given charge on a given uninsulated conductor S is an equipotential surface of S when S is freely electrified and alone in the field.*

Here we have two conductors, viz. S and an infinitesimal conductor at a point P.

In the first state (Fig. (i)) a given charge e_1 is to be induced on S at zero potential by a unit charge at P, so that

$$e_1 \text{ is given, } e_2=1,\ \phi_1=0 \text{ and } \phi_2 \text{ is unknown.}$$

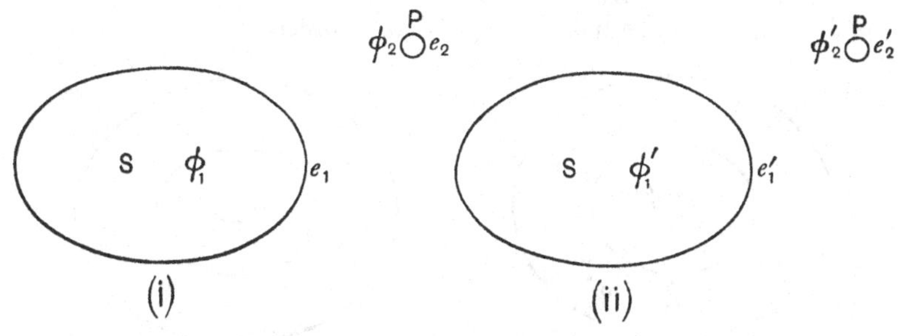

In the second state (Fig. (ii)) S is freely electrified and alone in the field, i.e. there is no charge at P and the conductor at P is regarded as too small to disturb by its presence the electrification on S; in this state therefore e_1' and ϕ_1' are definite constants, e_2' is zero and ϕ_2' is unknown. But by the Reciprocal Theorem

$$e_1\phi_1'+e_2\phi_2'=e_1'\phi_1+e_2'\phi_2,$$

so that, by substituting the particular values above, we get

$$e_1\phi_1'+\phi_2'=0; \quad \text{or} \quad \phi_2'=-e_1\phi_1'.$$

But e_1 is a given charge and ϕ_1' is the potential of S when freely electrified and therefore a constant, so that ϕ_2' is constant.

But ϕ_2' is the potential at P in the second state; hence the required locus of P is an equipotential surface of S when S is freely electrified and alone in the field.

(ii) *Two closed equipotential surfaces ϕ_1 and ϕ_2 are such that ϕ_2 encloses ϕ_1, and ϕ_P is the potential at a point P between them. If now a charge e be placed at P and the equipotential surfaces are replaced by conducting surfaces at zero potential, prove that the charges induced on these surfaces are given by*

$$\frac{e_1}{\phi_P-\phi_2}=\frac{e_2}{\phi_1-\phi_P}=\frac{e}{\phi_2-\phi_1}.$$

The distribution of potential in a field in air is not affected by placing a thin conducting sheet in the position of any equipotential surface, for the lines of force will still continue to cross the surface at right angles. Hence we may suppose the thin conducting surfaces to exist in both the fields described in the problem.

In order to complete the specification of the first field (Fig. (i)) we must suppose it to be produced by a charged conductor, having a charge e_0 and potential ϕ_0 say, inside the given equipotential surfaces, and let this be the only charge in the field. We can suppose also that there is a small conductor at P which carries the charge e in the second field (Fig. (ii)).

Then in the first field we have

charges e_0, 0, 0, 0

and potentials ϕ_0, ϕ_1, ϕ_P, ϕ_2 in order.

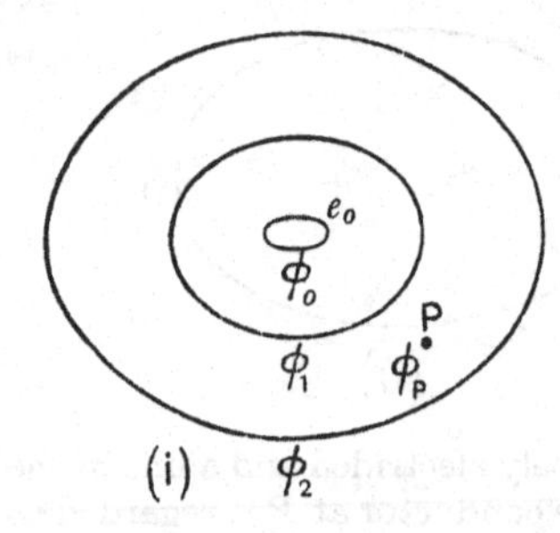

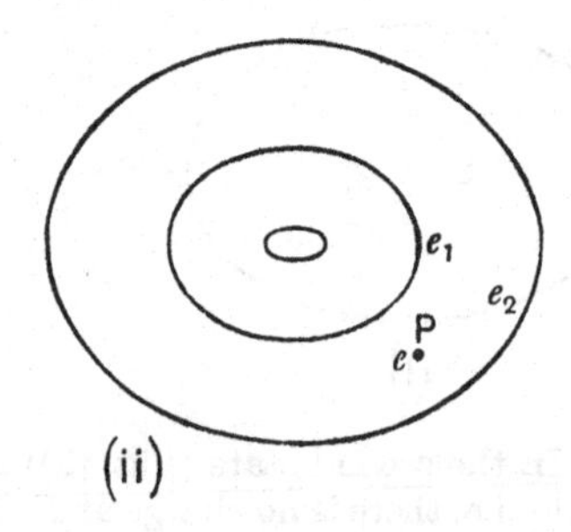

And in the second field, when there is no charge on the innermost conductor, and the other conductors save that at P are at zero potential, we have

charges 0, e_1, e, e_2

and potentials 0, 0, ϕ_P', 0 in order,

where ϕ_P' denotes the potential at P in the second field, and the innermost conductor is at zero potential because it is uncharged and lies inside a conductor at zero potential.

Applying the Reciprocal Theorem we get

$$e_1\phi_1 + e\phi_P + e_2\phi_2 = 0 \quad \text{.........................(1).}$$

Also since in the second field all the tubes of force which start from the charge e end on one or other of the conducting surfaces, therefore

$$e_1 + e + e_2 = 0 \quad \text{.........................(2);}$$

and from (1) and (2) we obtain the required result.

4·5. Energy as a quadratic function of charges or potentials. Taking the expression for the energy from **4·4**, viz.

$$W = \tfrac{1}{2}\Sigma(e\phi),$$

and substituting from **4·3** (1) for the potentials in terms of the charges, we get a quadratic expression for W in terms of the charges, which we may denote by W_e and write

$$W_e = \tfrac{1}{2}p_{11}e_1^2 + p_{12}e_1e_2 + \tfrac{1}{2}p_{22}e_2^2 + p_{13}e_1e_3 + \dots \quad \dots(1).$$

If instead we substitute from **4·3** (2) for the charges in terms of the potentials, we get a quadratic expression for W in terms of the potentials which we may write

$$W_\phi = \tfrac{1}{2}q_{11}\phi_1^2 + q_{12}\phi_1\phi_2 + \tfrac{1}{2}q_{22}\phi_2^2 + q_{13}\phi_1\phi_3 + \dots \quad \dots(2);$$

where, of course, $\qquad W_e = W_\phi = W.$

We observe that, differentiating partially,

$$\frac{\partial W_e}{\partial e_r} = p_{1r}e_1 + p_{2r}e_2 + \dots + p_{nr}e_n = \phi_r \quad \dots\dots\dots(3)$$

and

$$\frac{\partial W_\phi}{\partial \phi_r} = q_{1r}\phi_1 + q_{2r}\phi_2 + \dots + q_{nr}\phi_n = e_r \quad \dots\dots\dots(4).$$

From the foregoing quadratic expressions various theorems about the p's and q's may be proved. Thus taking the uncharged state of the conductors as the state of zero energy, W_e is essentially positive, therefore $p_{11}, p_{22}, p_{33}, \dots$ must all be positive. Then by taking all the e's to be zero except e_1 and e_2, it follows that $p_{11}p_{22} - p_{12}^2$ must be positive, and by similar arguments each of the successive discriminants

$$p_{11}, \quad \begin{vmatrix} p_{11} & p_{21} \\ p_{12} & p_{22} \end{vmatrix}, \quad \begin{vmatrix} p_{11} & p_{21} & p_{31} \\ p_{12} & p_{22} & p_{32} \\ p_{13} & p_{23} & p_{33} \end{vmatrix}, \dots$$

must be positive.

4·6. To deduce the mechanical forces between the conductors from the energy. First suppose that the charges on the conductors are given and remain constant. We may suppose that the relative positions of the conductors are determined by the values of certain co-ordinates of position $x_1, x_2, x_3, \dots$. The coefficients of potential, the p's, then depend on the x's. A change in any co-ordinate x will in general alter some or all of the p's and therefore alter the energy W_e. Let X be the force which tends to produce a displacement δx. The work that would be done in such a displacement is $X\,\delta x$ and this could only be done at the expense of the store of electrical energy and is therefore equal to the corresponding decrease in W_e, i.e.

$$X\delta x = -\delta W_e$$

or
$$X = -\frac{\partial W_e}{\partial x} \qquad \text{(1)};$$

and in this expression the e's are constant and the differentiation is performed on the p's which alone are functions of x.

Now we observe that although $W_\phi = W_e$ it is not open to us to substitute W_ϕ for W_e in (1), because of our hypothesis that the charges remain constant during the displacement.

If we take as an alternative hypothesis that the potentials are kept constant while the conductors are displaced, this introduces another physical consideration, viz. that any alteration of position of the conductors would alter the potentials and in that case the potentials can only be kept constant by connecting each conductor with a battery, i.e. with a source of energy on which it can draw, and we cannot in this case argue that the mechanical work done in a small displacement is the exact equivalent of the loss of energy, because we are assuming that there is a limitless store of energy available for doing work.

To deal with this case we proceed by an independent argument to prove that $\frac{\partial W_e}{\partial x} = -\frac{\partial W_\phi}{\partial x}$, thus:

We have
$$W_e + W_\phi = 2W = \sum_{r=1}^{n} (e_r \phi_r).$$

Consider a small displacement δx without any restriction on charges and potentials, i.e. suppose that charges, potentials and positions of conductors all undergo small changes.

Then
$$\delta W_e + \delta W_\phi = \delta \sum_{r=1}^{n} (e_r \phi_r),$$

or
$$\sum_{r=1}^{n} \frac{\partial W_e}{\partial e_r} \delta e_r + \frac{\partial W_e}{\partial x} \delta x + \sum_{r=1}^{n} \frac{\partial W_\phi}{\partial \phi_r} \delta \phi_r + \frac{\partial W_\phi}{\partial x} \delta x$$
$$= \sum_{r=1}^{n} (e_r \delta \phi_r) + \sum_{r=1}^{n} (\phi_r \delta e_r).$$

But from **4·5** (3) $\frac{\partial W_e}{\partial e_r} = \phi_r$, so that the terms in δe_r cancel out, and from **4·5** (4) $\frac{\partial W_\phi}{\partial \phi_r} = e_r$, so that the terms in $\delta\phi_r$ cancel out, and there remains

$$\frac{\partial W_e}{\partial x} + \frac{\partial W_\phi}{\partial x} = 0 \quad \text{.....................(2).}$$

By combining (1) and (2) we may therefore also express the force X tending to produce a displacement δx by the formula

$$X = \frac{\partial W_\phi}{\partial x} \quad \text{.......................(3).}$$

Hence the force is, by (1), the rate of decrease of the energy when expressed in terms of constant charges, or, by (3), it is the rate of increase of the energy when expressed in terms of constant potentials, the increase being supplied by the batteries which are used to maintain the constant potentials.

In a finite displacement which only changes the co-ordinate x, in which the *potentials are kept constant*, the work done by the mechanical forces is

$$\int X\,dx = \int \frac{\partial W_\phi}{\partial x}\,dx = W_\phi' - W_\phi$$

i.e. $\qquad = \text{the gain in electrical energy.}$

Thus we have mechanical work done and an equal gain in electrical energy, so that when the potentials are kept constant during a displacement, the batteries must supply an amount of energy equal to twice the mechanical work done.

4·61. Example. We shall again use the parallel plate condenser to illustrate the theory of the preceding article.

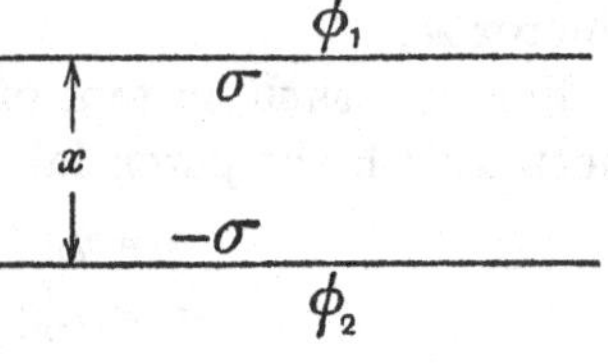

Let there be two parallel plates of area A with charges of surface density σ, $-\sigma$ and potentials ϕ_1, ϕ_2 at a distance x apart, and neglect the effect of the edges and consider only the field between the plates.

Then $\qquad \phi_1 - \phi_2 = 4\pi\sigma x \quad$ **(3·31)**,

and as in **3·71** or **4·4** the energy is

$$W = \tfrac{1}{2}A\sigma(\phi_1 - \phi_2);$$

so that $\qquad W_e = 2\pi A\sigma^2 x$ and $W_\phi = \dfrac{A(\phi_1 - \phi_2)^2}{8\pi x}$.

Then the force tending to increase x is either

$$-\frac{\partial W_e}{\partial x}=-2\pi A\sigma^2 \text{ or } +\frac{\partial W_\phi}{\partial x}=-\frac{A(\phi_1-\phi_2)^2}{8\pi x^2}=-2\pi A\sigma^2.$$

So that either formula gives a force per unit area $2\pi\sigma^2$, as in **3·1**, pulling the plates together.

Now suppose that the plates approach one another so that their distance becomes x'.

(i) If the charges are kept constant, the work done by the attraction is

$$\int_x^{x'} -2\pi A\sigma^2 dx = 2\pi A\sigma^2(x-x') = W_e - W_e'$$
$$= \text{loss of electrical energy.}$$

(ii) But if the potentials are kept constant, the work done by the attraction is

$$\int_x^{x'} -\frac{A(\phi_1-\phi_2)^2}{8\pi x^2}dx = \frac{A(\phi_1-\phi_2)^2}{8\pi}\left(\frac{1}{x'}-\frac{1}{x}\right)$$
$$= W_\phi' - W_\phi$$
$$= \text{gain in electrical energy;}$$

and in this case the battery to whose poles the plates are attached in order to maintain the constant potentials would have to supply both the gain in the electrical energy and an equal amount of energy to represent the mechanical work done by the force of attraction.

4·7. Electric screens. We saw in **2·31** (iii) that it is possible to have two independent electric fields inside and outside a conductor. We shall now shew that a conductor A_1 if enclosed in a second conductor A_2 is thereby screened from the effects of an external conductor A_3.

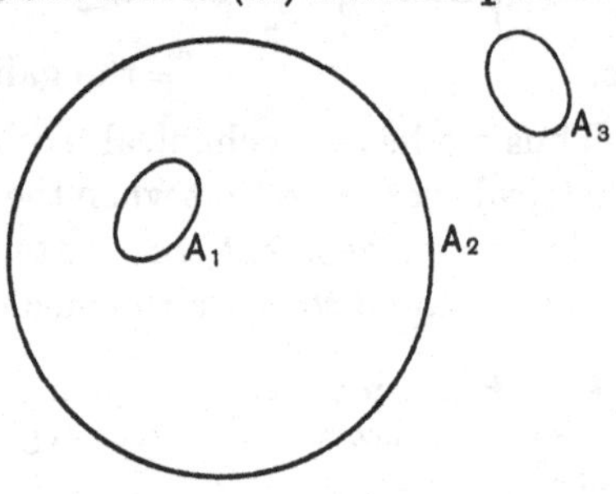

For any such system of conductors the charges are connected with the potentials by the relations

$$e_1 = q_{11}\phi_1 + q_{21}\phi_2 + q_{31}\phi_3 \quad \text{..............(1),}$$
$$e_2 = q_{12}\phi_1 + q_{22}\phi_2 + q_{32}\phi_3 \quad \text{..............(2),}$$
$$e_3 = q_{13}\phi_1 + q_{23}\phi_2 + q_{33}\phi_3 \quad \text{..............(3),}$$

where the q's are independent of the e's and ϕ's. In order to find out what we can about the q's take a special case: let A_2 be at zero potential and A_1 without charge. Then since A_2 contains no charge the potential is constant (i.e. zero) inside it

(2·44 (i)). Hence in equation (1) we have $e_1=0$, $\phi_1=\phi_2=0$, so that the equation reduces to

$$0=q_{31}\phi_3;$$

but ϕ_3 is unrestricted in value, therefore $q_{31}=0$.

Hence A_3 may be raised to any potential without affecting A_1 and *vice versa*; so that the conductor A_2 screens A_1 from the external field.

4·8. The quadrant electrometer. This is an instrument invented by Sir William Thomson (Lord Kelvin*) for measuring differences of potential. It consists of a metal short cylindrical box divided into four quadrants, shown in plan in the figure. Each quadrant is supported on a separate insulating stand but opposite quadrants are connected by wires. Inside the box a flat piece of aluminium C is suspended by a silk fibre so that it can turn in a horizontal position about the axis of the cylinder. From the lower surface of C along its axis hangs a fine metal wire which connects C with the inner surface of a condenser, which is maintained at a constant high potential. The opposing pairs of quadrants A, A' and B, B' are connected to the two bodies the difference of whose potentials is to be found, one of which might be the earth.

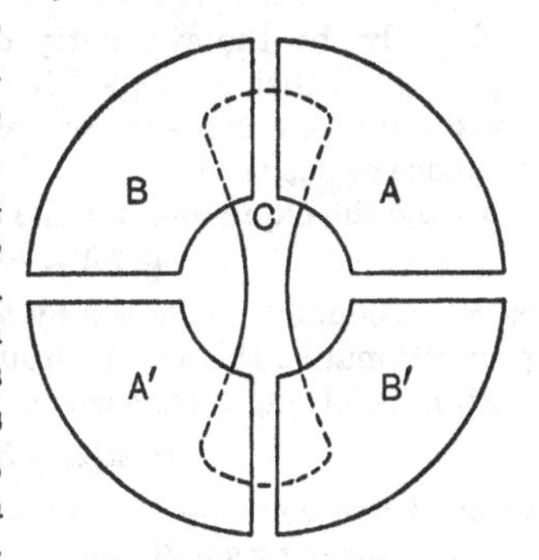

When A, A' and B, B' have the same potential, the needle C will be symmetrically placed with regard to them; but when the potentials of A, A' and B, B' differ, C will rotate and take up a position of equilibrium in which the couple on it produced by the electric field inside the box is balanced by the torsion of the silk thread. We have now to determine the relation between the potentials and the angle turned through by the needle C.

It will be convenient to denote the potentials of the pairs of quadrants A, A' and B, B' and the needle C by A, B and C respectively. Then the expression for the energy in terms of the potentials is

$$W_\phi=\tfrac{1}{2}q_{11}A^2+\tfrac{1}{2}q_{22}B^2+\tfrac{1}{2}q_{33}C^2+q_{23}BC+q_{31}CA+q_{12}AB,$$

where the q's depend in general on the position of the needle. Hence if θ denotes the angular displacement of the needle in the position of equilibrium, the couple tending to increase θ is

$$\frac{\partial W_\phi}{\partial\theta}=\frac{1}{2}\frac{\partial q_{11}}{\partial\theta}A^2+\frac{1}{2}\frac{\partial q_{22}}{\partial\theta}B^2+\frac{1}{2}\frac{\partial q_{33}}{\partial\theta}C^2+\frac{\partial q_{23}}{\partial\theta}BC+\frac{\partial q_{31}}{\partial\theta}CA+\frac{\partial q_{12}}{\partial\theta}AB.$$

But by hypothesis there is no couple when $A=B$, no matter what

* William Thomson, Lord Kelvin (1824–1907).

value C may have; so that, when $A=B$, the coefficients of the different powers of C in the last expression must vanish, i.e.

$$\frac{1}{2}\frac{\partial q_{11}}{\partial\theta}+\frac{1}{2}\frac{\partial q_{22}}{\partial\theta}+\frac{\partial q_{12}}{\partial\theta}=0,\quad \frac{\partial q_{23}}{\partial\theta}+\frac{\partial q_{31}}{\partial\theta}=0 \quad\text{and}\quad \frac{\partial q_{33}}{\partial\theta}=0.$$

Therefore, when A and B are unequal,

$$\frac{\partial W_\phi}{\partial\theta}=\tfrac{1}{2}(A-B)\left(A\frac{\partial q_{11}}{\partial\theta}-B\frac{\partial q_{22}}{\partial\theta}+2C\frac{\partial q_{31}}{\partial\theta}\right).$$

Also it is evident that if A, B and C are all increased by the same amount so that the potential differences inside the box are unaltered, the couple will be unaltered. Therefore

$$\frac{\partial q_{11}}{\partial\theta}-\frac{\partial q_{22}}{\partial\theta}+2\frac{\partial q_{31}}{\partial\theta}=0;$$

and lastly, by the symmetry of the arrangement, it is clear that if $q_{11}=q_0+\alpha\theta$, then $q_{22}=q_0-\alpha\theta$, because an increase in the area of the needle opposite the surfaces of A, A' is accompanied by an equal decrease opposite B, B'.

Hence the expression for the couple reduces to

$$\tfrac{1}{2}\alpha(A-B)\{C-\tfrac{1}{2}(A+B)\}.$$

But this couple is balanced by the torsion of the silk fibre, and that is proportional to the angle through which it is twisted from the equilibrium position, so that we have a relation

$$\theta=k(A-B)\{C-\tfrac{1}{2}(A+B)\},$$

where k is a constant.

For measuring small potential differences, if the condenser is highly charged, so that C is large compared to $\frac{1}{2}(A+B)$, then the deflection θ is proportional to the difference of potential $A-B$.

For measuring a large difference of potential it is usual to connect the condenser and therefore the needle with one pair of quadrants—say A, A'. Then we have $C=A$, and

$$\theta=\tfrac{1}{2}k(A-B)^2,$$

so that the deflection is proportional to the square of the difference of potential.

4·9. Examples. (i) *If q_{11}, q_{22}, q_{33} are the coefficients of capacity of three conductors A, B and C respectively, of which C encloses A and B, and if q_{12}, q_{13}, q_{23} are the coefficients of induction, then the capacity of C when alone in space is $q_{33}-(q_{11}+2q_{12}+q_{22})$.* [M. T. 1907]

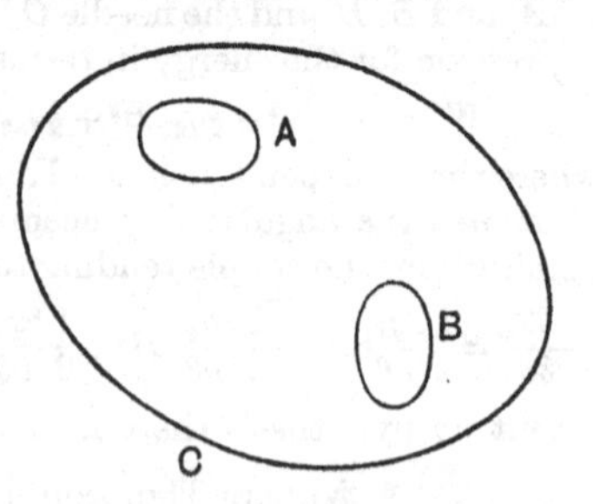

The relations between charges and potentials are

$$e_1=q_{11}\phi_1+q_{21}\phi_2+q_{31}\phi_3,$$
$$e_2=q_{12}\phi_1+q_{22}\phi_2+q_{32}\phi_3,$$
$$e_3=q_{13}\phi_1+q_{23}\phi_2+q_{33}\phi_3.$$

Let A, B be without charge and let C be raised to unit potential by a charge e_3, then since A, B are uncharged their potentials are also unity (2·44 (i)). Therefore the above relations become

$$0=q_{11}+q_{21}+q_{31},$$
$$0=q_{12}+q_{22}+q_{32},$$
$$e_3=q_{13}+q_{23}+q_{33};$$

whence we get $$e_3=q_{33}-(q_{11}+2q_{12}+q_{22}),$$
but this is the charge which will raise C to unit potential when alone in space, i.e. its capacity.

(ii) *Shew that, if a new insulated conductor be brought into the field of any system of conductors, their coefficients of potential of the type p_{rr} are in general diminished, and if p_{rr}, p_{rs}, p_{ss} be three coefficients of potential before the introduction of the new conductor, p_{rr}', p_{rs}', p_{ss}' the same coefficients afterwards $(p_{rr}p_{ss}-p_{rr}'p_{ss}')$ is not less than $(p_{rs}-p_{rs}')^2$.*

[M. T. 1896]

When the new conductor is introduced there will be a redistribution of the charges in the field, though the charge on each conductor remains the same, and positive work will be done by the electric forces so that there will be a diminution of the electric energy. Hence the difference

$$W-W'=\tfrac{1}{2}(p_{11}-p_{11}')\,e_1^2+(p_{12}-p_{12}')\,e_1e_2+\tfrac{1}{2}(p_{22}-p_{22}')^2\,e_2^2+\ldots$$

is positive for all values of the charges.

Hence $$p_{rr}-p_{rr}'\geqslant 0 \quad \text{and} \quad p_{ss}-p_{ss}'\geqslant 0$$

and $$(p_{rs}-p_{rs}')^2\leqslant(p_{rr}-p_{rr}')(p_{ss}-p_{ss}')$$

i.e. $$\leqslant p_{rr}p_{ss}+p_{rr}'p_{ss}'-p_{rr}p_{ss}'-p_{rr}'p_{ss}.$$

Therefore $$(p_{rs}-p_{rs}')^2\leqslant p_{rr}p_{ss}+p_{rr}'p_{ss}'-p_{rr}'p_{ss}'-p_{rr}'p_{ss}'$$

i.e. $$\leqslant p_{rr}p_{ss}-p_{rr}'p_{ss}'.$$

(iii) *Three insulated spherical conductors of the same size are placed at the corners of an equilateral triangle whose sides are large in comparison with the radii of the spheres. If the conductors are then touched in turn by another small charged spherical conductor, shew that the charges they receive are in geometrical progression.*

[I. 1899]

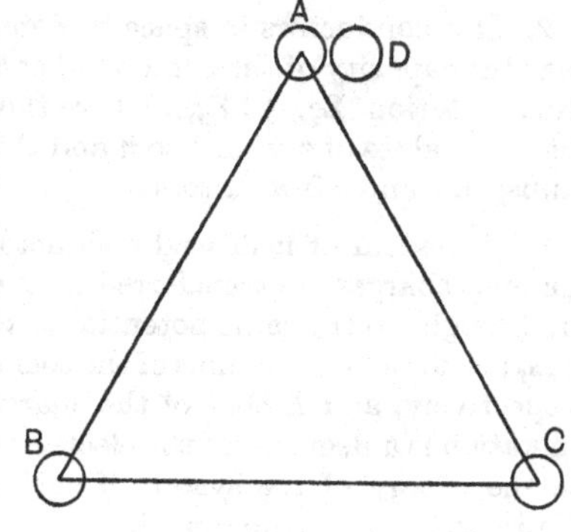

Let A, B, C be the fixed and D the movable conductor. Let e be the initial charge on D, and e_1, e_2, e_3 the charges which A, B, C in turn receive from it.

When A and D are in contact their common potential ϕ_1 satisfies relations

$$\phi_1 = p_{11}e_1 + p_{12}(e-e_1),$$
$$\phi_1 = p_{12}e_1 + p_{22}(e-e_1),$$

so that, by subtraction,

$$0 = (p_{11}-p_{12})e_1 + (p_{12}-p_{22})(e-e_1) \quad\text{..............(1).}$$

Next, when D is in contact with B, the charges on B and D are e_2 and $e-e_1-e_2$ and the coefficients of potential of B and D are by symmetry the same as the coefficients of potential of A and D when D was in contact with A, save that the potentials of B and D will both contain a term due to the charge e_1 on A, which we may take to be e_1/r, where r is a side of the triangle regarded as large compared to the dimensions of the conductors. Hence we have for the common potential of B and D

$$\phi_2 = p_{11}e_2 + p_{12}(e-e_1-e_2) + e_1/r,$$
$$\phi_2 = p_{12}e_2 + p_{22}(e-e_1-e_2) + e_1/r,$$

so that, by subtraction,

$$0 = (p_{11}-p_{12})e_2 + (p_{12}-p_{22})(e-e_1-e_2) \quad\text{...........(2).}$$

Similarly, for the common potential of C and D when in contact,

$$\phi_3 = p_{11}e_3 + p_{12}(e-e_1-e_2-e_3) + (e_1+e_2)/r,$$
$$\phi_3 = p_{12}e_3 + p_{22}(e-e_1-e_2-e_3) + (e_1+e_2)/r,$$

giving

$$0 = (p_{11}-p_{12})e_3 + (p_{12}-p_{22})(e-e_1-e_2-e_3) \quad\text{.........(3).}$$

From (1), (2) and (3) we get

$$\frac{e-e_1}{e_1} = \frac{e-e_1-e_2}{e_2} = \frac{e-e_1-e_2-e_3}{e_3},$$

which give

$$e_2^2 = e_1 e_3.$$

EXAMPLES

1. If the radii of two concentric spheres be a, b $(b>a)$, and if each sphere be electrified with a positive charge e, shew that the energy of the system will be $e^2(b+3a)/2ab$. [I. 1912]

2. If n conductors in space be connected together by wires, prove that the capacity of the compound conductor thus formed is, with the usual notation, $\Sigma q_{rr} + 2\Sigma q_{rs}$, where the first summation is taken for all integral values of r from 1 to n and the second for all unequal integral values of r and s from 1 to n. [I. 1899]

3. A system of insulated conductors having been charged in any manner, charges are transferred from one conductor to another till all are brought to the same potential ϕ. Shew that $\phi = E/(s_1+2s_2)$, where s_1, s_2 are the algebraic sums of the coefficients of capacity and induction respectively, and E that of the charges. Prove that the effect of the operation is a decrease of the electrostatic energy equal to what would be the energy of the system if each of the original potentials were diminished by the amount ϕ. [M. T. 1901]

4. If there are two conductors, A, B, and when B is uncharged the operation of charging A to unit potential raises B to potential ϕ, prove that when A is kept at zero potential a unit charge on B will induce on A a charge $-\phi$. [M. T. 1904]

5. If one of n conductors entirely surrounds the others, shew that $2n-1$ coefficients of potential are equal; and that if they are each equal to p, shew that the loss of energy when the outside conductor is connected with the earth is $\frac{1}{2}pe^2$, e being the quantity of electricity that passes to earth. [I. 1915]

6. If a conductor could be made larger without change of shape, would its capacity be increased or diminished? Give a reason for your answer.

If a conductor carrying a given charge is completely surrounded by another uncharged conductor, is the energy of the system increased or diminished? [St John's Coll. 1906]

7. Two equal uncharged conducting spheres are placed so that the coefficient of capacity for each is a and the coefficient of mutual induction is b. Each encloses a charged conductor, the charges being e_1, e_2 respectively. Prove that when the spheres are connected by a wire the loss of energy is $(e_1-e_2)^2/4(a-b)$. [I. 1914]

8. Shew that if a given charge of electricity be distributed between two conductors of the same shape and size, and symmetrically placed with respect to one another and neighbouring conductors, the maximum energy obtainable is when the whole charge is given to one, and the minimum when it is equally divided between them. [I. 1891]

9. Shew that if a given charge is distributed over a number of conductors so that the potential energy of the system when in electrical equilibrium is least, the conductors are at the same potential.

[M. T. 1897]

10. If conductors A, B, C, D, ..., where C completely surrounds A and B only, are charged in any manner and A and C are connected by a thin wire, shew that the charge remaining on A bears a constant ratio to the charge on B. [M. T. 1908]

11. Two equal and similar condensers, each consisting of a conducting sphere surrounded by a concentric spherical conducting shell of negligible thickness, are insulated and placed at a great distance r from one another, and charges e, e' are given to the inner surfaces of the condensers. Prove that, if the outer surfaces are connected by a fine wire, the loss of electric energy is approximately $\frac{1}{2}(e-e')^2\left(\frac{1}{b}-\frac{1}{r}\right)$, where b is the radius of the outer surface of either condenser.

[M. T. 1904]

12. Shew that if two conductors, for which

$$\phi_1 = p_{11} E_1 + p_{12} E_2,$$
$$\phi_2 = p_{12} E_1 + p_{22} E_2,$$

be connected by a fine wire, the new charges will be $E_1 + x$, $E_2 - x$, where x is such as to make the expression

$$p_{11}(E_1+x)^2 + 2p_{12}(E_1+x)(E_2-x) + p_{22}(E_2-x)^2$$

a minimum, and determine the apparent loss of energy.

What happens to this energy? [St John's Coll. 1910]

13. A system of conductors consists of three, C_1, C_2, C_3; and C_1 completely surrounds C_2. If C_2 were annihilated, the coefficients of potential of C_1 and C_3 would be P_{11}, P_{13}, P_{33}; and if C_3 were annihilated, the coefficients of potential of C_1 and C_2 would be ϖ_{11}, ϖ_{12}, ϖ_{22}. Shew that the actual coefficient of potential of C_2 on itself is

$$P_{11} - \varpi_{11} + \varpi_{22},$$

and write down the remaining coefficients of potential of the system. [M. T. 1909]

14. Two insulated fixed conductors are at given potentials when alone in the electric field and charged with quantities e_1, e_2 of electricity. Their coefficients of potential are p_{11}, p_{22}, p_{12}. But if they are surrounded by a large spherical conductor at potential zero with its centre near them, the two conductors require charges e_1', e_2' to produce the given potentials. Prove that, neglecting terms involving the inverse square of the radius of the enclosing conductor,

$$e_1' - e_1 : e_2' - e_2 = p_{22} - p_{12} : p_{11} - p_{12}.$$ [M. T. 1895]

15. Two conductors of capacities C_1, C_2, when each is at an infinite distance from any other body, are brought to a distance R apart where R is very great compared with their dimensions. Prove that the coefficient of mutual induction is $-\frac{C_1 C_2}{R}$ and the coefficients of capacity $C_1\left(1+\frac{C_1 C_2}{R^2}\right)$, $C_2\left(1+\frac{C_1 C_2}{R^2}\right)$. Verify the results by shewing that if the charges be Q_1, Q_2 the increase of electrical energy on the approach of the bodies is $\frac{Q_1 Q_2}{R}$. [I. 1892]

16. There are two concentric conducting spheres of radii a and b, they are insulated and the inner (a) has a charge e, but the outer, which is to be regarded as a shell of negligible thickness, has no charge. Shew that if the outer is connected by a thin wire with an insulated and uncharged conducting sphere of radius c at a great distance d, the loss of electric energy is

$$\frac{1}{2}\frac{e^2 c}{b(b+c)}\left\{1 - \frac{2b^2}{d(b+c)}\right\}.$$ [I. 1903]

17. The equipotential surface ϕ_1, due to the charges $A_1, A_2, \ldots$, encloses all those charges, and the equipotential surface ϕ_2, due to the charges $B_1, B_2, \ldots$, encloses all those charges, the two systems being too far apart to influence one another appreciably. If each of the surfaces is now made a conducting surface, without resultant charge, and if the two surfaces are joined by a fine conducting wire, shew that their common potential will be

$$\phi_1\phi_2\frac{\Sigma A+\Sigma B}{\phi_2\Sigma A+\phi_1\Sigma B}.$$ [M. T. 1907]

18. Two conductors A, B of a system are connected by a wire of negligible capacity so as to form a single conductor A'. Shew that the coefficient of potential of a third conductor C is diminished by

$$(p_{13}-p_{23})^2/(p_{11}+p_{22}-2p_{12}),$$

where $p_{11}, p_{12}, \ldots$ denote as usual the coefficients of potential of the original system, and the suffixes 1, 2, 3 refer to the conductors A, B, C. [I. 1906]

19. If a system of charged conductors is surrounded by a very large spherical sheet at zero potential with its centre in the neighbourhood of the conductors, obtain an approximate expression for the loss of energy due to the presence of the sphere in terms of its radius and the given charges, neglecting squares and higher powers of the ratios of the linear dimensions of the system to the radius of the sphere.

How is the result modified if the spherical sheet is insulated without charge instead of being at zero potential? [I. 1904]

20. A system consists of five conductors of which the second surrounds the first, the third surrounds the second and the fifth surrounds the fourth. Determine which of the coefficients of induction are zero. Find, independently, the relations which exist between the coefficients of potential and shew that only six of them are independent. [I. 1908]

21. A condenser is formed of two spheres, one inside the other, with their centres at a distance x from one another. Assuming the capacity to be a known function $f(x)$ of x, write down a formula for the force tending to increase x, when the charges on the spheres are e and $-e$. [St John's Coll. 1906]

22. Three equal spheres are placed at the corners of an equilateral triangle. When their potentials are ϕ, 0, 0 their charges are E, E', E' respectively. Shew that, when each sphere is at a potential ϕ', each has a charge $(2E'+E)\phi'/\phi$.

Find the potentials when the charges are E'', 0, 0. [M. T. 1934]

23. Prove that in a small displacement accompanied by small changes (δE) in the charges, the identity

$$\Sigma E\delta\phi - \Sigma\phi\delta E = \delta p_{11} E_1^2 + 2\delta p_{12} E_1 E_2 + \ldots$$

holds good. Deduce the equation

$$\delta q_{11} \phi_1^2 + 2\delta q_{12} \phi_1 \phi_2 + \ldots = -(\delta p_{11} E_1^2 + 2\delta p_{12} E_1 E_2 + \ldots)$$

and explain the relation of each of these expressions to the mechanical work done by the forces of the system in the displacement.

[St John's Coll. 1911]

24. Four perfectly equal uncharged and insulated conductors A_1, A_2, A_3, A_4 are such that when they are placed at the angular points of a square, they are symmetrically situated with respect to each other and the centre of the square. Another charged conductor is moved so as to touch each in succession in the same way, beginning with A_1; and e_1, e_2, e_3, e_4 are the charges on the conductors after they have been each touched once. Shew that

$$(e_1 - e_2)(e_1 e_3 - e_2^2) = e_1(e_2 e_3 - e_1 e_4). \qquad \text{[M. T. 1898]}$$

25. An insulated sphere of radius 25 cm. is charged and afterwards connected to an electrometer by a long fine wire, the deflection being 75 divisions. The system is then joined to a distant insulated sphere of radius 12 cm. and the deflection falls to 53 divisions. Calculate the capacity of the electrometer. [M. T. 1918]

26. Three concentric spherical conductors, radii a, b, c ($a<b<c$), have charges E_1, E_2, E_3 respectively. Shew that, if the inner conductor is now connected with the earth, the potentials of the conductors are diminished by amounts inversely proportional to their radii, and that the loss of energy is

$$\frac{a}{2}\left(\frac{E_1}{a} + \frac{E_2}{b} + \frac{E_3}{c}\right)^2. \qquad \text{[M. T. 1912]}$$

27. Three concentric spherical conductors, radii a, b, c ($a<b<c$), have charges e_1, e_2, e_3 respectively and potentials V_1, V_2, V_3. Write down the potentials in terms of the charges and shew that the electrostatic energy is

$$\frac{1}{2}\left\{\left(\frac{1}{a} - \frac{1}{b}\right) e_1^2 + bV_2^2 + \left(\frac{1}{c} - \frac{b}{c^2}\right) e_3^2\right\}. \qquad \text{[M. T. 1915]}$$

28. A system consists of three concentric spheres (a, b, c). The middle sphere has a charge e and the other two are first put to earth and then insulated, find the charges induced on the first and third spheres.

Calculate what would be the loss of energy (1) if the two innermost spheres were joined by a wire, (2) if the two outermost spheres were joined. [St John's Coll. 1911]

29. Use the theorem, that if a set of conductors suffer given small displacements the loss of electrical energy when the charges are kept constant is equal to the gain of electrical energy in the same displacements when the potentials are kept constant, to find the force X per

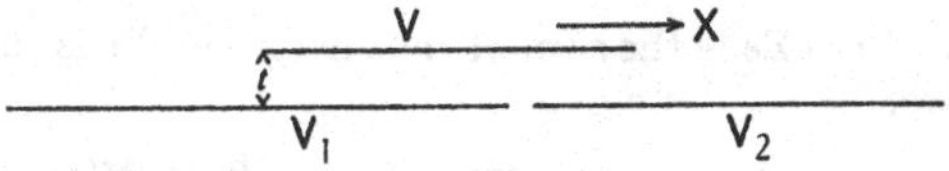

$V \longrightarrow X$, l

V_1 V_2

unit length acting on the upper plate of a parallel plate condenser arranged as shewn, edge on, with the potentials of the plates, in the figure. [M. T. 1929]

30. If three equal spherical conductors are placed at the corners of an equilateral triangle and raised to potentials ϕ_1, ϕ_2, ϕ_3, prove that, when they are put into electric communication with one another, the energy of the system is reduced by

$$\tfrac{1}{6}\{(\phi_1-\phi_2)^2+(\phi_2-\phi_3)^2+(\phi_3-\phi_1)^2\}\,W,$$

where W is the energy of the system, when the potentials are 1, -1, and 0 respectively. [I. 1912]

31. A spherical conductor of radius a has a charge E and is surrounded by three insulated spherical conducting shells, the internal radii of the shells being $2a$, $4a$ and $6a$, and the external radii $3a$, $5a$, $7a$, and their charges are E_1, E_2, E_3. Find the change in the energy of the system when the first and second shells are connected by a wire. [I. 1909]

32. If one conductor of a system of charged conductors contains all the others and there be $n+1$ in all, shew that there are n relations between either the coefficients of potential or the coefficients of induction, and if the potential of the largest be ϕ_0 and that of the others $\phi_1, \phi_2, \dots \phi_n$, then the most general expression for the energy is $\tfrac{1}{2}C\phi_0^2$ increased by a quadratic function of $\phi_1-\phi_0, \phi_2-\phi_0, \dots \phi_n-\phi_0$, where C is a definite constant for all positions of the inner conductors. [M. T. 1905]

33. If three condensers of capacities C_1, C_2, C_3 are joined in cascade and charged with a charge Q, and if then this connection is broken and the condensers are joined in parallel, shew that the loss of energy is

$$\tfrac{1}{2}Q^2\frac{C_1(C_2-C_3)^2+C_2(C_3-C_1)^2+C_3(C_1-C_2)^2}{C_1C_2C_3(C_1+C_2+C_3)}.$$

[M. T. 1907]

34. Recent experiments lead to the inferences that in C.G.S. electrostatic units the charge on an electron is $3{\cdot}4\times10^{-10}$, and that the mass of an electron is $6{\cdot}1\times10^{-28}$.

If electrons pass between two metal plates at a potential difference of 2000 volts, and the whole of their lost potential energy is converted into kinetic energy, calculate the velocity with which they strike the second plate, on the above data. [M. T. 1913]

ANSWERS

12. $\frac{1}{2}\{E_1(p_{12}-p_{11})+E_2(p_{22}-p_{12})\}^2/(p_{11}+p_{22}-2p_{12})$.

19. $(\Sigma e)^2/2r$, where Σe is the sum of the charges and r is the radius of the sphere.

20. $q_{13}, q_{14}, q_{15}, q_{24}, q_{25}, q_{34}$ are zero.

21. $e^2 f'(x)/2\{f(x)\}^2$.

22. $(E+E')E''/(E^2+EE'-2E'^2)$, $-E'E''/(E^2+EE'-2E'^2)$.

25. 3·9 abs. units.

28. $-\dfrac{a(c-b)}{b(c-a)}e$, $-\dfrac{c(b-a)}{b(c-a)}e$; $\dfrac{1}{2}\dfrac{e^2 a(b-a)(c-b)^2}{b^3(c-a)^2}$, $\dfrac{1}{2}\dfrac{e^2 c(b-a)^2(c-b)}{b^3(c-a)^2}$.

29. $(2V-V_1-V_2)(V_2-V_1)/8\pi t$.

31. $(E+E_1)^2/24a$.

34. $1{\cdot}92\times 10^9$ cm. per sec.

Chapter V

DIELECTRICS

5·1. Thus far we have assumed that the electric fields under consideration are produced by charged conductors in air. Faraday found that when some non-conducting substance other than air fills the space between the plates of a condenser its capacity is altered, being always increased in a definite ratio when the same substance is used. The ratio, usually denoted either by K or by ϵ, is called the **specific inductive capacity** of the substance or its **dielectric constant.**

As compared with a vacuum the specific inductive capacity of air has been found to be 1·000585, so that it is immaterial whether we take air or vacuum as the standard for comparison provided that the air is at constant temperature and pressure.

Taking the specific inductive capacity of air to be unity, those of other gases do not differ much from unity, but they may vary with the pressure. The specific inductive capacities of solids and liquids vary considerably, e.g. solid paraffin 2·29, sulphur 3·97, glass 3·2 to 7·6, distilled water (a semi-conductor) 76.

5·11. Inference drawn from parallel plate condensers. Consider two parallel plate condensers A and B which are exactly similar save that in A the medium between the plates is air while in B it is a dielectric of specific inductive capacity K.

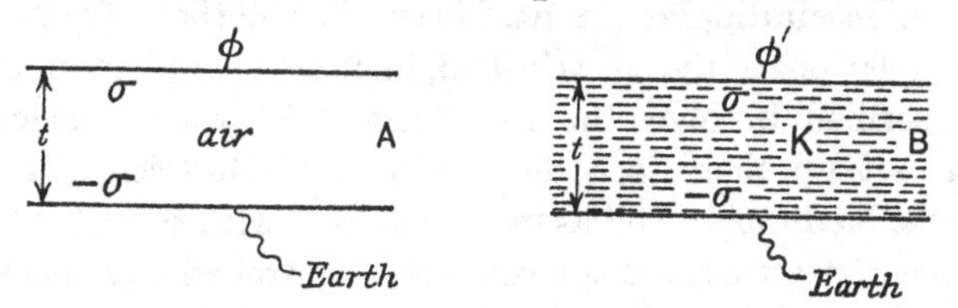

Let C, C' be the capacities per unit area of A and B, then, by definition of K,

$$C' = KC \qquad \text{.........................(1).}$$

Let the upper plates of the condensers have equal charges σ per unit area, let their potentials be ϕ, ϕ'; and let the lower plates be earth connected.

Then, by definition of capacity,

$$C=\sigma/\phi \text{ and } C'=\sigma/\phi' \quad \text{..............(2).}$$

Hence from (1) and (2)

$$\phi'=\phi/K \quad \text{.......................(3);}$$

or, for the same charges, the drop of potential in the dielectric is $1/K$ of what it is in air under like conditions.

Neglecting edge effects, in both condensers the tubes of force are straight, and if E, E' denote the electric intensities in the two fields and t the distance between the plates of the condensers, we have

$$Et=\phi \text{ and } E't=\phi',$$

so that

$$E'=E/K \quad \text{.......................(4);}$$

or the gradient of potential in the dielectric is $1/K$ of what it is in air under like conditions; and though we have only demonstrated this by a comparison of fields in which the gradients of potential are constants, yet we shall base our theory on the hypothesis that it is true for condensers of any shape.

We assume therefore that if a point charge e is embedded in a medium of uniform specific inductive capacity K the intensity of the field at distance r is $\frac{e}{Kr^2}$, and that the force between two small charged bodies with charges e, e' at a distance r apart is ee'/Kr^2.

5·2. Electric displacement. Maxwell's* method of explaining the relations of the electric field, including the phenomenon just described, involves the assumption that when an electric field is set up there is a displacement of electricity in the dielectric medium as well as in conductors; with the difference that in a dielectric the displacement is controlled and checked by some kind of elastic force which does not exist in conductors.

* James Clerk Maxwell (1831–1879), Cavendish Professor of Experimental Physics at Cambridge.

Thus if the plates A, B of an uncharged condenser are connected by a wire and by the action of an electromotive force, a quantity Q of electricity is transferred along the wire from the plate B to the plate A; then, at the same time, the electrification of the plates produces a certain electromotive force from A towards B in the dielectric between the plates and this causes a displacement of electricity in the dielectric. This displacement is not a continuing flow, but a displacement which takes place throughout the dielectric while the electrostatic field is in process of creation, the amount of electricity which is displaced across any surface drawn between the plates being Q, the quantity which passes across the wire; so that if S be a closed surface which surrounds the plate B, a quantity Q passes out of S along the wire and simultaneously a quantity Q enters S by the displacement which takes place at all points of the dielectric. In this respect electric displacement resembles the displacement of an incompressible fluid, in that the amount of electricity within a closed surface remains constant; if a charge Q is brought into the region bounded by a surface S, there is simultaneously a displacement of electricity of amount Q outwards across S in the dielectric medium. The direction of displacement is from a positive charge into the dielectric medium.

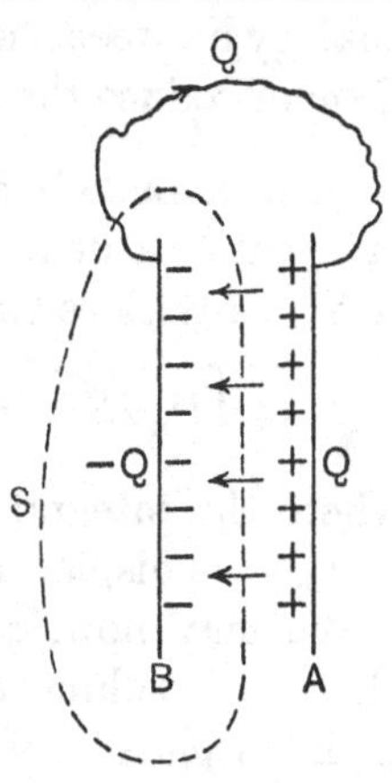

In view of what was stated in **2·11** about the nature of electricity, it may be as well to remark that the above comparison of electricity with an incompressible fluid is a highly artificial one and the resemblance is only to be regarded as holding good in this one particular of 'displacement'.

In Maxwell's theory, now commonly spoken of as the 'Classical Theory' of the electromagnetic field, the vector called the electric displacement plays an essential part. It was defined by Maxwell as the quantity of electricity displaced across a unit of area placed perpendicularly to the direction in which electricity is displaced. It is thus a measurable directed quantity. It is denoted by the symbol **D**, but in accordance with modern usage we shall define **D** as 4π times the quantity of electricity displaced as above. The reason for so doing is that, unless we define and adopt a system of 'rational units', we cannot exclude the factor 4π from many of our formulae, and we get more uniformity in the formulae and a closer analogy between the formulae of magnetism and electrostatics if we introduce the factor 4π at this stage.

5·21. Generalized form of Gauss's Theorem. In accordance with the definition of **D** and the explanation given in **5·2**, if S be any closed surface in an electric field

$$\int D_n dS = 4\pi \text{ (total charge within the surface)} \quad \ldots(1),$$

where the integral on the left represents the total outward flux of the displacement vector across the surface.

We may now, as in **2·33**, apply this theorem to a small element of volume δv enclosing a point P, and shew that, if **D** is the displacement and ρ the density of electricity at P, then

$$\operatorname{div} \mathbf{D} = 4\pi\rho \quad \ldots\ldots\ldots\ldots\ldots\ldots\ldots\ldots(2),$$

with the consequence that at all points in the field at which there is no volume density of electricity

$$\operatorname{div} \mathbf{D} = 0 \quad \ldots\ldots\ldots\ldots\ldots\ldots\ldots\ldots(3).$$

Vectors whose divergences vanish through a region are called *solenoidal* or *streaming vectors*, or are said to satisfy *the solenoidal condition*. This condition is also satisfied by the velocity of an incompressible fluid.

Next, if there is a surface of discontinuity of **D**, by applying the modified form of Gauss's Theorem above to a short cylinder extending across the surface, we can shew, exactly as in **2·34**, that if $\mathbf{D}_1$, $\mathbf{D}_2$ denote the displacement on opposite sides of the surface, the normal components measured away from the surface satisfy the relation

$$(D_1)_n + (D_2)_n = 4\pi\sigma \quad \ldots\ldots\ldots\ldots\ldots\ldots(4),$$

where σ is the surface density of the charge on the surface; with the consequence that across every *uncharged* interface between two dielectric substances the normal component of the displacement is continuous, i.e.

$$(D_1)_n = (D_2)_n \quad \text{.......................(5)}$$

when both are measured in the same sense.

5·22. Relations between D and E. There are thus two vectors **D** and **E** whose principal characteristics are that **D** is a solenoidal or streaming vector through uncharged space, with a continuous normal component across any uncharged interface, and that **E** is the gradient of a potential function. The two vectors are interrelated because **D** only exists when **E** exists; but the relation between them depends on the dielectric medium.

It is seen at once that if we assume that *in air* (or vacuum)

$$\mathbf{D} = \mathbf{E} \quad \text{..........................(1),}$$

then the relations **5·21** (1), (2), (3), (4) are merely Gauss's Theorem in its integral and local forms as set forth in **2·3**, **2·33** and **2·34**.

And if further we assume that in an isotropic* dielectric

$$\mathbf{D} = K\mathbf{E} \quad \text{.......................(2),}$$

where K is the dielectric constant, this makes the value of **E** in a dielectric $1/K$ of what it would be in air as required in **5·11**.

5·23. Recapitulation. In the previous chapters we limited our considerations to electric fields produced by charged conductors in air, but in the present chapter we broaden the basis of our considerations by admitting the possibility of the existence in the field of masses of dielectric substances. This has led to the introduction of a new vector **D** called the displacement; but in a static field displacement does not refer to a continuing process but to something which happened in the creation of the field. In varying electromagnetic fields, however, 'displacement' undergoes a time-rate of change, and 'displacement currents' play an essential part in the theory of the propagation through space of electrical effects. In

* A substance is *isotropic* if any spherical portion of the substance possesses no directional property, in contrast with a crystalline substance, for which the relation between **D** and **E** would be more complicated.

a static field the vector **D** satisfies certain fundamental relations specified in **5·21**, and this theory does not invalidate any of the previous discussion of fields in air but includes them as a special case in which $K=1$, or $\mathbf{D}=\mathbf{E}$.

5·24. Objections. The discussion of the so-called *displacement* in **5·2** is based on the theory explained in Maxwell's *Treatise on Electricity and Magnetism*,* but it is open to the objection that there is no experimental evidence that in dielectrics there is an actual displacement of electricity of the kind described. According to modern views it is preferable to define the vector **D** solely in terms of the conditions which it has to satisfy; and to this end we assert that a complete theory of the electrostatic field requires

(i) the existence of a single-valued potential function ϕ;

and, in addition to the electric intensity **E** which is the negative gradient of ϕ, a vector **D** which satisfies

(ii) the generalized form of Gauss's Theorem, and

(iii) the relation $\mathbf{D}=K\mathbf{E}$;

these conditions being sufficient to determine the field of any given charges. There is then no need for any further definition of **D** apart from the fact that it satisfies these relations.

5·25. Modification of Coulomb's Theorem. Suppose that a conductor on which there is a charge of surface density σ is in contact with a dielectric medium of constant K. Then since there is no field inside the conductor, in **5·21** (4) we may put $(D_2)_n=0$, so that

$$(D_1)_n=4\pi\sigma,$$

where $(D_1)_n$ is the normal component of displacement directed from the surface of the conductor into the dielectric. Then since $\mathbf{D}=K\mathbf{E}$, we have

$$KE_n=4\pi\sigma,$$

or

$$E_n=4\pi\sigma/K \quad\quad (1);$$

shewing again that the electric vector is $1/K$ of what it would be in air.

* 3rd edition 1892, pp. 65–70.

5·26. Refraction of the lines of force. When the lines of force cross an *uncharged* interface between two dielectrics, they undergo a refraction in accordance with a definite law. It was proved in **2·411** that in crossing the common surface of two media the tangential component of the electric intensity is continuous, and the argument of **2·411** is valid whatever the media. It was also proved in **5·21** that at an *uncharged* interface between two dielectrics the normal component of displacement is continuous.

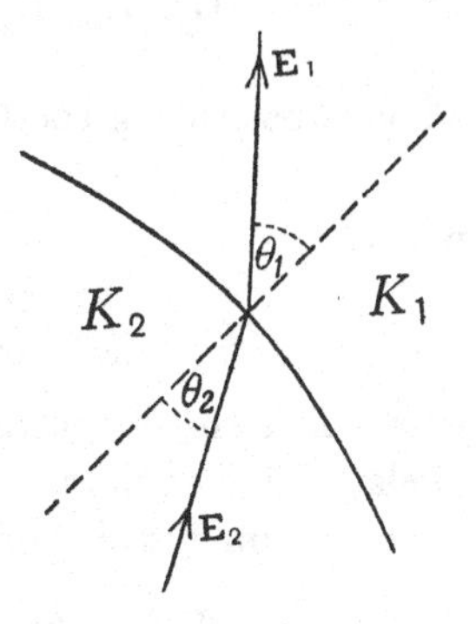

Let K_1, K_2 be the dielectric constants and let the electric intensities at points close to the interface and to one another in the two media be $\mathbf{E}_1$, $\mathbf{E}_2$, making angles θ_1, θ_2 with the normal to the interface. Then from the continuity of the tangential component of the intensity, we have

$$E_1 \sin\theta_1 = E_2 \sin\theta_2;$$

and from the continuity of the normal displacement we have

$$K_1 E_1 \cos\theta_1 = K_2 E_2 \cos\theta_2;$$

from which it follows that

$$K_1 \cot\theta_1 = K_2 \cot\theta_2,$$

and this is the law of refraction of the lines of force. It applies of course to the case of lines of force entering or leaving a block of dielectric substance surrounded by air, in which case one of the K's is unity.

5·3. The potential. The equations to be satisfied by the potential in a field which contains dielectric bodies may now be deduced from **5·21** and **5·22**.

In every electrostatic field we have

$$\mathbf{E} = -\mathbf{grad}\,\phi \qquad \text{.......................(1)}$$

with rectangular components

$$E_x,\ E_y,\ E_z = -\frac{\partial\phi}{\partial x},\ -\frac{\partial\phi}{\partial y},\ -\frac{\partial\phi}{\partial z} \qquad \text{.........(2):}$$

so that, if as in **5·22** we put $\mathbf{D}=K\mathbf{E}$, we have

$$D_x,\ D_y,\ D_z=-K\frac{\partial\phi}{\partial x},\ -K\frac{\partial\phi}{\partial y},\ -K\frac{\partial\phi}{\partial z}\ \text{......(3)}.$$

Then substituting these values in **5·21** (2), viz.

$$\operatorname{div}\mathbf{D}=4\pi\rho,$$

we get

$$\frac{\partial}{\partial x}\left(K\frac{\partial\phi}{\partial x}\right)+\frac{\partial}{\partial y}\left(K\frac{\partial\phi}{\partial y}\right)+\frac{\partial}{\partial z}\left(K\frac{\partial\phi}{\partial z}\right)=-4\pi\rho\ \text{...(4)},$$

where K is the dielectric constant and ρ is the volume density of electricity at the point (x, y, z); and at points at which there is no volume density of electricity the equation for ϕ is

$$\frac{\partial}{\partial x}\left(K\frac{\partial\phi}{\partial x}\right)+\frac{\partial}{\partial y}\left(K\frac{\partial\phi}{\partial y}\right)+\frac{\partial}{\partial z}\left(K\frac{\partial\phi}{\partial z}\right)=0\ \text{......(5)}.$$

We observe that equations (4) and (5) include the case in which the dielectric constant K varies from point to point of the region. Also, that in a homogeneous dielectric since K is constant it may be removed from (5); so that (5) reduces to $\nabla^2\phi=0$; hence in a homogeneous dielectric where there are no charges the potential satisfies Laplace's equation, just as in empty space.

Using **5·21** (4), at a surface of discontinuity of the medium, if σ is the surface density on the interface and ϕ_1, ϕ_2 the potential functions in the dielectrics whose constants are K_1, K_2 on opposite sides of the interface, we have

$$K_1\frac{\partial\phi_1}{\partial n_1}+K_2\frac{\partial\phi_2}{\partial n_2}=-4\pi\sigma\ \text{..............(6)},$$

where ∂n_1, ∂n_2 are elements of the normal drawn from the interface into the medium on each side.

5·4. Comparisons. When fields containing dielectrics are compared with fields in air, the following considerations sometimes lead to simple results. Suppose that the equipotential surfaces are known for a given distribution of charges in air. Let the space between any two of these surfaces be filled with homogeneous dielectric of constant K. The new field will retain the same surfaces as its equipotential surfaces, though the values of the potential over the surfaces will not be the same as before. The conditions to be satisfied in the new field are (i) that the potential shall be continuous; and (ii) that the normal displacement shall be continuous across the interfaces; i.e. that D_n or KE_n shall be continuous. Both conditions can be satisfied by making the gradient of potential in the air regions in the new field the same as in the original, and the gradient of potential in the dielectric region $1/K$ of what it was in the original field.

We shall illustrate this method by the following example.

5·41. Example. *A conductor with a charge e is surrounded by equipotential surfaces. If the space between the surfaces of potentials ϕ_1, ϕ_2 is filled with homogeneous dielectric of constant K, shew that the energy is reduced by*

$$\tfrac{1}{2}e(\phi_1-\phi_2)\left(1-\frac{1}{K}\right).$$

Let ϕ be the potential of the conductor A before the dielectric is introduced, as in fig. (i), so that the energy is $\frac{1}{2}e\phi$. After the introduction of the dielectric the charge of the conductor is unaltered, but its potential becomes ϕ', and the potentials of the given equipotentials B, C become ϕ_1', ϕ_2', as in fig. (ii).

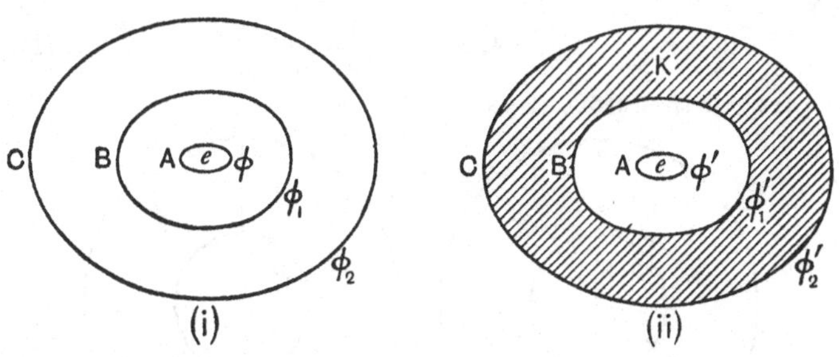

Then using the facts that the potential drop between A and B in air is unaltered; that the drop between B and C in the dielectric is $1/K$ of what it is in air; and that the drop between C and infinity in air is unaltered, we have

$$\phi'-\phi_1'=\phi-\phi_1,$$

$$\phi_1'-\phi_2'=\frac{1}{K}(\phi_1-\phi_2),$$

and

$$\phi_2'-0=\phi_2-0.$$

Therefore by addition

$$\phi'=\phi-(\phi_1-\phi_2)\left(1-\frac{1}{K}\right).$$

But the energy of the new field is $\frac{1}{2}e\phi'$, therefore the loss of energy

$$=\tfrac{1}{2}e(\phi-\phi')=\tfrac{1}{2}e(\phi_1-\phi_2)\left(1-\frac{1}{K}\right).$$

5·42. Again, suppose that the solution of a condenser problem is known when air is the medium between the conducting surfaces, and that we imagine the whole of a tube of force to be filled with a homogeneous dielectric of constant K. The equipotential surfaces will remain the same for they must still cut the lines of force at right angles, and the conditions of the problem require the potential drop along all the tubes of force to be the same. Hence, if the potentials of the conductors are kept constant, the charge at the end of the dielectric tube must be K times what it was in air; for otherwise the potential drop in this tube would fall to $1/K$ of what it was before.

5·43. Example. *The space between two concentric conducting spherical shells of radii a and b, $a<b$, is divided into two parts by a diametral plane, one part being filled with a dielectric of specific inductive capacity K_1 and the other part with a dielectric of specific inductive capacity K_2. Shew that the equipotentials are spheres concentric with the two conducting shells, and find the capacity of the condenser which is formed of the two conducting shells.* [M. T. 1934]

Let ϕ_1, ϕ_2 be the potentials of the conductors and σ the surface density on the sphere of radius a when the medium between the conductors is air. Then as in **3·4**

$$\phi_1-\phi_2=4\pi a^2\sigma\left(\frac{1}{a}-\frac{1}{b}\right) \quad \text{.....................(1)},$$

since $4\pi a^2\sigma$ is the total charge.

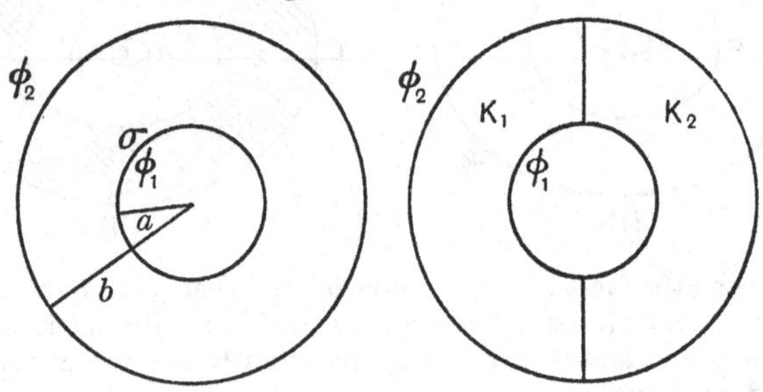

When the dielectrics are inserted, the equipotential surfaces between the conductors will still be concentric spheres since the lines of force will be radial; and in order to maintain the same potential drop as before along all the tubes of force, the surface density over one half of the inner sphere will need to be $K_1\sigma$ and over the other half $K_2\sigma$, making the total charge $2\pi a^2\sigma(K_1+K_2)$, and the capacity being the ratio of this to the potential difference $\phi_1-\phi_2$ is, from (1),

$$\tfrac{1}{2}(K_1+K_2)\,ab/(b-a).$$

5·5. Examples. (i) *A condenser of capacity one microfarad is to be formed by piling $2n$ square sheets of metallic foil each of* 10 *cm. edge, interleaved with sheets of insulating paper* 0·1 *mm. thick of dielectric constant* 3, *the sheets of foil being attached alternately to the two terminals. Find the necessary value of n, given that one microfarad is 9×10^5 electrostatic units.* [M. T. 1932]

Let ϕ_1, ϕ_2 be the potentials of the terminals, and let the sheets numbered 1, 3, 5, ... have positive charges and be connected to the terminal of potential ϕ_1. Neglecting edge effects, the surface densities on opposing faces of the metal sheets must be equal and opposite and may be denoted as in the figure.

Denoting the distance between the sheets by t, and the dielectric constant by K, and considering the field between the sheets 1 and 2, we have

$$\phi_1 - \phi_2 = tE_1 = \frac{4\pi}{K}\sigma_1 t,$$

where E_1 is the electric intensity.

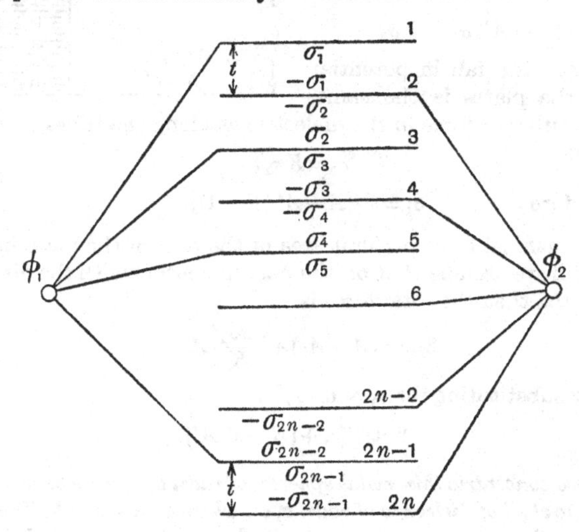

But there is a similar relation for each successive pair of sheets, viz.

$$\phi_1 - \phi_2 = tE_2 = \frac{4\pi}{K}\sigma_2 t,$$

....

$$\phi_1 - \phi_2 = tE_{2n-1} = \frac{4\pi}{K}\sigma_{2n-1} t.$$

Therefore, by addition,

$$(2n-1)(\phi_1 - \phi_2) = \frac{4\pi t}{K}(\sigma_1 + \sigma_2 + \dots + \sigma_{2n-1}).$$

But the capacity $= \dfrac{\text{area of a sheet} \times \Sigma\sigma}{\phi_1 - \phi_2} = \dfrac{(2n-1)K \times \text{area}}{4\pi t}$,

where $t = 0{\cdot}01$ cm., $K = 3$ and the area $= 100$ sq. cm.

Therefore the capacity $= (2n-1)\,300/0{\cdot}04\pi$ electrostatic units; and for this to be as much as 9×10^5, we find that n will need to be 189.

(ii) *Two large plane conducting plates each of area A are placed parallel to each other in air at a distance d apart. One is given a charge Q and the other is connected to earth. If a cylindrical piece of dielectric of specific inductive capacity K, thickness d, area of section A' $(A' < A)$, is placed between them with its generating lines perpendicular to the plates, shew that, neglecting edge effects, the mechanical force acting on each plate is*

$$2\pi Q^2/\{A + (K-1)A'\}. \qquad \text{[M. T. 1924]}$$

Let σ_1 be the surface density on the part of the positively charged plate in air, and σ_2 that on the part in contact with the dielectric. Then

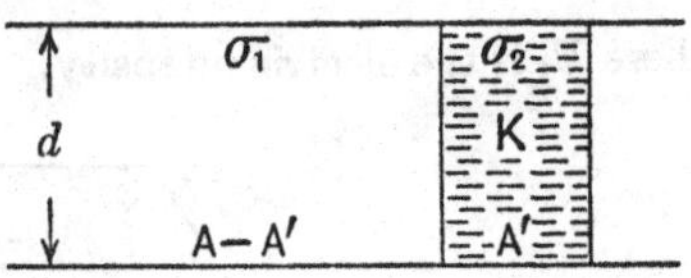

$$Q = (A - A')\,\sigma_1 + A'\sigma_2.$$

But since the fall in potential between the plates is the same along the tubes of force in the dielectric as along the tubes in air, we must have

$$\sigma_2 = K\sigma_1,$$

and therefore $$\sigma_1 = Q/\{A + A'(K-1)\}.$$

The mechanical force per unit area of the part of the plate in air is $2\pi\sigma_1^2$ and if we assume that on the part in contact with the dielectric it is $2\pi\sigma_2^2/K$, then the total force is

$$2\pi\sigma_1^2(A - A_1) + \frac{2\pi\sigma_2^2}{K}A',$$

which, by substituting for σ_1 and σ_2,

$$= 2\pi Q^2/\{A + (K-1)A'\}.$$

(iii) *Two concentric thin metal spheres, of radii a, b, have between them a concentric shell of dielectric of radii c, c′, where $a < c < c' < b$. The inner sphere is earth connected and the outer is given a charge e. Find the ratio in which this charge is divided between the inner and outer surfaces of the sphere, and shew that the potential at a point in the dielectric at a distance r from the centre is*

$$\frac{e}{b}\left\{\frac{1}{a} - \frac{1}{c} + \frac{1}{K}\left(\frac{1}{c} - \frac{1}{r}\right)\right\} \Big/ \left\{\frac{1}{a} - \frac{1}{c} + \frac{1}{K}\left(\frac{1}{c} - \frac{1}{c'}\right) + \frac{1}{c'} - \frac{1}{b}\right\},$$

where K is the dielectric constant. [I. 1926]

Of the charge e let the portion e' reside on the inner surface of the sphere of radius b and the portion $e - e'$ on its outer surface. By consideration of the tubes of force which start from the charge e', we see that the charge on the sphere of radius a is $-e'$.

Then remembering that a uniform spherical distribution produces no field inside itself, the field in the space between the spheres taken along the radius in the sense of r increasing has the following values:

$$\text{for } a < r < c,\ \frac{-e'}{r^2};\ \text{for } c < r < c',\ \frac{-e'}{Kr^2};\ \text{and for } c' < r < b,\ \frac{-e'}{r^2}.$$

Let ϕ_1 be the potential of the outer sphere, then since the inner is earth connected, we have, for the potential difference,

$$0-\phi_1=-\int_a^c \frac{e'}{r^2}dr-\int_c^{c'}\frac{e'}{Kr^2}dr-\int_{c'}^b \frac{e'}{r^2}dr$$

or

$$\phi_1=e'\left\{\frac{1}{a}-\frac{1}{c}+\frac{1}{K}\left(\frac{1}{c}-\frac{1}{c'}\right)+\frac{1}{c'}-\frac{1}{b}\right\}.$$

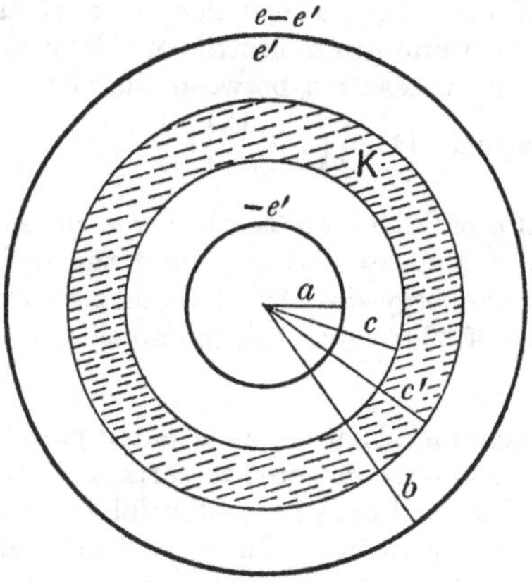

Again, the field for $r>b$ is e/r^2, so that we also have

$$\phi_1=\int_b^\infty \frac{e}{r^2}dr=\frac{e}{b}.$$

Therefore

$$\frac{e}{b}=e'\left\{\frac{1}{a}-\frac{1}{c}+\frac{1}{K}\left(\frac{1}{c}-\frac{1}{c'}\right)+\frac{1}{c'}-\frac{1}{b}\right\},$$

which determines the ratio $e':e-e'$.

Again, if ϕ denotes the potential at a point in the dielectric at a distance r from the centre, and we take the fall in potential between the sphere of radius a and this point, we have

$$0-\phi=-\int_a^c \frac{e'}{r^2}dr-\int_c^r \frac{e'}{Kr^2}dr$$

or

$$\phi=e'\left\{\frac{1}{a}-\frac{1}{c}+\frac{1}{K}\left(\frac{1}{c}-\frac{1}{r}\right)\right\},$$

and then by substituting for e' in terms of e we obtain the required result.

EXAMPLES

1. Two large parallel plates at distance d are maintained at potentials ϕ_1, ϕ_2; find the force per unit area on either of them, and find the effect of interposing a slab of dielectric of specific inductive capacity K and thickness t between the plates. [I. 1903]

2. The plates of a parallel plate condenser are at distance h apart. Prove that, if a slab of uniform dielectric, of thickness t and having a dielectric constant K, is inserted between the plates, the capacity is increased in the ratio $1:\left\{1-\frac{t}{h}\left(1-\frac{1}{K}\right)\right\}$. [M. T. 1924]

3. If the area of a parallel plate condenser is one square metre, the distance between the plates 0·01 cm., and the dielectric constant $K=7$, find in ergs the energy stored in the condenser when charged to a potential difference of 300 volts, that is, one electrostatic unit. [M. T. 1934]

4. Two large plane metal sheets are situated parallel to each other at distance a apart and are connected to earth. A third sheet is placed midway between them and raised to potential V. Two slabs of homogeneous dielectric substances of thicknesses t, t' and dielectric constants K, K' are placed parallel to the sheets one in each gap. Shew that the central sheet is attracted towards the former or latter slab according as

$$\frac{t'}{t} \text{ is less or greater than } \frac{K'(K-1)}{K(K'-1)}.$$

Find by how much the central plate must be moved to be in equilibrium. [M. T. 1933]

5. If two parallel conducting plates be separated by layers of two dielectrics of thickness t_1 and t_2 and specific inductive capacities K_1 and K_2 and the plates have potentials ϕ_1 and ϕ_2, prove that the force on the first plate is $\frac{SK_1K_2^2}{8\pi}\frac{(\phi_1-\phi_2)^2}{(K_1t_2+K_2t_1)^2}$, where S is the area of either plate. [I. 1892]

6. Two large parallel conducting plates are maintained at potentials ϕ_1 and ϕ_2 and the space between them is filled up by slabs of dielectric whose S.I.C.'s are K_1 and K_2, whose thicknesses are d_1 and d_2, and whose common face is parallel to the plates. Find the potential at any point between the plates.

Shew that the potential everywhere is the same as if the dielectrics were replaced by an insulated conducting sheet along their common face, and the charge on this plate per unit of area were

$$\frac{(K_1-K_2)(\phi_1-\phi_2)}{4\pi(d_1K_2+d_2K_1)}.$$ [M. T. 1909]

7. Find the capacity of the condenser C, D described in the following problem, making it clear in what units your answer is expressed. A condenser consists of two plates A, B at a distance a cm. apart. A second condenser consists of two plates C, D at a distance b cm., with a slab of paraffin of thickness c cm. and specific inductive capacity K between them. The slab and plates are all of area S sq. cm. The plates A, C are insulated and B, D are put to earth. A is first raised to potential V in electrostatic units, then connected to C by a fine wire and finally the slab is withdrawn.

Compare the attractions between the plates A, B in the three stages and prove that the work done against electrical forces in withdrawing the slab is

$$\frac{ScV^2}{8\pi(a+b)} \frac{(K-1)}{K(a+b-c)+c} \text{ ergs.}$$

[It may be assumed in the above that the lines of force are perpendicular to the slab and plates and that the condensers are too far apart to affect one another by induction.] [M. T. 1913]

8. Find the capacity of a spherical conductor of radius a, closely surrounded by a concentric spherical shell of dielectric whose inductive capacity is K and outer radius b. Find the energy when the charge is e. [St John's Coll. 1905]

9. If a spherical conductor, of radius a, with no other conductor in the neighbourhood, is coated with a uniform thickness d of shellac of which K is the specific inductive capacity, shew that the capacity of the conductor is increased in the ratio $K(a+d) : Ka+d$. [I. 1902]

10. If the space between concentric conducting spheres of radii a and b $(a < b)$ is filled with two dielectrics of specific inductive capacities K, K', their common surface being a concentric sphere of radius c, find the capacity of the condenser. [I. 1904]

11. Evaluate the coefficients of capacity and induction for two concentric spherical conductors of radii a, b, (i) when the intervening medium is air, (ii) when the inner conductor is closely surrounded by a concentric shell of a dielectric whose S.I.C. is K, the outer radius of the shell being $b-t$. [St John's Coll. 1906]

12. A spherical condenser consists of two concentric conducting spheres of radii a, b $(b > a)$. A spherical shell of dielectric K extends from the inner sphere to a distance c $(< b)$ from the centre. The inner sphere is insulated and receives a charge e and the outer is earthed. Prove that the potential inside the dielectric is

$$\frac{e}{Kr} - \frac{e}{Kc} + \frac{e}{c} - \frac{e}{b},$$

where r is the distance of a point from the centre. Find also the capacity of the condenser. [M. T. 1935]

13. A spherical condenser of radii a, d is modified by the insertion of a concentric spherical shell of specific inductive capacity K, radii b and c $(d>c>b>a)$. Find its capacity as modified. [M. T. 1928]

14. A metallic sphere is surrounded by a thin concentric conducting shell formed by two hemispheres with their rims in contact, the space between the sphere and shell being filled with a dielectric of specific inductive capacity K. If charges e, e' be given to the shell and sphere, shew that if the halves of the shell remain in contact the charges must be of opposite sign and the ratio of their magnitudes must lie between the limits $1 \pm 1/\sqrt{K}$. [M. T. 1908]

15. If half the space between two concentric conducting spheres be filled with solid dielectric of specific inductive capacity K, the dividing surface between the solid and the air being a plane through the centre of the spheres, shew that the capacity of the condenser will be the same as if the whole dielectric were of uniform specific inductive capacity $\frac{1}{2}(1+K)$. [M. T. 1894]

16. The space between two concentric conducting spheres is filled on one side of a diametral plane with dielectric of S.I.C. K and on the other side with dielectric of S.I.C. K'. The inner sphere is of radius a and has a charge e; shew that the force on it perpendicular to this diametral plane is

$$\frac{1}{2}\frac{K-K'}{(K+K')^2}\frac{e^2}{a^2}. \qquad \text{[I. 1901]}$$

17. The space between two concentric conducting spheres of radii a and A is filled with n concentric layers of dielectric of S.I.C. K_1, K_2, ... K_n. The radii from a to A inclusive are a series in harmonical progression.

Prove that C, the capacity of the condenser, is given by

$$1 = C\left(\frac{1}{K_1}+\frac{1}{K_2}+\dots+\frac{1}{K_n}\right)\left(\frac{1}{a}-\frac{1}{A}\right)\frac{1}{n}. \qquad \text{[I. 1902]}$$

18. Three thin conducting sheets are in the form of concentric spheres of radii $a+d$, a, $a-c$ respectively. The dielectric between the outer and middle sheet is of specific inductive capacity K, and that between the middle and inner sheets is air. At first the outer sheet is uninsulated, the inner sheet is uncharged and insulated, the middle coating is charged to potential V and insulated. The inner sheet is now uninsulated without connection with the middle sheet. Prove that the potential of the middle sheet falls to

$$KVc(a+d)/\{Kc(a+d)+d(a-c)\}. \qquad \text{[M. T. 1895]}$$

19. A curve is drawn on the surface of a freely electrified oval conductor surrounded by air, dividing it into two parts on which the charges are E and E' respectively. The space, enclosed by the portion of the surface of the conductor on which the charge is E' and by the lines of force drawn from every point of the curve to infinity, is filled with a dielectric of S.I.C. K. Prove that the potential of the conductor is now diminished in the ratio $E+E' : E+KE'$. [M. T. 1906]

20. A very long condenser is formed of a dielectric of given specific inductive capacity bounded by two coaxial cylinders of given radii: if the two surfaces are coated with tinfoil, the outer being uninsulated and the other raised to a given potential, determine the energy per unit length. [M. T. 1903]

ANSWERS

1. (i) $(\phi_1-\phi_2)/8\pi d^2$; (ii) $(\phi_1-\phi_2)^2/8\pi(d-t+t/K)^2$.

3. $2{\cdot}78\times 10^5$.

4. $\frac{1}{2}t\left(1-\frac{1}{K}\right)-\frac{1}{2}t'\left(1-\frac{1}{K'}\right)$.

6. $0<x<d_1$, $\phi=\phi_1+\dfrac{(\phi_2-\phi_1)K_2x}{K_1d_2+K_2d_1}$;

$d_1<x<d_2$, $\phi=\phi_2-\dfrac{(\phi_2-\phi_1)K_1(d_1+d_2-x)}{K_1d_2+K_2d_1}$.

7. $S/4\pi(b-c+c/K)$ cm. The attractions are as

$$1:(b-c+c/K)^2/(a+b-c+c/K)^2:b^2/(a+b)^2.$$

8. $\dfrac{Kab}{b-a(1-K)}$; $\frac{1}{2}e^2\left\{\dfrac{1}{K}\left(\dfrac{1}{a}-\dfrac{1}{b}\right)+\dfrac{1}{b}\right\}$.

10. $KK'abc/\{K'b(c-a)+Ka(b-c)\}$.

11. (i) $q_{11}=-q_{12}=q_{22}-b=ab/(b-a)$.
(ii) $q_{11}=-q_{12}=q_{22}-b=Kab(b-t)/\{b(b-t-a)+Kat\}$.

13. $Kabcd/\{Kcd(b-a)+Kab(d-c)+ad(c-b)\}$.

20. $K\phi^2/\{4\log(b/a)\}$, where ϕ is the given potential and a, b the inner and outer radii.

Chapter VI

ELECTRICAL IMAGES

6·1. Theorem of the equivalent layer. Let S be a closed equipotential surface surrounding certain charges $\sum e$ in a field produced by the charges $\sum e$ and by certain other charges $\sum e'$ external to S. The lines of force cut S at right angles everywhere, and this would still be the case if we supposed a thin conducting sheet to occupy the position defined by S. But, from considerations of tubes of force, this conductor would now have a charge $-\sum e$ on its inner surface; and, since we suppose its total charge to be zero, the charge on its outer surface would be $\sum e$. In the substance of the conducting sheet the lines of force cease to exist but otherwise the original field is unaltered. Also, as explained in **2·31** (iii), the fields inside and outside S are now independent of one another, and the field inside S may be removed while the field outside S remains unaltered. But the field outside S is now due to a charge $\sum e$ spread over the surface together with the same external charges $\sum e'$ as before; and, since S is an equipotential surface, the charge $\sum e$ is in equilibrium on the surface.

The surface density of the charge $\sum e$ is given by Coulomb's Law, viz.

$$4\pi\sigma = E_n = -\partial\phi/\partial n,$$

where ϕ is the potential due to $\sum e$ and $\sum e'$ jointly.

Hence *if S be an equipotential surface in a field due to given charges, the charges inside S can be distributed over S (regarded as a conductor) in an equilibrium layer without affecting the external field.*

Again, as regards the field inside S, in fig. (i) it is a variable field; but in fig. (ii) there is no charge inside S and S is an equipotential surface, therefore the potential is constant inside S and the field is zero. But in fig. (ii) the field inside S is due to the joint effects of the surface layer $\sum e$ and the external charges

$\sum e'$, and since these together produce a zero field, therefore the part of the field inside S in fig. (i) due to the external charges $\sum e'$ could equally well be produced by a charge $-\sum e$ spread over the surface.

Hence *if S be an equipotential surface in a field due to charges $\sum e$ inside S and $\sum e'$ outside S, the field inside S will be unaltered if the external charges $\sum e'$ are abolished and a charge $-\sum e$ is distributed over S (regarded as a conductor) in an equilibrium layer.*

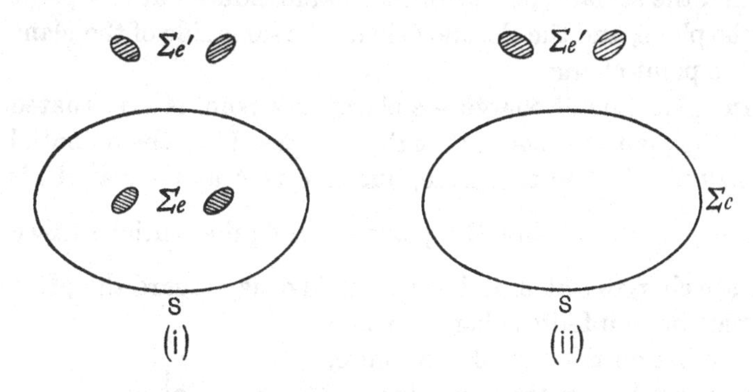

At first glance it may look as though this result is independent of what the charges $\sum e'$ may be, but this is not so, because S must be an equipotential in the field due to *all* the charges, and the surface density of the layer $4\pi\dfrac{\partial\phi}{\partial n}$ depends on the potential due to all the charges. In fact through the remainder of this chapter we shall be concerned mainly with varieties of the problem—given S and $\sum e'$, to determine $\sum e$.

The theorems of this article are known as Green's Theorem of the Equivalent Layer. They can also be proved analytically.

6·2. Images.* *Definition.* An electrical image is a point charge or set of point charges on one side of a conducting surface which would produce on the other side of the surface

* The theory of electrical images was the discovery of William Thomson, afterwards Lord Kelvin, who communicated his first ideas on the subject by letter to M. Liouville in 1845, the year in which he graduated as Second Wrangler.

the same electric field as is produced by the actual electrification of the surface.

6·21. Field produced by a point charge in the presence of a conductor at zero potential with an infinite plane face. Let there be a point charge e at a point A at a distance a from an infinite conducting plane at zero potential which cuts the plane of the paper at right angles in the line XX'. The tubes of force which start from e fall on the plane and it is required to find the surface density of the charge induced at any point of the plane, and the electric field on the same side of the plane as the point charge e.

Imagine a point charge $-e$ placed at a point A' so situated that the plane bisects AA' at right angles. Then the potential at any point P on the plane, due to e at A and $-e$ at A', is $\frac{e}{AP}-\frac{e}{A'P}=0$; so that the plane is an equipotential surface for the charges e at A and $-e$ at A'. We may regard the plane as part of an infinite sphere S whose centre lies on the left of the figure, thus reproducing the conditions of **6·1**, viz. a closed equipotential surface S with a charge e outside it and a charge $-e$ inside it; and by the theorem of the equivalent layer the internal charge $-e$ may be spread over S in a state of equilibrium without affecting the external field. Moreover, by the uniqueness theorem **4·21** (i), there is only one way in which a charge $-e$ can be distributed over the plane in equilibrium in the presence of the external point charge e. Hence so far as the field in air is concerned, i.e. in the region on the right of XX' in the figure, the electricity induced on the plane produces exactly the same effect as would a point charge $-e$ at A'.

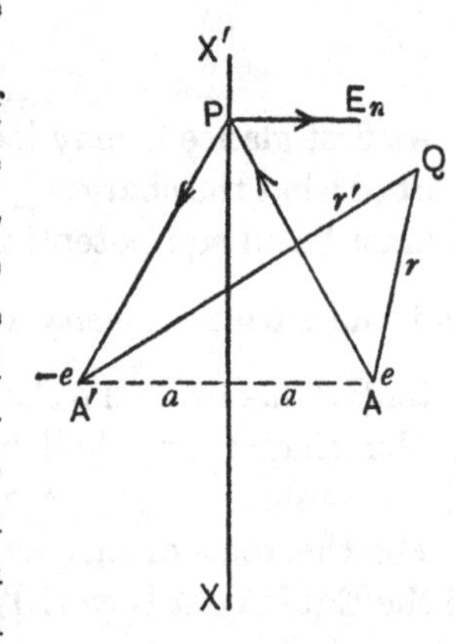

The point charge $-e$ at A' is commonly described as the image, in the plane, of e at A, by analogy from optical images, but it is well to bear in mind that the charge $-e$ at A' produces the same effect at a point on the right of the plane as the whole surface charge induced on the plane.

If Q is any point in air at distances r, r' from A, A', the potential of the field at Q is

$$\phi = \frac{e}{r} - \frac{e}{r'} \quad \text{.......................(1)},$$

whereof the first term is directly due to e at A, and the second term gives the effect of the induced charge.

To find the surface density induced at a point P on the plane, we note that the electric vector at P is the resultant of a repulsion e/AP^2 along AP and an attraction e/PA'^2 along PA'. So that if E_n denotes the normal component, by Coulomb's Theorem

$$4\pi\sigma = E_n$$
$$= -\frac{e}{AP^2}\cdot\frac{a}{AP} - \frac{e}{PA'^2}\cdot\frac{a}{PA'}$$
$$= -2ea/AP^3,$$

so that $$\sigma = -ea/2\pi AP^3,$$

or the surface density is inversely proportional to the cube of the distance from A.

6·3. Sphere and point charge. *To find the electric field outside a conducting sphere, due to an external point charge and the charge which it induces on the sphere; and to determine the surface density of the induced charge.*

(i) *When the sphere is at zero potential.* Let O be the centre of the sphere and a its radius: and let there be an external point charge e at A, where $OA = f$.

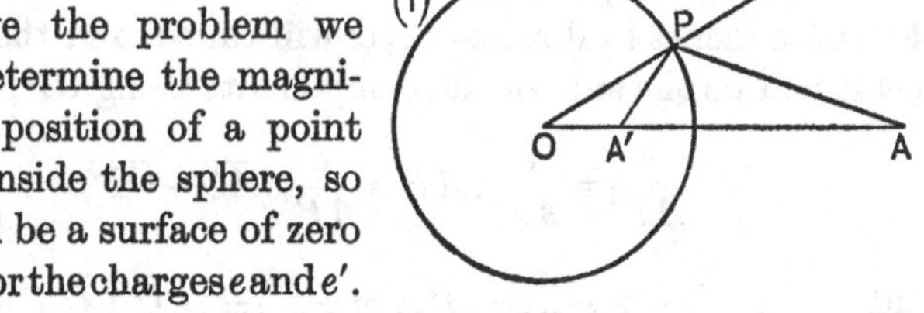

To solve the problem we have to determine the magnitude and position of a point charge e' inside the sphere, so that it will be a surface of zero potential for the charges e and e'.

Place a charge e' at A' the inverse of the point A, i.e. the point on OA such that $OA'.OA = a^2$. Then we can determine e' so that the potential is zero at every point of the sphere. This will be so, if for any point P on the sphere

$$\frac{e'}{A'P} + \frac{e}{AP} = 0,$$

i.e. if

$$e' = -e\frac{A'P}{AP} = -e\frac{OP}{OA},$$

by similar triangles, since $OA.OA' = OP^2$. Therefore

$$e' = -ea/f.$$

Hence the sphere is a surface of zero potential for charges e at A and $-ea/f$ at A', and therefore, by the theorem of the equivalent layer, for a charge e at A and a charge $-ea/f$ distributed in equilibrium over the surface; and since all conditions are satisfied this must be the solution of the problem proposed (4·21 (i)).

The total induced charge is therefore $-ea/f$ and this indicates the proportion of the tubes of force which start from e and fall on the sphere.

The potential at any external point Q is given by

$$\phi = \frac{e}{AQ} - \frac{ea}{f.A'Q} \quad \text{......................(1)}.$$

For the surface density at any point P on the sphere we have

$$4\pi\sigma = E_n \quad \text{........................(2)},$$

where E_n is the resultant of $\frac{e}{AP^2}$ along AP and $\frac{ea}{f.PA'^2}$ along PA'.

We know that at the surface of the conductor the resultant intensity is along the normal OP, so that if we resolve each of the foregoing components of E_n in the directions AO and OP the components in direction AO will cancel and the required result will be the sum of the components along OP. Thus

$$\frac{e}{AP^2} \equiv \frac{e}{AP^3}.\overline{AP} \equiv \frac{e}{AP^3}(\overline{AO} + \overline{OP})$$

and

$$\frac{ea}{f.PA'^2} \equiv \frac{ea}{f.PA'^3}.\overline{PA'} = \frac{ea}{f.PA'^3}(\overline{PO} + \overline{OA'}).$$

Therefore $$E_n = \overline{OP}\left(\frac{e}{AP^3} - \frac{ea}{f.PA'^3}\right)$$

$$= \frac{ea}{AP^3}\left\{1 - \frac{a}{f}\left(\frac{AP}{A'P}\right)^3\right\}.$$

But $$AP : A'P = f : a,$$

therefore $$\sigma = \frac{E_n}{4\pi} = -\frac{e(f^2 - a^2)}{4\pi a . AP^3} \quad \text{..................(3)}.$$

(ii) *When the sphere is insulated and without charge.* We remark that by the uniqueness theorem **4·21** (i) we have only to find *a* solution which satisfies the necessary conditions and then we know that it is *the* solution.

Since the sphere is now to be insulated and without charge, the total induced charge is zero, the effect of induction being to separate equal quantities of positive and negative electricity. But the sphere though no longer at zero potential must have a constant potential. So all the conditions of the problem will be satisfied if we superpose on the field of case (i) the field due to a charge $+ea/f$ uniformly distributed over the sphere, for this will make the total charge zero and leave the sphere an equipotential surface.

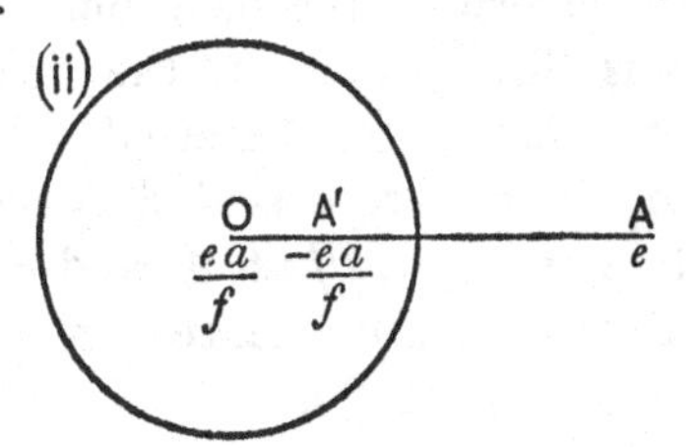

Since the external field due to a uniform spherical charge is the same as if the charge were collected at the centre of the sphere, we may now say that the external field is due to e at A, $-ea/f$ at A' and ea/f at O.

To find the surface density we have only to add to that given in (3) that due to the uniformly distributed charge ea/f, so that in this case

$$\sigma = -\frac{e(f^2 - a^2)}{4\pi a . AP^3} + \frac{e}{4\pi af} \quad \text{..............(4)}.$$

It is clear that at points such that $AP^3 = f(f^2 - a^2)$ there is no electrification. This defines a circle on the sphere separating positive from negative electrification.

(iii) *When the sphere is insulated and has a total charge Q.* For the solution in this case it is only necessary to superpose on the field (ii) that due to a charge Q uniformly distributed over the sphere. The external field may now be represented as due to e at A, $-ea/f$ at A' and $Q+ea/f$ at O; and the surface density at any point P is

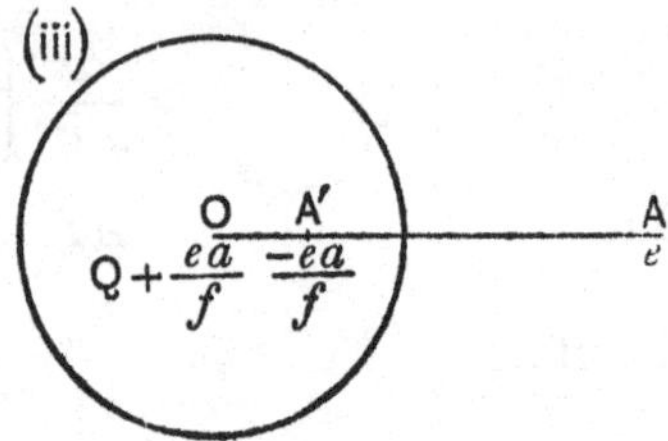

$$\sigma = -\frac{e(f^2-a^2)}{4\pi a . AP^3} + \frac{e}{4\pi af} + \frac{Q}{4\pi a^2} \quad \text{............(5).}$$

(iv) *When the sphere is maintained at a given potential ϕ.* Since a uniformly distributed charge ϕa on the surface of the sphere would give it a potential ϕ, it is clearly only necessary to superpose the field due to this distribution on the field of case (i). The external field may now be represented as due to e at A, $-ea/f$ at A' and ϕa at O; and the surface density at any point P is

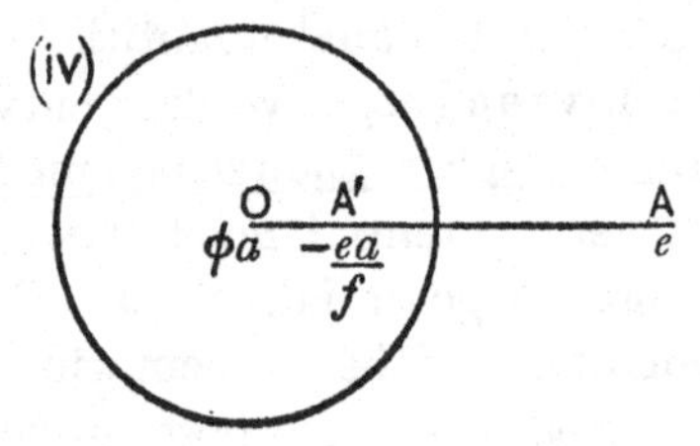

$$\sigma = -\frac{e(f^2-a^2)}{4\pi a . AP^3} + \frac{\phi}{4\pi a} \quad \text{..................(6).}$$

6·31. The force on the point charge e can be found in any of the foregoing cases by using the law of inverse squares. Thus in **6·3** (iii), the point charge e is repelled from a sphere having a total charge Q with a force

$$e\left(Q+\frac{ea}{f}\right)\Big/ OA^2 - \frac{e^2a}{f}\Big/ A'A^2,$$

where $A'A = f - a^2/f$. Therefore the force

$$= \frac{eQ}{f^2} + \frac{e^2a}{f^3} - \frac{e^2af}{(f^2-a^2)^2}.$$

Whether this is a repulsion or attraction depends on the relative values. It is clearly a repulsion if Q/e is large enough.

Put $f = a + x$, and let the distance x of e from the surface of the sphere be small. Then, if we expand the expression for the force on e, the principal terms are

$$\frac{eQ}{a^2}-\frac{e^2a^2}{4x^2},$$

so that the force is an attraction unless $Q > ea^4/4x^2$.

6·32. It is to be observed that electrical images are all 'virtual' or 'imaginary'. We cannot have a field on one side of a surface due to an *image* on the *same* side of the surface. In the cases considered so far the point charge e is a real charge and the other point charges are all imaginary charges put in to aid the solution of the problem in question.

6·33. Point charge *e* inside a sphere at zero potential. The external field in this case is zero since there is no charge outside the surface of zero potential. Let the conducting sphere be of internal radius a with a point charge e at a point A at a distance f ($<a$) from the centre O.

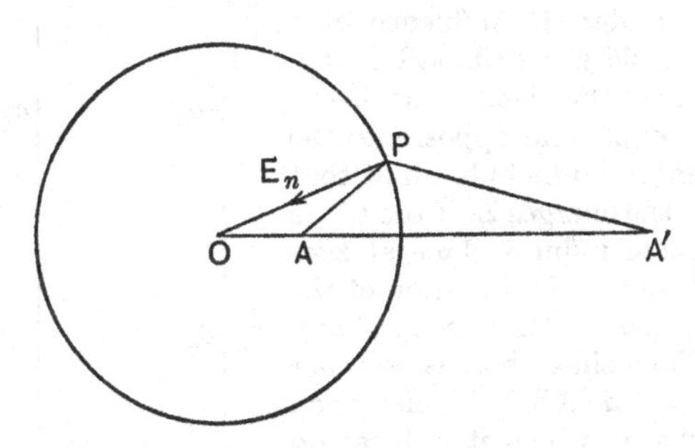

As in **6·3** (i), we can shew that a charge e' at the inverse point A' together with e at A will make the sphere a surface of zero potential if $e' = -ea/f$.

Then, as in the second part of the theorem of the equivalent layer (**6·1**), considering the field inside the sphere due to e at A and $-ea/f$ at A', the contribution due to the external charge $-ea/f$ can equally well be produced by a distribution $-e$ in equilibrium over the surface of the sphere. But the field which we are investigating is the field inside the sphere due to e at A and the charge $-e$ which it induces on the inner surface of the conducting sphere; so that, conversely, this field may be represented as due to e at A and an image $-ea/f$ at A'.

By an argument which is, step by step, the same as that of **6·3** (i) we can shew that the surface density at a point P is given by

$$\sigma = -\frac{e(a^2-f^2)}{4\pi AP^3} \quad \text{......................(1).}$$

The force on the point charge e is an attraction along AA' equal to $-ee'/AA'^2$ or $e^2af/(a^2-f^2)^2$.

6·34. Examples. (i) *A thin plane conducting lamina of any shape and size is under the influence of a fixed electrical distribution on one side of it. If σ_1 be the density of the induced charge at a point P on the side of the lamina facing the fixed distribution, and σ_2 that at the corresponding point on the other side, prove that $\sigma_1 - \sigma_2 = \sigma_0$, where σ_0 is the density at P of the distribution induced on an infinite plane conductor at zero potential coinciding with the lamina.*

Let L denote the lamina and R the remainder of the infinite plane of which L is a part. Denote the given fixed distribution by Σe. Consider an electric field which contains the lamina L insulated and uncharged, but under the influence of a hypothetical rigidly distributed charge on R, in which the density at every point of R is equal and opposite to the density which would be induced at that point of R by the charges Σe if the plane were a complete infinite plane at zero potential; so that in the notation of the question $-\sigma_0$ denotes the density at any point of R. Then since there is no other electricity in the field but this charge on R and the charge which it induces on the lamina, the field is symmetrical about the plane of the lamina, and tubes of force proceeding from the charge on R fall on both sides of the lamina in exactly the same way, so that at corresponding points on opposite sides the surface density has the same value, say σ_2. This field is represented in section in fig. (i).

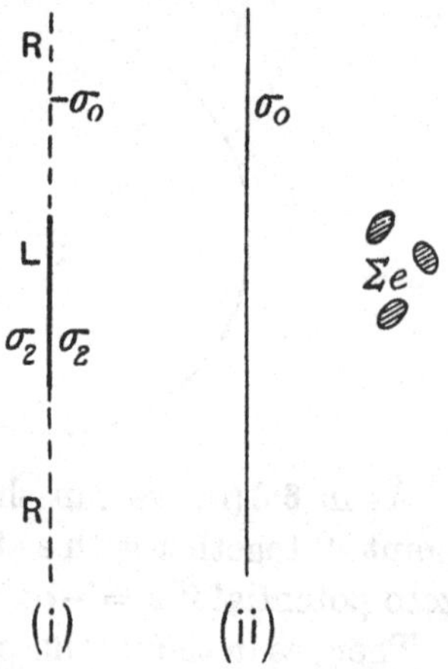

R
$-\sigma_0$
L
σ_2 σ_2
R
(i)

σ_0
Σe
(ii)

Consider a second field fig. (ii) in which the given fixed charges Σe induce a charge of density denoted by σ_0 on an infinite plane conductor, at zero potential, at the same distance from Σe as is the lamina.

Now superpose the fields by moving fig. (ii) to the left until the plane coincides with the plane in fig. (i). The charges on R annihilate one another, and there remain the charges Σe and the lamina L with a charge of density σ_2 on the side remote from Σe and a charge $\sigma_1 = \sigma_2 + \sigma_0$ at the corresponding point on its near side; so that we have proved that $\sigma_1 - \sigma_2 = \sigma_0$ as required.

(ii) *A straight wire of length 2l is charged with electricity of amount σ per unit length. It is placed in the presence of a conducting sphere of radius a and the sphere is earthed. The perpendicular distance from the centre of the sphere to the wire is c and the ends of the wire are equidistant from the centre of the sphere. Shew that the sphere receives a charge of amount*

$$-2\sigma a \sinh^{-1} l/c. \qquad \text{[M. T. 1934]}$$

It is assumed that the charge on the wire is 'rigid', i.e. that its distribution is unaffected by induction.

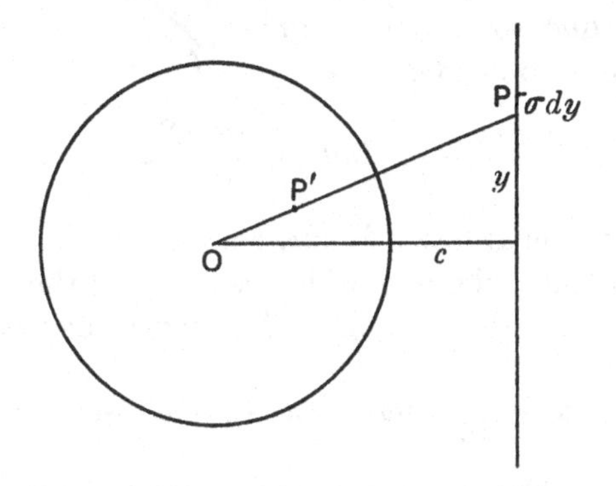

If σdy be the charge of an element dy of the wire at P, at a distance y from its middle point, there is an image at the inverse point P' with a charge

$$-\sigma dy . \frac{a}{OP} = -\frac{\sigma a\, dy}{\sqrt{(c^2+y^2)}}.$$

Hence the total induced charge, being the sum of all such image charges,

$$= -\int_{-l}^{l} \frac{\sigma a\, dy}{\sqrt{(c^2+y^2)}} = -2\sigma a \sinh^{-1} l/c.$$

6·4. Geometrical method of evaluating an induced charge. *If, in the solution of a problem, an electrostatic field in air is expressed as a field due to a number of point charges, then the charge induced on any portion of a conducting surface can be expressed in terms of the solid angles which this portion subtends at the point charges.*

Let dS be an element of area of the conducting surface at P subtending a small solid angle $d\omega$ at a point A, at which there is a point charge e; and suppose that it is required to find the contribution which this point charge makes to the induced charge on a finite portion of the surface, on the side remote from A.

If σ denotes the surface density, then the charge on dS is σdS, where $4\pi\sigma = E_n$, the normal intensity.

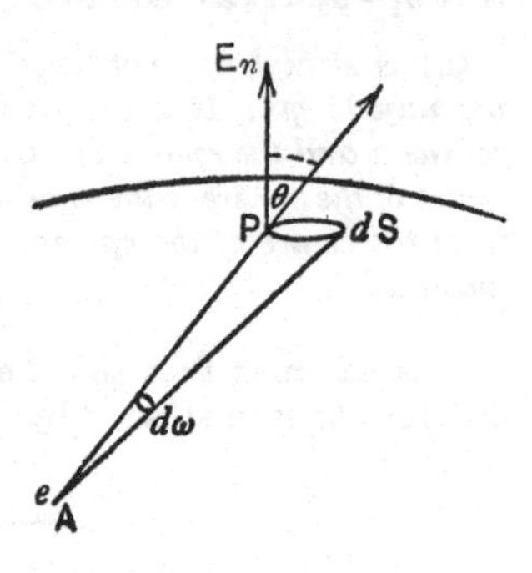

But the point charge e contributes to the intensity a component e/r^2 along AP, where $AP = r$. And if AP makes an angle θ with the normal, the normal component due to e is $e\cos\theta/r^2$, and its contribution to σdS is

$$\frac{e\cos\theta . dS}{4\pi r^2}; \quad \text{but} \quad \frac{\cos\theta . dS}{r^2} = d\omega,$$

therefore this contribution is $e d\omega/4\pi$.

If dS subtends solid angles $d\omega'$, $d\omega''$, ... at the other point charges e', e'', ..., they will make like contributions, so that

$$\sigma dS = \frac{1}{4\pi}(e d\omega + e' d\omega' + e'' d\omega'' + \ldots);$$

and by integrating we get, as the charge induced on any finite area S,

$$\frac{1}{4\pi}(e\omega + e'\omega' + e''\omega'' + \ldots) \quad \ldots\ldots\ldots\ldots\ldots(1),$$

where ω, ω', ω'', ... are the solid angles which S subtends at the charges.

It must be observed that we assumed that the induced charge lies on the side of the surface remote from the point charges, but in general some of the charges lie on one side of the surface and some on the other, and it is clearly necessary to change the sign of any term in (1) if the corresponding charge lies on the same side of the surface as the induced charge.

6·41. Examples. (i) *A point charge e is placed at distances a, b from two infinite conducting planes at zero potential which meet at right angles. Find the ratio in which the induced charge is divided between the planes.*

Let the plane of the paper cut the given planes at right angles in the lines Ox, Oy and contain the point charge e at A. If $ABCD$ is a rectangle with its sides parallel to Ox, Oy, it is easy to see that the two given planes would be at zero potential under the influence of point charges e at A and C and $-e$ at B and D.

The charge induced on the plane Ox can therefore be expressed in terms of the solid angles which it subtends at A, B, C and D. These are measured as in **1·33**. Thus the plane through Ox perpendicular to the plane of the paper, having a straight boundary through O and otherwise unlimited, subtends at A the solid angle between two planes, both at right angles to the plane of the paper and one passing through AO while the other is parallel to Ox; and the solid angle subtended at A being twice the angle between these planes is

$$2(\pi - AOx) \text{ or } 2(\pi - \tan^{-1} b/a).$$

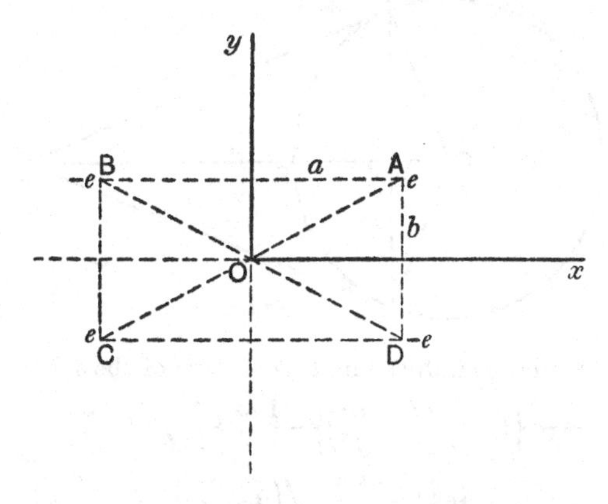

The plane clearly subtends an equal solid angle at D; and at B and C it subtends a solid angle $2OBA$ or $2\tan^{-1} b/a$.

Hence taking account of the fact that the charge induced on the plane is on the same side of it as the charges at A and B and on the opposite side to the charges at C and D, we have for the induced charge

$$-\frac{e}{4\pi}\cdot 2(\pi - \tan^{-1} b/a) + \frac{e}{4\pi}\cdot 2\tan^{-1}\frac{b}{a} + \frac{e}{4\pi}\cdot 2\tan^{-1}\frac{b}{a} - \frac{e}{4\pi}\cdot 2(\pi - \tan^{-1} b/a),$$

where we have written down the contributions of the charges at A, B, C, D in turn.

This makes the total charge induced on the plane through Ox equal to $-\frac{2e}{\pi}\tan^{-1}\frac{a}{b}$; similarly the charge induced on the plane through Oy is $-\frac{2e}{\pi}\tan^{-1}\frac{b}{a}$, the two together amounting to $-e$.

(ii) *A point charge e is placed at a distance f from the centre of an insulated uncharged sphere of radius a. Shew that the total charge on the smaller part of the sphere cut off by the polar plane of the point is*

$$-\tfrac{1}{2}e\left\{1+\frac{a^2}{f^2}-\sqrt{\left(1-\frac{a^2}{f^2}\right)}\right\}. \qquad \text{[M. T. 1919]}$$

In this case, as in **6·3** (ii), the point charges are e at A, $-ea/f$ at the inverse point A' which lies on the polar plane of A, and ea/f at the centre O. The solid angles which the spherical cap PBP' in the figure subtends at A, A' and O respectively are

$$2\pi(1-\cos OAP),\ 2\pi \text{ and } 2\pi(1-\cos POA)$$

or

$$2\pi\left\{1-\sqrt{\left(1-\frac{a^2}{f^2}\right)}\right\},\ 2\pi \text{ and } 2\pi\left(1-\frac{a}{f}\right).$$

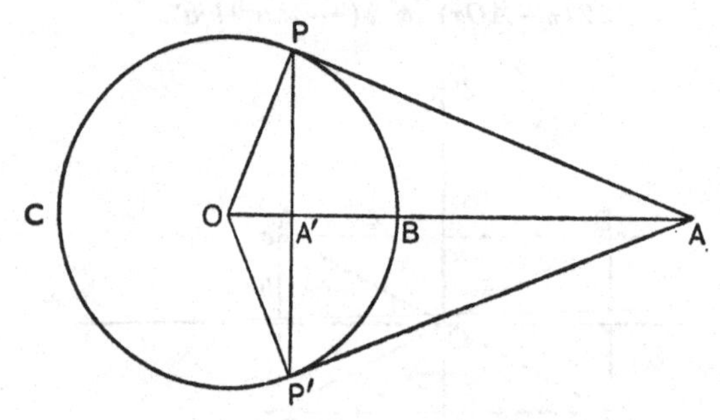

Therefore the charge induced on this portion of the sphere is

$$-\tfrac{1}{2}e\left\{1-\sqrt{\left(1-\frac{a^2}{f^2}\right)}\right\}-\frac{1}{2}\frac{ea}{f}+\frac{1}{2}\frac{ea}{f}\left(1-\frac{a}{f}\right)$$

or

$$-\tfrac{1}{2}e\left\{1+\frac{a^2}{f^2}-\sqrt{\left(1-\frac{a^2}{f^2}\right)}\right\}.$$

In the same way the spherical cap PCP' subtends at A, A' and O solid angles $2\pi(1-\cos OAP)$, 2π and $2\pi(1+\cos POA)$, and taking account of the fact that the induced charge is on the side of this surface remote from all the point charges its amount is

$$+\tfrac{1}{2}e\left\{1-\sqrt{\left(1-\frac{a^2}{f^2}\right)}\right\}-\frac{1}{2}\frac{ea}{f}+\frac{1}{2}\frac{ea}{f}\left(1+\frac{a}{f}\right)$$

or

$$+\tfrac{1}{2}e\left\{1+\frac{a^2}{f^2}-\sqrt{\left(1-\frac{a^2}{f^2}\right)}\right\},$$

as is otherwise obvious since the total charge is zero.

6·42. Point charge and an infinite conducting plane with a hemispherical boss. Let XX' be an infinite conducting plane with a hemispherical boss DCD' of centre O and radius a, and let a charge e be at a point A on the axis of the boss, where $OA=f\,(>a)$, the plane and boss being at zero potential. Then if B be the inverse of A in the sphere and A', B' the geometrical images of A, B in the plane, point charges e at A, $-e$ at A', $-ea/f$ at B and ea/f at B' will have the plane and sphere as a surface of zero potential. Therefore the field in air, i.e. on the right of the conductor, due to the charge e at A and the induced electricity will be the same as the field of the four point charges.

From **6·3** the surface density at a point P on the boss is given by

$$\sigma = -\frac{e(f^2-a^2)}{4\pi a . AP^3} + \frac{e(f^2-a^2)}{4\pi a . A'P^3} \quad \text{..................(1);}$$

and from **6·21** the surface density at a point Q on the plane is given by

$$\sigma = -\frac{ea}{2\pi . AQ^3} + \frac{(ea/f)\, OB}{2\pi . BQ^3},$$

or

$$\sigma = -\frac{ea}{2\pi . AQ^3} + \frac{ea^3}{2\pi f^2 . BQ^3} \quad \text{..................(2).}$$

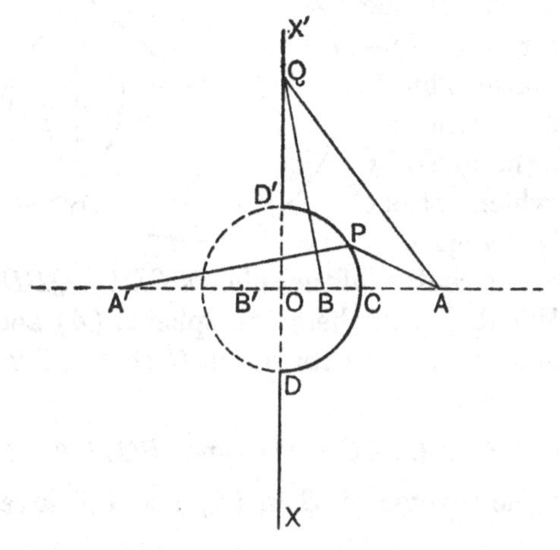

Applying the method of **6·4** to find the total charge induced on the boss, the solid angles subtended by the hemisphere DCD' at both A and A' are

$$2\pi(1-\cos OAD) \text{ or } 2\pi\{1-f/\sqrt{(a^2+f^2)}\};$$

at B the solid angle is

$$2\pi(1+\cos OBD) \text{ or } 2\pi\{1+a/\sqrt{(a^2+f^2)}\},$$

since $OB=a^2/f$; and at B' the solid angle is

$$2\pi(1-\cos OB'D) \text{ or } 2\pi\{1-a/\sqrt{(a^2+f^2)}\}.$$

Hence taking account of the signs of the charges and the side of the surface on which each lies, we get for the induced charge

$$-\frac{e}{2}\left\{1-\frac{f}{\sqrt{(a^2+f^2)}}\right\} - \frac{ea}{2f}\left\{1+\frac{a}{\sqrt{(a^2+f^2)}}\right\} + \frac{ea}{2f}\left\{1-\frac{a}{\sqrt{(a^2+f^2)}}\right\}$$
$$-\frac{e}{2}\left\{1-\frac{f}{\sqrt{(a^2+f^2)}}\right\},$$

the four terms being the contributions arising from the charges at A, B, B', A' in order, and the result is

$$-e\left\{1-\frac{f^2-a^2}{f\sqrt{(f^2+a^2)}}\right\}.$$

Since the total induced charge is $-e$, that on the plane must be

$$-e(f^2-a^2)/f\sqrt{(f^2+a^2)}.$$

6·5. Conductor bounded by two orthogonal spheres. The distribution of an electric charge on the surface of any conductor and the capacity of the conductor can be found, if we can determine a set of point charges which will have the surface of the conductor for an equipotential surface. This is well exemplified by the case of a conductor bounded by two spheres which cut one another at right angles.

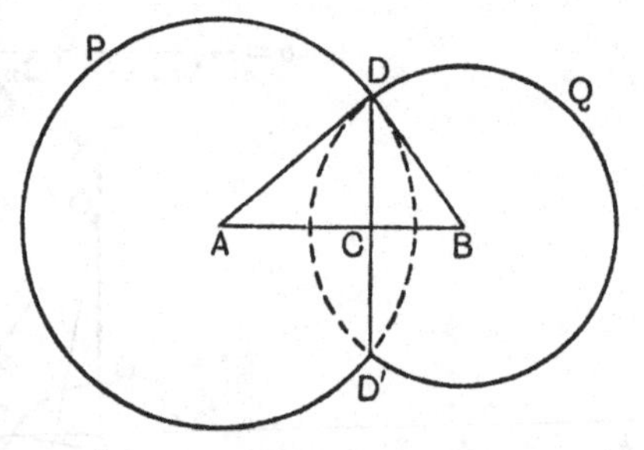

Let A, B be the centres of the spheres PDD', QDD' and a, b their radii. We shall call them the spheres (A) and (B). If their plane of intersection cuts AB in C, then ADB is a right angle, and we have

$$AB^2=a^2+b^2,\quad AC.AB=AD^2 \text{ and } BC.BA=BD^2;$$

so that C is the inverse of B in (A) and the inverse of A in (B).

Let a charge a be placed at A, its image in (B) is a charge $-ab/\sqrt{(a^2+b^2)}$ at C, and these charges together have (B) as a surface of zero potential. Similarly a charge b at B and the same image $-ab/\sqrt{(a^2+b^2)}$ at C would have (A) as a surface of zero potential. But the charge a alone would raise (A) to unit potential, and b alone would raise (B) to unit potential. Hence for a set of three point charges a at A, $-ab/\sqrt{(a^2+b^2)}$ at C and b at B the two spheres would be at unit potential; and by the theorem of the equivalent layer, if the total charge

$$a+b-ab/\sqrt{(a^2+b^2)}$$

were distributed over the surface $PDQD'$, the external field would be unaltered, and the potential of the surface would still be unity, so that the capacity of the conductor is

$$a+b-ab/\sqrt{(a^2+b^2)}.$$

Again the surface density at a point P on (A) due to the charge a at A is $a/4\pi a^2$, and due to b at B and its image at C is $-\dfrac{b\,(AB^2-a^2)}{4\pi a\,.\,BP^3}$ (**6·3** (i) (3)), so that the whole surface density at P is $\dfrac{1}{4\pi a}\left(1-\dfrac{b^3}{BP^3}\right)$, and in like manner the surface density at a point Q on (B) is $\dfrac{1}{4\pi b}\left(1-\dfrac{a^3}{AQ^3}\right)$.

The whole charge on the portion DPD' of the surface may be found by the method of **6·4**. Thus the solid angles which it subtends at A, C and B are respectively

$$2\pi\,(1+\cos CAD),\ \ 2\pi \text{ and } 2\pi\,(1-\cos CBD)$$

or

$$2\pi\left\{1+\frac{a}{\sqrt{(a^2+b^2)}}\right\},\ \ 2\pi \text{ and } 2\pi\left\{1-\frac{b}{\sqrt{(a^2+b^2)}}\right\};$$

so that the charge on DPD' is

$$\tfrac{1}{2}a\left\{1+\frac{a}{\sqrt{(a^2+b^2)}}\right\}-\frac{ab}{2\sqrt{(a^2+b^2)}}+\tfrac{1}{2}b\left\{1-\frac{b}{\sqrt{(a^2+b^2)}}\right\}$$

or

$$\frac{1}{2}\left\{a+b+\frac{a^2-ab-b^2}{\sqrt{(a^2+b^2)}}\right\};$$

and similarly for the charge on DQD'.

6·51. Series of images. The fact that C was a common inverse point in **6·5** enabled us to get a solution with only three point charges, and in any case in which the spheres cut at an angle which is a submultiple of π the problem could be solved by a finite number of point charges. But otherwise, as in the case of two non-intersecting spheres (A) and (B) of radii a and b, of which (A) is at unit potential and (B) is at zero potential, if we begin by placing a charge a at the centre of (A) its image must be taken in (B) and then the image in (A) of this image must be taken, and so on indefinitely, so that there would be an infinite series of images. The sum of the charges inside (A) would be its coefficient of capacity q_{11}, and the sum of the charges inside (B) would be the coefficient of mutual induction q_{12}. In some cases the distance between the conductors is large compared with their radii, so that approximate results may be found by taking a few terms of the series of images.

6·52. Example. *The outer of two concentric spherical conductors of radii a, b $(a<b)$ is earth connected and the inner is charged to potential ϕ. A point charge Q is then placed in the space between the spheres at a distance c from the centre. Prove that the charges on the spheres are as*

$$Q/c-Q/b-\phi : Q/a-Q/c+\phi.$$

[I. 1896]

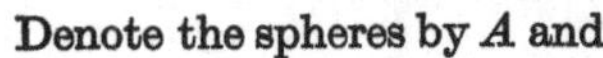

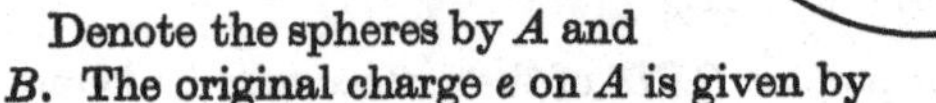

Denote the spheres by A and B. The original charge e on A is given by

$$\phi=\frac{e}{a}-\frac{e}{b}, \text{ or } e=\frac{ab\phi}{b-a} \quad \text{.....................(1).}$$

Let O be the centre and P the position of the charge Q; $OP=c$.
Let P_1 be the inverse of P in A,
Q_1 be the inverse of P_1 in B,
P_2 be the inverse of Q_1 in A,
Q_2 be the inverse of P_2 in B and so on;

and denote image charges by square brackets. Then we have

$$OP_1=\frac{a^2}{c},\ [P_1]=-Q\frac{a}{c}; \qquad OQ_1=\frac{b^2}{OP_1}=\frac{b^2c}{a^2},\ [Q_1]=-[P_1]\frac{b}{OP_1}=Q\frac{b}{a};$$

$$OP_2=\frac{a^2}{OQ_1}=\frac{a^4}{b^2c},\ [P_2]=-[Q_1]\frac{a}{OQ_1}=-Q\frac{a^2}{bc}; \qquad OQ_2=\frac{b^2}{OP_2}=\frac{b^4c}{a^4},\ [Q_2]=-[P_2]\frac{b}{OP_2}=Q\frac{b^2}{a^2};$$

$$OP_3=\frac{a^2}{OQ_2}=\frac{a^6}{b^4c},\ [P_3]=-[Q_2]\frac{a}{OQ_2}=-Q\frac{a^3}{b^2c}; \text{ and so on.}$$

Similarly, let Q_1' be the inverse of P in B,
P_1' be the inverse of Q_1' in A,
Q_2' be the inverse of P_1' in B and so on;

then in the same way we may shew that

$$[Q_1']=-Q\frac{b}{c}; \quad [P_1']=Q\frac{a}{b},$$

$$[Q_2']=-Q\frac{b^2}{ac}; \quad [P_2']=Q\frac{a^2}{b^2},$$

$$[Q_3']=-Q\frac{b^3}{a^2c}; \text{ and so on.}$$

Then the total charge on the sphere A is

$$e+[P_1]+[P_2]+[P_3]+\dots+[P_1']+[P_2']+\dots$$

$$=\frac{ab\phi}{b-a}-Q\frac{a}{c}\left(1+\frac{a}{b}+\frac{a^2}{b^2}+\dots\right)+Q\frac{a}{b}\left(1+\frac{a}{b}+\frac{a^2}{b^2}+\dots\right)$$

$$=\frac{ab}{a-b}\left\{\frac{Q}{c}-\frac{Q}{b}-\phi\right\}.$$

And since there is no external field the total charge on B is minus the sum of the charges it contains, i.e.

$$-Q-\frac{ab}{a-b}\left\{\frac{Q}{c}-\frac{Q}{b}-\phi\right\}, \text{ or } \frac{ab}{a-b}\left\{\frac{Q}{a}-\frac{Q}{c}+\phi\right\};$$

and hence the result.

6·6. Field due to an electric doublet. An arrangement of two equal and opposite point charges $-e$ and e at an infinitesimal distance δs apart constitutes an **electric doublet**, or **bipole**, the magnitude of the charges being such that the product $e\delta s$ has a finite limit as e increases and δs decreases.

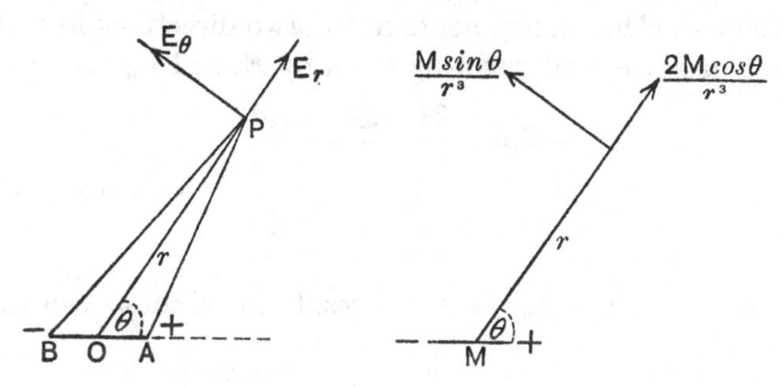

This limit is called the **moment of the doublet** and denoted by M, and the line joining the charges in the sense from the negative to the positive charge is called the **axis of the doublet.**

In the chapter on magnetism we shall prove that M may be used as a vector to denote both the magnitude of the moment and the direction of the axis of the doublet and that doublets can be compounded and resolved by applying the vector law of addition to their moments.

Let a doublet consist of charges $-e$ at B and e at A, where $AB=\delta s$, and the limit of $e\delta s$ as e increases and δs decreases is finite and $=M$.

Let O be the middle point of BA, and let P be a point (r, θ) referred to O as origin and the axis BA as axis of co-ordinates. The field is clearly symmetrical about the axis of the doublet, and the potential at P is given by

$$\phi = \frac{e}{AP} - \frac{e}{BP}$$

$$= \frac{e}{r - \frac{1}{2}\delta s \cos\theta} - \frac{e}{r + \frac{1}{2}\delta s \cos\theta}$$

$$= \frac{e}{r}\left(1 + \frac{\delta s \cos\theta}{2r} \dots\right) - \frac{e}{r}\left(1 - \frac{\delta s \cos\theta}{2r} \dots\right)$$

$$= \frac{e\,\delta s \cos\theta}{r^2} \text{ to the first order in } \delta s$$

or $$\phi = \frac{M\cos\theta}{r^2} \quad \dots\dots\dots\dots(1).$$

Since ϕ is a function of two variables r, θ, therefore the electric intensity at P has components in the two directions in which r and θ increase; and denoting them by E_r and E_θ, we have

$$\left.\begin{aligned} E_r &= -\frac{\partial\phi}{\partial r} = \frac{2M\cos\theta}{r^3} \\ \text{and} \qquad E_\theta &= -\frac{1}{r}\frac{\partial\phi}{\partial\theta} = \frac{M\sin\theta}{r^3} \end{aligned}\right\} \quad \dots\dots\dots(2);$$

and the electric vector $\mathbf{E}$ is the resultant of these two components.

6·61. Insulated conducting sphere in a uniform field of force. If in 6·3 (ii) we imagine the point charge e outside the insulated sphere to move away to an infinite distance and at the same time to increase in such a way that e/f^2 remains finite and equal to F, the lines of force in the finite part of the field will be lines radiating from a point at infinity, i.e. parallel lines and the undisturbed field will be of uniform intensity F. At the same time, as the point A in the figure of 6·3 (ii) moves to infinity the point A' moves up to O, and the charges $-ea/f$ at A' and ea/f at O, being at a distance a^2/f apart, form a doublet of moment ea^3/f^2, or Fa^3, whose axis is in the direction of the undisturbed field F.

It follows that when an insulated sphere of radius a is placed in a uniform field of intensity F the induced charge produces the same external field as would a doublet of moment Fa^3 placed at the centre of the sphere with its axis in the direction of the field.

Take the undisturbed field F from left to right as indicated in the figure; then the total field at a point (r, θ) outside the sphere is made up of three components, viz. the given field F and the two components of intensity due to the doublet Fa^3, i.e. $\dfrac{2Fa^3 \cos\theta}{r^3}$ radially and $\dfrac{Fa^3 \sin\theta}{r^3}$ transversely in the sense in which θ increases. It is easy to see that at a point on the

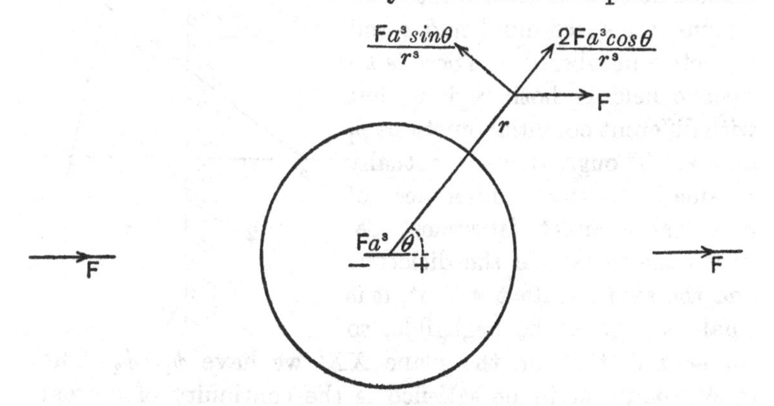

surface of the sphere, these three are components of a single resultant $3F \cos\theta$ along the normal. Therefore the surface density induced on the sphere is given by

$$4\pi\sigma = 3F \cos\theta \quad \text{..................(1).}$$

It is positive over one half of the sphere and negative over the other half, the circle $\theta = \frac{1}{2}\pi$ being a line of no electrification.

The total positive charge on the hemisphere can be expressed as an integral, viz.

$$\int_0^{\frac{1}{2}\pi} 2\pi a^2 \sigma \sin\theta \, d\theta = \tfrac{3}{4} F a^2 \quad \text{...............(2),}$$

and there is an equal and opposite negative charge.

It should be noted that if we take an axis of x in the direction of the uniform field F, the potential of the field must be $-Fx$

(in order that the negative gradient may be F), and the potential of the doublet is by **6·6** $Fa^3 \cos\theta/r^2$, so that the whole potential of the external field is

$$\phi = -Fx + \frac{Fa^3 \cos\theta}{r^2} \quad \text{.................(3).}$$

6·7. Dielectric problems. A point charge in front of a block of uniform dielectric with an infinite plane face. Let e be the charge at A, XX' the plane face of the block of dielectric and A' the point in the dielectric such that XX' bisects AA' at right angles. The plane XX' divides space into two regions, air space numbered 1 and dielectric numbered 2. There is an electric field in both regions, but with different potential functions ϕ_1 and ϕ_2. Though there is actually a small constant difference of potential (contact difference) between the surface of the dielectric and the air in contact with it, it is small enough to be negligible, so we assume that on the plane XX' we have $\phi_1 = \phi_2$. The other condition to be satisfied is the continuity of normal displacement (**5·21** (5)), since there is no charge on the interface.

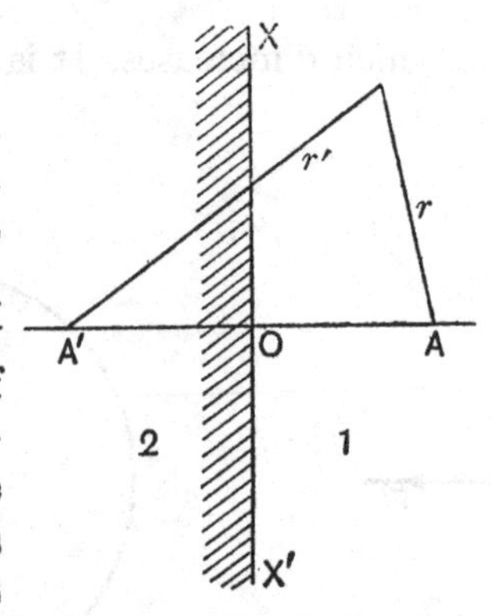

We will shew that we can satisfy these conditions by means of potentials, in air

$$\phi_1 = \frac{e}{r} + \frac{e'}{r'} \quad \text{.......................(1),}$$

where r, r' denote distances from A, A'; and in dielectric

$$\phi_2 = \frac{e''}{r} \quad \text{..........................(2).}$$

We remark that we cannot have a term of the form e'''/r' in ϕ_2, because it would make the potential infinite at A', and there is no singularity in the dielectric which could account for such an infinity in ϕ_2.

Then since $\phi_1=\phi_2$ on XX' (where $r=r'$), we must have

$$e+e'=e'' \quad \text{.......................(3).}$$

And since the normal displacement is got by multiplying the normal intensity by the dielectric constant, the condition of continuity of normal displacement requires that

$$e-e'=Ke'' \quad \text{.......................(4),}$$

where K is the specific inductive capacity of the dielectric, the minus sign arising from the fact that e' is on the opposite side of the interface to e and e''.

Thus all conditions are satisfied by

$$e'=\frac{1-K}{1+K}e \text{ and } e''=\frac{2e}{1+K} \quad \text{...........(5).}$$

That is to say, the field in air is identical with a field created by the given charge e at A together with a charge $\frac{1-K}{1+K}e$ at A'; and the field in the dielectric is such as would be produced by a single charge $\frac{2e}{1+K}$ at A. It follows that the lines of force in the dielectric are straight lines radiating from A as shewn in the figure.

The force on the charge e is ee'/AA'^2 or $ee'/4a^2$, if a is the distance of the charge from the block. This means that the charge is attracted to the dielectric block, or *vice versa* by a force $\frac{K-1}{K+1}\frac{e^2}{4a^2}$, representing a tension in the lines of force.

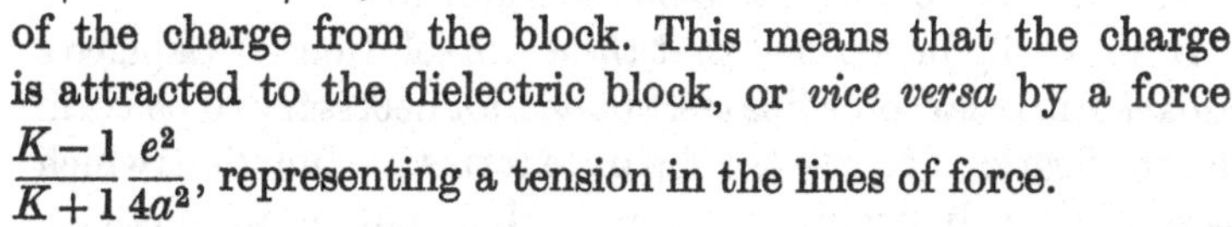

6·71. Dielectric sphere in a uniform field of force. Let a sphere of radius a of homogeneous dielectric of constant K be placed in a uniform field of electric intensity F. To find how the field is modified by the presence of the sphere. The surface of the sphere divides space into two regions—air, numbered 1 and dielectric numbered 2.

Take an axis Ox through the centre O of the sphere in the direction of the undisturbed field F. Since the negative gradient of the potential is F, the potential of the undisturbed field must be given by

$$\phi = -Fx = -Fr\cos\theta \quad \text{..................(1).}$$

To solve the problem we have to find potential functions, for regions 1 and 2, i.e. solutions of Laplace's equation, continuous across the surface $r=a$, making the normal displacement continuous at $r=a$, and tending to the form (1) at a great distance from the sphere.

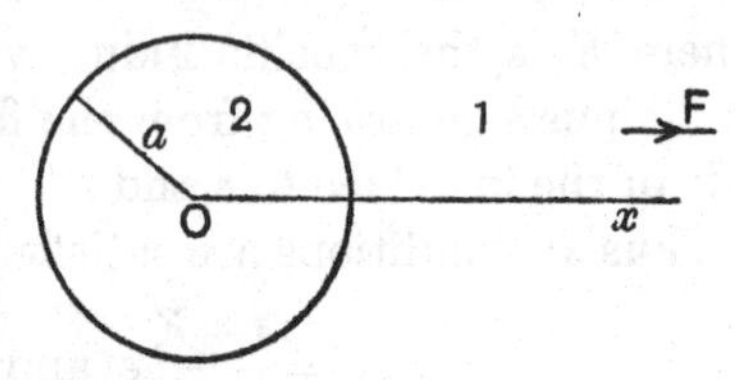

Further we notice that in **6·7** the field in the dielectric is a definite multiple of what the field would be if the dielectric were not present; and the field in air due to the presence of the dielectric is a definite multiple of what the field would be if the dielectric were a conductor. So we endeavour to obtain a solution of this problem in analogous fashion, remembering that if the sphere were a conductor it would produce in region 1 the same effect as a doublet at its centre.

We assume therefore that the potentials in regions 1 and 2 are given by

$$\phi_1 = -Fr\cos\theta + \frac{A\cos\theta}{r^2} \quad \text{..............(2)}$$

and

$$\phi_2 = Br\cos\theta \quad \text{........................(3),}$$

where A, B are constants to be determined.

It can easily be verified that these are solutions of Laplace's equation in polar co-ordinates, but it is not necessary to perform the verification because the separate terms are functions which have previously occurred in our work as potential functions.

As r increases, ϕ_1 tends to the form $-Fr\cos\theta$, which is one of the requisite conditions to be satisfied. Also, when $r=a$, we have $\phi_1=\phi_2$, so that

$$-Fa + \frac{A}{a^2} = Ba \quad \text{....................(4);}$$

and $\dfrac{\partial\phi_1}{\partial r} = K\dfrac{\partial\phi_2}{\partial r}$, so that

$$-F - \frac{2A}{a^3} = KB \quad \text{....................(5).}$$

Therefore all the conditions are satisfied by (2) and (3) provided that

$$A=\frac{K-1}{K+2}Fa^3 \text{ and } B=-\frac{3F}{K+2} \quad \text{.........(6);}$$

i.e. the field in the dielectric is uniform and $3/(K+2)$ of what it would be if the dielectric were not present. But we must remember that it is D_n or KE_n which is continuous, so that the

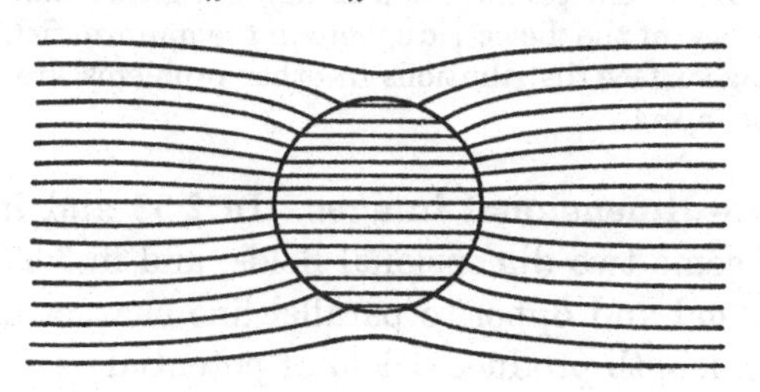

number of unit tubes of force which cross a plane area perpendicular to the field in the dielectric is $3K/(K+2)$ of what it would be if the dielectric were not present; this means that, for large values of K, there are roughly three times as many tubes of force in the dielectric as there would be in air.

Also, in air the dielectric sphere produces the same effect as would a doublet of moment $\frac{K-1}{K+2}Fa^3$ placed at its centre.

6·72. Converse problems. We may use the case just considered to illustrate a converse problem: *What distribution of electricity in air will produce a given field of potential?*

The solution is found from the fact that at every point of space at which there is a volume density ρ, it is determined by Poisson's equation **2·41** (3)

$$\nabla^2\phi=-4\pi\rho \quad \text{...........................(1)}$$

and on every surface on which there is a surface density σ by **2·411** it is given by

$$\frac{\partial\phi_1}{\partial n_1}+\frac{\partial\phi_2}{\partial n_2}=-4\pi\sigma \quad \text{.......................(2),}$$

where ∂n_1, ∂n_2 are directed away from the surface.

Now apply this process to the field given by

$$\phi_1=-Fr\cos\theta+\frac{K-1}{K+2}\frac{Fa^3\cos\theta}{r^2} \quad (r>a) \quad \text{...........(3),}$$

$$\phi_2=-\frac{3Fr\cos\theta}{K+2} \quad (r<a) \quad \text{........(4).}$$

It is easy to verify that $\nabla^2\phi_1=0$ and $\nabla^2\phi_2=0$, so that there is no volume density of electricity anywhere.

Also $\partial n_1 = \partial r$ and $\partial n_2 = -\partial r$, so that by differentiating and putting $r=a$ in (2),

$$-4\pi\sigma = -F\cos\theta - 2\frac{K-1}{K+2}F\cos\theta + \frac{3F\cos\theta}{K+2}$$

or

$$\sigma = \frac{3}{4\pi}\frac{K-1}{K+2}F\cos\theta \quad \text{......................(5).}$$

Hence a surface distribution of density σ, given by (5), on a sphere of radius a in air, would produce the same field inside and outside as exists in the case of the dielectric sphere in the uniform field. This and corresponding surface distributions in other problems are sometimes called '*fictive layers*'.

6·8. Two-dimensional images. In **2·53** and in **3·43** we considered some two-dimensional fields, and in **2·532** we saw that two equal and opposite parallel line charges of amount $\pm e$ per unit length produce a field of potential

$$\phi = C - 2e\log r + 2e\log r' \quad \text{..............(1),}$$

where r, r' denote distances from the lines.

It follows that the equipotential curves in the two-dimensional field are the family of coaxial circles with limiting points A, A' at the charges.

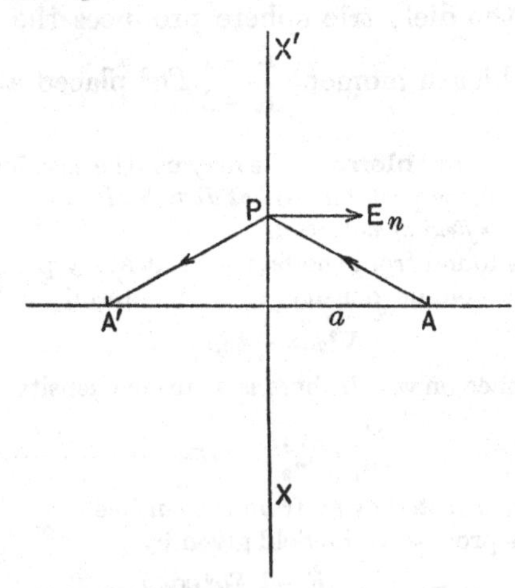

The plane XX' bisecting AA' at right angles is also an equipotential surface, and will be the surface $\phi = 0$ if we take $C = 0$.

Hence the image of a line charge through A with regard to a parallel conducting plane XX' at zero potential is an equal and opposite line charge through A'.

The surface density induced at any point P on the plane is given by

$$4\pi\sigma = E_n = -\frac{2e}{AP}\cdot\frac{a}{AP} - \frac{2e}{PA'}\cdot\frac{a}{PA'}$$

or

$$\sigma = -ea/\pi AP^2 \quad \text{.....................(2)},$$

where a is the distance of A from XX'.

6·81. Image of a uniform line charge in a parallel circular cylinder. Continuing the same argument, if O be the centre of one of the coaxial circles which are equipotential curves surrounding the point A', the limiting points A, A' are inverse points with regard to this circle. Let a be the radius of the circle and let $OA = f$.

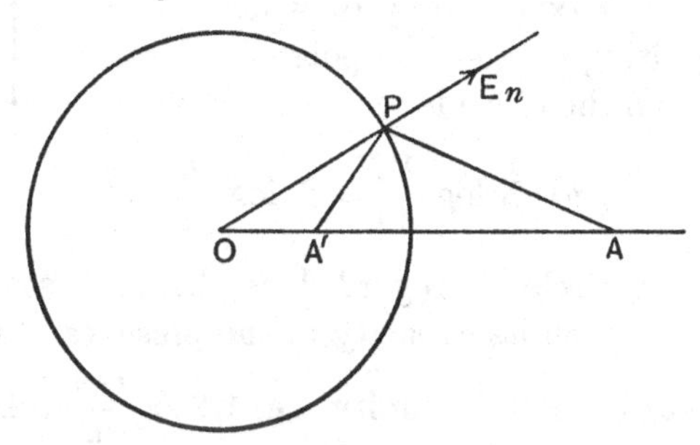

The potential due to line charges $\pm e$ through A and A' is

$$\phi = C - 2e \log r + 2e \log r',$$

and at a point P on the circle this takes the form

$$C - 2e \log \frac{AP}{A'P} = C - 2e \log \frac{f}{a};$$

so that the cylinder of which the circle is a cross-section will be at zero potential if we take $C = 2e \log (f/a)$. We can then shew as in **6·3** (i) that the surface density at P is given by

$$\sigma = -\frac{e(f^2 - a^2)}{2\pi a . AP^2} \quad \text{.....................(1)}.$$

To deduce the case of the cylinder insulated and uncharged

since the charge per unit length when the potential is zero is the image charge $-e$, we must superpose the field due to a charge e uniformly distributed, and that is equivalent to a line charge e through O.

6·82. Capacity of a telegraph wire. Reverting to **6·8**, let XX' be the surface of the earth and consider the field of horizontal line charges e, $-e$ through A, A', where AA' is bisected at right angles by XX'.

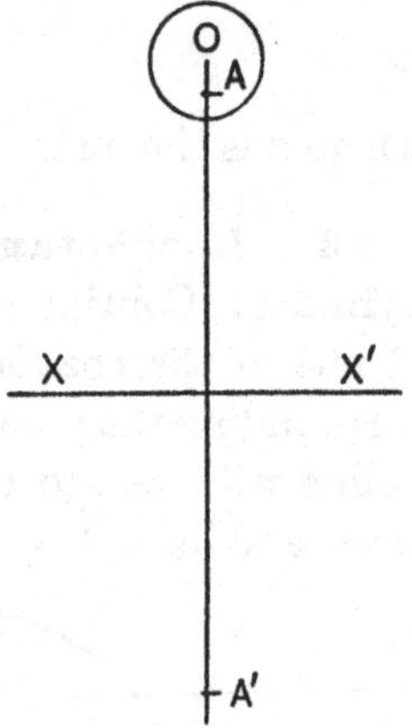

The potential

$$\phi = C - 2e \log r + 2e \log r'$$

will be such that XX' is a surface of zero potential if $C=0$. Take a small circle of centre O and radius a and let it be one of the coaxial family of which A, A' are limiting points. The potential at a point P on the circle is

$$\phi = 2e \log \frac{A'P}{AP} = 2e \log \frac{OA'}{a}.$$

Now let the circle be regarded as the cross-section of a telegraph wire, then its capacity in the presence of the earth is the value of e when ϕ is unity, i.e. $1/2 \log \dfrac{OA'}{a}$. But if h is the height of O, $OA + OA' = 2h$ and $OA \,.\, OA' = a^2$, so that $OA' = h + \sqrt{(h^2 - a^2)}$; and the capacity

$$= 1/2 \log \left\{ \frac{h + \sqrt{(h^2 - a^2)}}{a} \right\}.$$

Since a is small compared to h, the value is approximately $1/2 \log (2h/a)$.

EXAMPLES

1. A positive point charge at A is fixed in front of a large conducting plane which is earthed. Shew that the tubes of force, which start on the point charge and make an angle ϕ or less with the normal from A to the plane (not with the normal produced), end on a circular region of the plane whose radius subtends an angle θ at A, where

$$\sin \tfrac{1}{2}\phi = \sqrt{2} \sin \tfrac{1}{2}\theta.$$ [M. T. 1930]

2. Two equal charges are condensed at points distant a and b from an infinite conducting plate which is to earth: the charges are on the same side and on the same line perpendicular to the plate. Shew that the lines of force from the nearer charge fall on a circular area of the plate of radius r, where

$$a/\sqrt{(a^2+r^2)} + b/\sqrt{(b^2+r^2)} = 1.$$ [I. 1907]

3. Two equal point charges of opposite signs are placed at points A, B whose distance apart is $2a$, in front of an infinite plane conductor, the distance of each from the plane conductor being a. Shew that the force required to maintain either charge in its position is inclined to the plane at an angle of 45°, and is less than it would be if the plane were absent in the ratio $(2\sqrt{2}-1):2$. [St John's Coll. 1906]

4. Two equal charges are placed at a distance a from an infinite conducting plane at zero potential, and at a distance $3a/2$ from each other. Shew that, if these charges have the same sign, the resultant mechanical force experienced by either charge makes an angle $\tan^{-1} \frac{224}{243}$ with the normal to the plane. [M. T. 1915]

5. A positive charge e is placed at distances a and b from two semi-infinite planes at zero potential intersecting at right angles. Find the surface densities of electrification on the nearest point of each plane. [M. T. 1927]

6. Two point charges A, B, each of amount Q, are placed at equal distances from each other and from an infinite conducting plane sheet at potential zero. Find the charges induced on the portions of the sheet cut off by the plane through A perpendicular to AB. Sketch the lines of force in the plane through AB perpendicular to the sheet. [M. T. 1924]

7. A conductor is formed of two planes inclined to one another at an angle $\frac{1}{3}\pi$; the planes are bounded by their line of intersection and are of very great area; the conductor is maintained at zero potential and is influenced by a charge e at a point P in the space between the two planes, such that the perpendicular from P to the common edge makes angles α, $\frac{\pi}{3}-\alpha$ with the planes. Prove that the charges induced on the two planes are $e\left(-1+\frac{3\alpha}{\pi}\right)$, $-e.\frac{3\alpha}{\pi}$ respectively. [I. 1902]

8. Shew that, if the inducing charge is at a distance from the centre of a sphere at zero potential equal to double the radius, the surface densities at the nearest and the most remote points of the sphere are in the ratio 27 : 1. [St John's Coll. 1907]

9. A point charge e is held at a distance f from the centre of an insulated spherical conductor of radius a, which carries a charge Q. Prove that the surface density at the point of the sphere most remote from the charge e will be zero if

$$Q = -ea^2(3f+a)/\{f(f+a)^2\}.$$

[St John's Coll. 1906]

10. An insulated conducting sphere of radius a is under the influence of a point charge e at a distance f ($>a$) from its centre. What is the least positive charge that must be given to the sphere in order that the surface density may be everywhere positive. [M. T. 1893]

11. A charge e is placed at a distance c from the centre of an insulated uncharged sphere of radius a, c being greater than a. Shew that the part of the sphere which is positively charged is separated from that which is negatively charged by a circle of points at distance from e given by

$$r^3 = c(c-a)(c+a).$$

[M. T. 1933]

12. A point charge e is placed at a point A at a distance f from the centre of an insulated conducting sphere of radius a ($<f$). Determine the charge that must be placed on the sphere in order that the line of no electrification may be the intersection of the sphere by the polar plane of A; and shew that the numerical ratio of the greatest negative and positive surface densities is in this case $\{(f+a)/(f-a)\}^{\frac{3}{2}}$. [I. 1932]

13. A sphere is insulated and carries a charge equal to that placed at an external point; prove that if the ratio of the distance of the point from the centre to the radius is greater than 2·62 : 1, the surface density on the sphere will be nowhere negative. [St John's Coll, 1909]

14. The centre of an *insulated uncharged* conducting sphere of radius a is at the middle point of the straight line joining two equal point charges of electricity, which are a distance $2c$ apart. If a/c is small, shew that the force on either charge would be increased in the approximate ratio

$$1 + \frac{24a^5}{c^5} : 1,$$

if the sphere were removed. [M. T. 1931]

15. An insulated uncharged conducting sphere is placed centrally between two charges of equal magnitude. Shew that if they are of like signs the repulsion between them is diminished; but if of unlike signs the attraction between them is increased. [M. T. 1921]

16. If a sphere of radius a is earthed and positive charges e, e' are placed on opposite sides of the sphere, at distances $2a$, $4a$ respectively from the centre and in a straight line with it, shew that the charge e' is repelled from the sphere if

$$e' < 25e/144.$$ [M. T. 1935]

17. A field is produced by a point charge e in the presence of an uninsulated spherical conductor of radius a whose centre is at a distance c from the charge. Find the force with which the charge is attracted to the sphere; and shew that the force is changed to a repulsion by insulating the sphere and connecting it to a large distant conductor at potential ϕ provided that

$$\phi > \frac{ec^3}{(c^2 - a^2)^2}.$$ [St John's Coll. 1915]

18. A conducting sphere of radius a at zero potential is under the influence of a charge at an external point O. Shew that the fraction of the induced charge which lies within a right circular cone with vertex O and axis through the centre of the sphere is $(a-c)/a$, where $2c$ is the intercept made by the sphere on any generator of the cone, and both intercepts made by the cone on the sphere are included.

[I. 1911]

19. A point charge e is placed at a distance $6a$ from the centre of an insulated sphere of radius a, which has a charge Q. Shew that if there is a point of equilibrium in the field, whose distance from the centre is $4a$ taken towards e, then $Q = \frac{447}{3174} e$.

Give a rough drawing to indicate the nature of the equipotential surfaces. [M. T. 1915]

20. Prove that, if a small charged conductor be at a distance r from the centre in a space bounded by a spherical conducting surface of radius a at zero potential, the force repelling it from the centre is $are^2/(a^2 - r^2)^2$, where e is the charge. [M. T. 1923]

21. Within a spherical hollow, in a conductor connected to earth, equal point charges e are placed at equal distances f from the centre, on the same diameter. Shew that each is acted on by a force equal to

$$e^2\left[\frac{4a^3f^3}{(a^4 - f^4)^2} + \frac{1}{4f^2}\right].$$ [St John's Coll. 1905]

22. A point charge e is held at a distance f from the centre of an insulated uncharged spherical conductor of radius a; prove that the force required to hold the point charge in position is $e^2af/(f^2 - a^2)^2$ or $e^2a^3(a^2 - 2f^2)/f^3(f^2 - a^2)^2$, according as f is less or greater than a.

[St John's Coll. 1907]

23. In a spherical cavity of radius a within a conductor a charge e is placed at a point P distant f from the centre C.

Shew that a line of force from e making initially an angle α with PC

will meet the surface of the cavity at a distance $(a^2-f^2)/(a-f\cos\alpha)$ from P. [I. 1906]

24. A charge e of electricity is situated at a distance c from the centre of a conducting sphere of radius a. If the sphere is insulated and uncharged, find the rise in its potential due to the presence of the charge, and shew that there is an attractive force between the sphere and the charge equal to

$$e^2 \cdot \frac{a^3(2c^2-a^2)}{c^3(c^2-a^2)^2}.$$ [M. T. 1925]

25. A point charge e is placed inside a hemispherical hollow of radius a in an uninsulated conductor. The point charge is on the axis of symmetry at a distance f from the plane boundary. Shew that it is in equilibrium if f/a is equal to a root of the equation

$$x^8 - 8x^7 - 2x^4 - 8x^3 + 1 = 0.$$

Shew also that this equation has one and only one root between 0 and 1. [M. T. 1932]

26. An infinite plane conductor at zero potential with a hemispherical boss of radius a is under the influence of a point charge $+e$ at a point on the axis of symmetry distant $2a$ from the plane. Find the potential at any point and sketch the lines of force. Determine the slope at the point charge of the critical line of force separating lines of force going to the boss and to the plane. [M. T. 1928]

27. An infinite plane conducting sheet has upon it a hemispherical conducting boss of radius a and the whole is at potential zero. A charge e is placed at a distance c from the centre on the radius of the sphere which is normal to the plane. Shew that the density at any point of the plane is

$$\sigma = -\frac{e}{2\pi c^2}\left(\frac{c^3}{r^3} - \frac{a^3}{r'^3}\right),$$

where r, r' are the distances of the point from the charge e and its image in the sphere.

If a/c is small, shew that σ is greatest at a distance $a^{\frac{3}{5}} c^{\frac{2}{5}}$ approximately from the centre. [M. T. 1934]

28. An infinite plane has a hemispherical boss, and a unit point charge is placed in front of the boss on the common normal to the plane and the hemisphere. Shew that the charge induced on the part of the plane external to the boss is $-\cos^2\theta\sec\theta$, where θ is the angle subtended at the charge by any radius of the base of the hemisphere.

[M. T. 1916]

29. A dome of conducting material is built in the form of a hemisphere on the ground; a small charged conductor is situated midway between the dome and the ground, on the vertical through the centre of the dome. Obtain the system of images, and prove that the mechanical force on the small conductor is upwards, and equal to $\frac{47}{225}e^2/a^2$, where e is the charge on the conductor and a is the radius of the dome. [M. T. 1914]

30. A point charge e at P is at a distance $PO=c$ from an infinite plane conductor, which is uninsulated. Shew that, when a conducting hemisphere of radius a ($<c$) is placed on the plane with its centre at O and its vertex towards P, the number of unit tubes of force which fall on the plane outside a circle of centre O and radius r ($>a$) is decreased by $ea^3/c\,(a^4+c^2r^2)^{\frac{1}{2}}$. [M. T. 1909]

31. A hollow conductor has the form of a quarter of a sphere bounded by two perpendicular diametral planes. Find the image of a charge placed at any point inside it. [M. T. 1897]

32. A conducting surface consists of two infinite planes which meet at right angles and a quarter of a sphere of radius a fitted into the right angle. If the conductor is at zero potential and a point charge e is symmetrically placed, with regard to the planes and the spherical surface, at a great distance f from the centre, shew that the charge induced on the spherical portion is approximately $-5ea^3/\pi f^3$. [M. T. 1903]

33. A thundercloud, which may be regarded as an electric doublet with its axis vertical, is moving uniformly along a horizontal straight line and directly approaching an observer on the ground who is recording the electric intensity close to the surface. Shew that the rate of change of the electric intensity vanishes when the elevation of the cloud is $\tan^{-1}\frac{1}{2}$. [M. T. 1923]

34. A sphere of radius a has its centre at a distance c from an infinite plane conductor at zero potential. Shew that its capacity is, approximately, $a+a^2/2c$, where a/c is regarded as small. [M. T. 1911]

35. A condenser is formed of two concentric conducting spheres of radii a and b ($a<b$). If a point charge e is placed between the spheres, at a distance c from the centre, and the spheres are connected by a fine wire, the charge on the inner sphere after connection will be

$$-ea\,(b-c)/c\,(b-a). \qquad \text{[I. 1907]}$$

36. A charge e is placed midway between two equal spherical conductors which are kept at zero potential. Shew that the charge induced on each is

$$-e\left(m-\frac{m^2}{2}+\frac{m^3}{4}-\frac{3m^4}{8}\right),$$

neglecting higher powers of m, which is the ratio of the radius of a conductor to half the distance between their centres. [I. 1908]

37. If a particle charged with a quantity e of electricity be placed at the middle point of the line joining the centres of two equal spherical conductors kept at zero potential, shew that the charge induced on each sphere is

$$-2em(1-m+m^2-3m^3+4m^4)$$

neglecting higher powers of m, which is the ratio of the radius to the distance between the centres of the spheres. [I. 1893]

38. A spherical conductor of diameter a is kept at zero potential in the presence of a fine uniform wire, in the form of a circle of radius c in a tangent plane to the sphere with its centre at the point of contact, which has a charge e of electricity; prove that the electrical density induced on the sphere at a point whose direction from the centre of the ring makes an angle ψ with the normal to the plane is

$$-\frac{c^2 e \sec^3\psi}{4\pi^2 a}\int_0^{2\pi}(a^2+c^2\sec^2\psi-2ac\tan\psi\cos\theta)^{-\frac{3}{2}}\,d\theta.$$

[I. 1892]

39. A conductor is formed by the larger segments of two spheres cutting at right angles; A, B are the centres of the two spheres, C any external point; D, E are the images of C in the spheres with centres at B, A respectively and AD, BE meet in F. If the conductor be uninsulated and any charge be placed at C, shew that the charge induced on the sphere bears to the inducing charge the ratio

$$-\sqrt{(\Delta ABD)}-\sqrt{(\Delta ABE)}+\sqrt{(\Delta ABF)}:\sqrt{(\Delta ABC)},$$

where ΔABC represents the area of the triangle ABC. [I. 1893]

40. A point charge is placed in front of the plane face of an unlimited mass of dielectric; find the potential function, and sketch the lines of force. Shew that the number of unit tubes of induction which cross the face is $K/(K+1)$ of those emitted by the charge. [I. 1910]

41. Shew that the capacity of a spherical conductor of radius a is increased in the ratio $1:1+\dfrac{K-1}{K+1}\dfrac{a}{2b}$ by the presence of a large mass of dielectric with a plane face, at a distance b from the centre of the sphere, if a/b is so small that its square may be neglected.

[St John's Coll. 1907]

42. A long fine straight wire carrying a charge e per unit length is parallel to and at a distance f ($>a$) from the axis of an insulated uncharged long conducting cylinder whose cross-section is a circle of radius a. Shew that the attraction per unit length between the cylinder and the wire is $2e^2a^2/f(f^2-a^2)$. Also shew that the surface density on the cylinder vanishes at the points of contact of tangents drawn from points on the wire. [I. 1928]

43. A charged sphere of radius a, with its centre at height h_1 above the ground, large compared with a, is at potential V_1: determine approximately the potential to which it will attain, when it is raised to a height h_2. [M. T. 1917]

44. The polar equation of a surface of revolution is

$$(r^2+c^2-2cr\cos\theta)(r-b)^2=a^2r^2.$$

If a conductor be made of this form, insulated and electrified, find the law of distribution of electricity on the conductor and the potential due to it at any point of external space. [I. 1897]

ANSWERS

5. $-\frac{e}{2\pi a^2}\left\{1-\frac{a^3}{(a^2+4b^2)^{\frac{3}{2}}}\right\}, \ -\frac{e}{2\pi b^2}\left\{1-\frac{b^3}{(4a^2+b^2)^{\frac{3}{2}}}\right\}.$

6. $-\frac{3}{4}Q, \ -\frac{5}{4}Q.$

10. $ea^2(3f-a)/f(f-a)^2.$

12. $ea\left\{\frac{1}{\sqrt{(f^2-a^2)}}-\frac{1}{f}\right\}.$

17. $e^2ca/(c^2-a^2)^2.$

24. $e/c.$

26. $\cos^{-1}\{(3-\sqrt{5})/\sqrt{5}\}.$

43. $V_1\left\{1+\frac{a}{2h_1}-\frac{a}{2h_2}\right\}.$

44. Let A be the point $(c, 0)$ and r' denote distance from A. When the potential of the conductor is unity, the potential at any point of external space is $a/r'+b/r$; and the electrification at a point P on the conductor is given by $4\pi\sigma=$ resultant of a/r'^2 along AP and b/r^2 along OP.

Chapter VII

ELECTRIC CURRENTS

7·1. Current strength. When two conductors at different potentials are connected by a wire there is a gradient of potential along the wire and equilibrium no longer exists, but there is a transfer of electricity called a current in the wire from one conductor to the other. The **strength of the current** in the wire may be measured by the rate of increase of the positive charge on one of the conductors, say $C = dQ/dt$, where C is the current and Q the charge.

Such a current would be transient unless the conductors are connected with the terminals of a battery or 'cell' by which they are maintained at different potentials.

The presence of the current is manifested by heating of the wire and the creation of a magnetic field in its neighbourhood.

The field of a steady current is said to be a stationary field, because, although there is a flow of electricity, the conditions remain the same at all points as time progresses. We must point out however that directly the two conductors above are connected the field ceases to be electrostatic; we have therefore no longer any ground for asserting the existence of a single-valued potential function, and the fact that the field of a steady current is derivable from a potential function (not single-valued) is a new empirical assumption.

In the electron theory metals contain a number of electrons moving freely between the molecules. If there be no external electric field, the velocities of the free electrons are in random directions and there is no tendency for a collective motion in one direction rather than another. But an electric field directs the flow and establishes a drift of electrons in the direction of the field, though held in restraint by collision of electrons with one another and with metal molecules.

For our purposes it is sufficient to measure the current in the way indicated above; or, by considering a cross-section of the wire and stating that if, in a short time δt, n units of positive electricity cross the section in one direction and n' units of negative electricity cross it in the opposite direction, then the strength of the current is $(n+n')/\delta t$, so that it is immaterial whether we regard the current as a flow of positive electricity in one direction or a flow of negative electricity in the opposite direction, or partly one and partly the other.

7·11. The voltaic cell. The simplest form of cell consists of a plate of zinc and a plate of copper immersed in a vessel of dilute sulphuric acid, with copper wires attached to the plates. When the wires are joined the 'circuit is completed' and a current flows from the zinc to the copper in the liquid and from the copper to the zinc through the wire. A chemical action takes place in the cell; the acid attacks the zinc plate, zinc sulphate being formed and hydrogen liberated from the acid. But the tiny bubbles of hydrogen are carried across by the current and adhere to the copper plate where with their positive charges placed opposite to negative charges on the zinc plate they act like a parallel plate condenser and produce a field opposed to the current in the liquid and consequently reduce the effectiveness of the cell by increasing what is known as its resistance. This process is called *the polarization of the cell*; its effect is to reduce the potential difference of the terminals A, B, so that such a cell is of no use when a permanent potential difference is required, and more elaborate forms of cell have been devised to get rid of polarization.

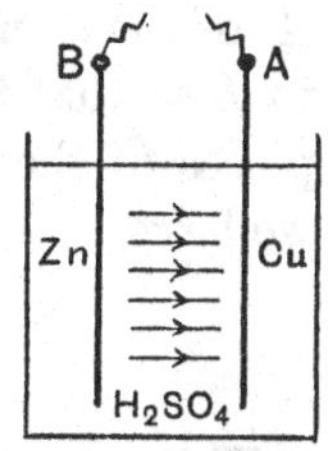

7·12. Electromotive force. We made reference in **2·411** to the possibility of a difference of potential between two substances in contact. Consider a closed ring—fig. (i)—formed of three metals A, B, C joined in sections. In equilibrium the potential is constant throughout each metal, but there may be differences at the contacts which may be denoted by ${}_A\phi_B$, ${}_B\phi_C$, ${}_C\phi_A$.

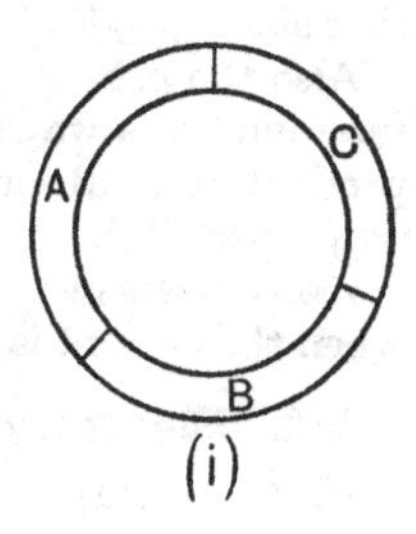

(i)

Such a discontinuity in the potential is associated in a static field with the existence of a 'double sheet' of electricity over the surfaces in contact, i.e. surface charges of opposite signs.

The contact differences will be such that in equilibrium

$$_A\phi_B = \phi_A - \phi_B,\ _B\phi_C = \phi_B - \phi_C \text{ and } _C\phi_A = \phi_C - \phi_A$$

and $$_A\phi_B + {}_B\phi_C + {}_C\phi_A = 0 \quad \text{..................(1)}.$$

But if the circuit is formed of two metals A, B, and an acid C as in a voltaic cell, there is not equilibrium but a flow of electricity. Consider this arrangement when A and B are separated by an air gap, and suppose that a small piece of metal A' of the same kind as A has been soldered on to B so that the gap is an air gap between two pieces of the same metal A, A' (fig. (ii)). Then we have

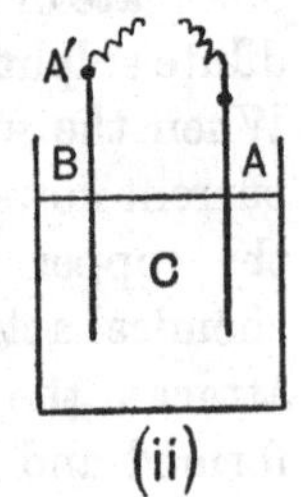

(ii)

$$\phi_{A'} - \phi_A = (\phi_{A'} - \phi_B) + (\phi_B - \phi_C) + (\phi_C - \phi_A)$$
$$= {}_A\phi_B + {}_B\phi_C + {}_C\phi_A \quad \text{........................(2)}.$$

This potential difference $\phi_{A'} - \phi_A$ can be tested by an electrometer. We define the *electromotive force of the cell* as this potential difference when the cell is 'open' and the terminals are of the same substance; and we see that it is the sum of the contact differences of potential of the elements which compose the cell.

7·121. It should be noticed that contact difference of potential depends upon temperature. Thus if we have a ring formed of two metals A and B and the junctions are kept at different temperatures t, t', there will be a current round the ring since $[_A\phi_B]_t \neq [-{}_B\phi_A]_{t'}$.

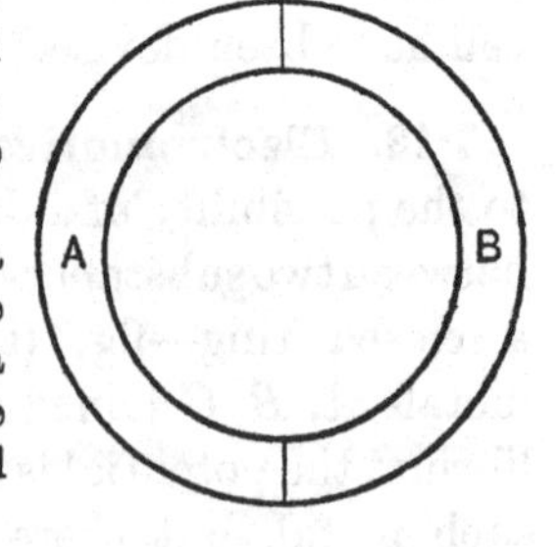

Also the passage of a current through a junction is always accompanied by the production or absorption of heat, in such a way that if heat is produced when the current passes in one way it will be absorbed when the current is reversed.

7·2. Electrolysis. In addition to the production of heat and of a magnetic field, the passage of a current through various substances in solution frequently results in a chemical decomposition and this process is called **electrolysis.**

When a current passes through a liquid from one metal plate to another, the plates are called **electrodes**, that of higher potential is called the **anode**, and that of lower potential the **cathode**. The substance which is decomposed is called an **electrolyte** and the constituents into which it is decomposed are called **ions**; that which is deposited at the **anode** is the **anion**, and that which is deposited at the **cathode** is the **cation**.

When a current passes between platinum electrodes through acidulated water, the water is decomposed into oxygen and hydrogen. In the process molecules of water in contact with the anode are decomposed, the oxygen atoms form into molecules which collect on the anode and the hydrogen atoms attack neighbouring molecules of water seizing and combining with their oxygen atoms and liberating hydrogen atoms, which in turn attack their neighbours. This process takes place across the whole space between the electrodes and in the end liberated hydrogen atoms assemble on the cathode.

The Laws of Electrolysis are due to Faraday who also originated the terms we have just defined. Faraday found that there is always a constant ratio between the amount of an ion deposited and the quantity of electricity which passes.

To every substance a number is attached called its **electrochemical equivalent.** It is *the number of grams of the substance deposited during the passage of a unit of electricity*. The Laws of Electrolysis may be summarized in the statement that *the number of grams of a substance deposited at an electrode is equal to the number of units of electricity which pass through the electrolyte multiplied by the electrochemical equivalent of the substance.* In the case of elementary ions the electrochemical equivalent is the atomic weight divided by the valency, and, in the case of compound ions, it is the molecular weight divided by the valency.

An apparatus in which the quantity of a cation deposited can be measured—e.g. the weight of silver deposited from a solution of silver nitrate—may be used to measure the quantity of electricity which passes in a given time. Such an apparatus is called a *voltameter*.

7·3. Ohm's Law. *The current between two points of a circuit varies directly as the electromotive force and inversely as the resistance.*

We have already defined the electromotive force of a cell as the difference of potential of its terminals on open circuit, and we may in the same way speak of the electromotive force between two points as the difference of potential between them. It also represents the work that would be done, or the energy used up, as a unit of electricity passes from one point to the other. Ohm's Law* is really a definition of the resistance of a conductor between two points; for it makes the resistance *the ratio of the electromotive force to the current which it produces.*

It must be observed that electromotive *force* is not a good name for what it is intended to connote, because its physical dimensions are not those of *force* but of *potential*; at the same time it is the physical entity which results in the motion of electricity.

If C is the current from P to Q, R the resistance and ϕ_P, ϕ_Q denote the potentials of P and Q, Ohm's Law may be expressed by the equation

$$\phi_P - \phi_Q = CR \qquad \text{.....................(1)},$$

assuming there to be no source of energy in the conductor between P and Q.

We observe that if an electric field has a single-valued potential function ϕ, then in any closed circuit for which Q coincides with P we have $\phi_Q = \phi_P$, so that from (1) there can be no current in the circuit. Hence the existence of a current implies a source of energy such as a cell or battery.

Let A, B be the terminals of a cell [fig. **7·11**], R the resistance of the wire joining A, B, r the resistance of the liquid in the cell and C the current. Then if E is the electromotive force of the cell, since it drives the current C through a total resistance $R+r$, we have

$$E = CR + Cr \qquad \text{.....................(2)}.$$

* G. S. Ohm (1787–1854), German physicist.

But from (1), when the circuit is closed, we have

$$CR = \phi_A - \phi_B \quad \text{..................(3)};$$

and from 7·12 E is the value of $\phi_A - \phi_B$ when the circuit is open. It follows from (2) that the potential difference of the terminals is less when the circuit is closed than when it is open by the amount Cr, needed to drive the current through the liquid in the cell.

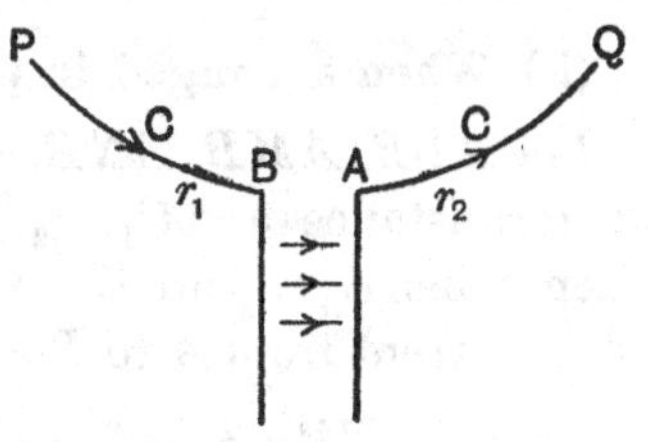

Now let P, Q denote any two points on the closed circuit and consider the flow from P to Q in the sense $PBAQ$. If r_1, r_2 denote the resistances of PB, AQ, we have

$$\phi_P - \phi_B = Cr_1,$$

$$\phi_B - \phi_A = -E + Cr, \text{ from (2) and (3)},$$

and $$\phi_A - \phi_Q = Cr_2;$$

so that $$\phi_P - \phi_Q + E = C(r_1 + r + r_2) \quad \text{..................(4)}.$$

Consequently we may express Ohm's Law in this rather more general form: *If P, Q are any two points connected by a wire; ϕ_P, ϕ_Q the potentials at P, Q; E_{PQ} the internal electromotive force in PQ, i.e. the electromotive force of any battery located in PQ and driving current in the sense from P to Q; C_{PQ} the current from P to Q, and R_{PQ} the total resistance of the wire including that of the battery, then*

$$\phi_P - \phi_Q + E_{PQ} = C_{PQ} R_{PQ} \quad \text{..............(5)}.$$

7·31. Resistance of a set of conductors. (i) When arranged in series.

Let $AB, BC, CD, \dots LM$ be the conductors, $r_1, r_2, \dots r_n$ their resistances and C the current flowing from A to M when the potentials of the ends of the conductors are $\phi_A, \phi_B, \dots \phi_M$; and let R be the whole resistance between A and M. Then by Ohm's Law

$$\phi_A - \phi_B = Cr_1, \; \phi_B - \phi_C = Cr_2, \dots \phi_L - \phi_M = Cr_n,$$

so that by addition

$$\phi_A - \phi_M = C(r_1 + r_2 + \ldots + r_n).$$

But $$\phi_A - \phi_M = CR,$$

therefore $$R = r_1 + r_2 + \ldots + r_n \quad \ldots\ldots\ldots\ldots\ldots\ldots(1).$$

(ii) When arranged in parallel.

Let ALB, AMB, ANB, ... be the conductors, $r_1, r_2, \ldots r_n$ their resistances and $C_1, C_2, \ldots C_n$ the currents passing through them when ϕ_A, ϕ_B are the potentials at A, B. Then if C is the total current from A to B and R the total resistance,

$$CR = \phi_A - \phi_B = C_1 r_1 = C_2 r_2 = \ldots = C_n r_n.$$

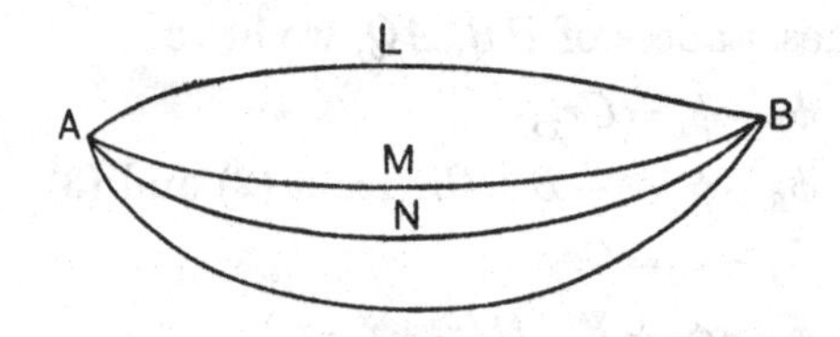

But $$C = C_1 + C_2 + \ldots + C_n,$$

therefore $$\frac{1}{R} = \frac{1}{r_1} + \frac{1}{r_2} + \ldots + \frac{1}{r_n} \quad \ldots\ldots\ldots\ldots\ldots\ldots(2).$$

It follows that if n conductors of the same resistance r are joined in series the resistance of the compound conductor is nr, but if they are joined in parallel the resistance of the compound conductor is r/n.

7·32. Specific resistance and conductivity. The **specific resistance** *of a substance is the resistance of a unit cube of the substance to a current passing parallel to one of its edges.*

If s denotes the specific resistance of a substance, the resistance of a wire of length l whose cross-section is the unit of area is ls, by **7·31** (i). And if the cross-section be α units of area, the wire may be regarded as made up of α wires of unit cross-section arranged in parallel, so that, by **7·31** (ii), the resistance is ls/α.

Conductivity. *The reciprocal of the specific resistance of a substance is called its conductivity.*

7·33. Units. In the form in which we have stated it, Ohm's Law requires that a unit of electromotive force should produce a unit current through a unit resistance. The names of the practical units are the **volt**, the **ampère*** and the **ohm**. Thus an electromotive force of one volt produces a current of one ampère through a resistance of one ohm.

In terms of the *absolute electrostatic units* defined in **3·9**:

1 volt $= \frac{1}{300}$ absolute *electrostatic* C.G.S. unit of E.M.F.,

1 ampère $= 3 \times 10^9$ absolute *electrostatic* C.G.S. units of current,

and 1 ohm $= 3^{-2} \times 10^{-11}$ absolute *electrostatic* C.G.S. unit of resistance.

But the reader must note that these physical quantities occur most often in connection with electromagnetic fields so that it is important to know their values in terms of *absolute electromagnetic units*, the basis of which we have not yet explained. The values are as follows:

1 volt $= 10^8$ absolute *electromagnetic* C.G.S. units of E.M.F.,

1 ampère $= 10^{-1}$ absolute *electromagnetic* C.G.S. unit of current,

and 1 ohm $= 10^9$ absolute *electromagnetic* C.G.S. units of resistance.

7·34. Arrangement of cells. Cells are said to be *arranged in series* when the zinc of the first is joined to the copper of the second, the zinc of the second to the copper of the third and so on. The electromotive force of such an arrangement being the sum of all the contact differences of potential is clearly the sum of the electromotive forces of the separate cells; and as in **7·31** the resistance is the sum of the separate resistances.

If on the other hand the cells are *joined in parallel*, i.e. if all the zinc plates of a number of like cells are joined and also all the copper plates, this is merely equivalent to increasing the size of the plates without altering the nature of the cell and thus will not affect its electromotive force; but by **7·31** if there are n cells their resistance in parallel will be reduced to $\frac{1}{n}$th of that of a single cell.

Let mn equal cells be arranged in m parallel sets each containing n cells in series. Let r be the resistance of a cell, E its electromotive force and R the external resistance which completes the circuit.

* After A. M. Ampère (1775–1836), French physicist.

The E.M.F. of each series of n cells is nE and its resistance nr, and when m such series are joined in parallel the E.M.F. remains nE and the resistance becomes nr/m. Hence if C be the total current, we have

$$nE = C\left(R + \frac{nr}{m}\right), \text{ or } C = \frac{E}{\frac{r}{m} + \frac{R}{n}}.$$

This may be written

$$C = \frac{E}{\left(\sqrt{\frac{r}{m}} - \sqrt{\frac{R}{n}}\right)^2 + 2\sqrt{\left(\frac{Rr}{mn}\right)}}.$$

Hence the current will be a maximum when $\frac{r}{m} = \frac{R}{n}$, or when $nr/m = R$, i.e. when the internal resistance is equal to the external resistance, provided this arrangement is possible; and the maximum value of the current is seen to be

$$C_{\max} = En/2R.$$

7·4. Joule's Law.* Let R be the resistance of a length AB of wire, E the electromotive force from A to B and C the current. Then in time δt an amount $C\delta t$ of electricity is driven from A to B by the potential difference E, so that an amount of work $EC\delta t$ is done, or $RC^2\delta t$ units of work. This work is transformed into heat and is made manifest in the heating of the wire. The rate of heat production is therefore, in mechanical units, RC^2 per unit time; and this is true for any portion of the circuit provided that R denotes its resistance and C the current.

7·41. The heating effect, to which reference was made in **7·121**, when a current passes through the junction of two metals is known as the *Peltier effect*.† In this case the quantity of heat emitted or absorbed is *directly proportional to the current*, and not to the square of the current as in the Joule effect.

7·5. Kirchhoff's Laws ‡ for steady currents in a network of wires.

1. *In a state of steady flow the algebraical sum of the currents which meet at a junction of wires is zero.*

* J. P. Joule (1818–1889), English physicist.
† J. C. A. Peltier (1785–1845), French physicist.
‡ G. R. Kirchhoff (1824–1887), German physicist.

This law means that the term 'steady flow' implies that electricity does not accumulate at any point of the network.

2. *In any closed circuit in a network the algebraical sum of the products of the resistance and current in each conductor is equal to the algebraical sum of the electromotive forces of the batteries in the circuit.*

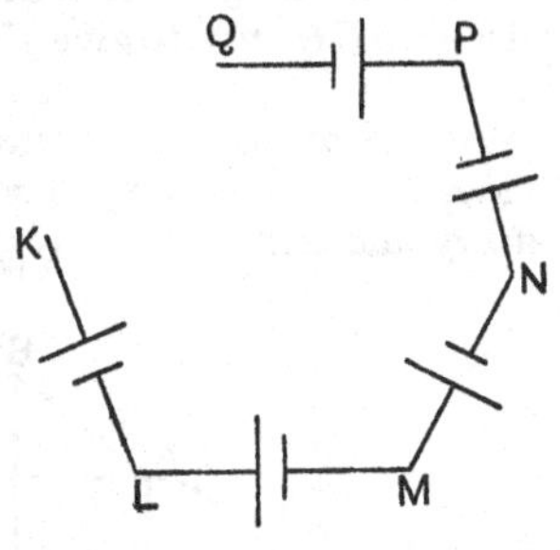

Let $L, M, N, \ldots K$ be the junctions in the circuit. Adopting the notation of **7·3**, viz. R_{PQ} denotes the resistance of the conductor PQ including that of any battery it may contain, C_{PQ} is the current from P to Q, E_{PQ} the electromotive force of the battery driving current from P to Q, and ϕ_P, ϕ_Q are the potentials at P and Q, we have, as in **7·3** (5)

$$C_{LM} R_{LM} = \phi_L - \phi_M + E_{LM},$$
$$C_{MN} R_{MN} = \phi_M - \phi_N + E_{MN},$$
$$\ldots\ldots\ldots\ldots\ldots\ldots\ldots\ldots$$
$$C_{KL} R_{KL} = \phi_K - \phi_L + E_{KL};$$

so that by addition

$$\Sigma CR = \Sigma E \quad \ldots\ldots\ldots\ldots\ldots\ldots\ldots(1).$$

It follows that, in any closed circuit in the network in which there is no battery, we must have

$$\Sigma CR = 0 \quad \ldots\ldots\ldots\ldots\ldots\ldots\ldots(2).$$

7·51. Examples. (i) *A system of conducting wires, all of equal resistance, form the edges of a cube. Find the currents carried by the wires if a battery is connected without extra resistance to two adjacent vertices of the cube.* [M. T. 1930]

Let E be the electromotive force and R the resistance of the battery, r the resistance of each edge of the cube. We have to find the currents in twelve wires, but the solution of the problem will become tedious if we use more unknown quantities than is necessary; so we make what use we can of Kirchhoff's first law from the outset and also make use of symmetry.

Referring to the diagram let the current enter the cube at A and let x denote the current in AF. By symmetry the currents in AB, AD must be equal, so we denote them by y. We also observe that if the battery were reversed, the currents would also be reversed, so that the currents in GF and KF will also be y. At B the current in AB divides, let z denote the current in BG, then $y-z$ must be the current in BC. In like manner the currents in DK, DC are z and $y-z$. At C the currents in BC, DC unite and give a current $2(y-z)$ in CH, which divides at H into $y-z$ in both HG and HK. Then the currents z in BG and $y-z$ in HG unite to give y in GF, and similarly y in KF as previously stated.

Also the total current through the battery is the current which divides at A, i.e. $x+2y$. Therefore, from the circuit formed of the battery and AF,

$$R(x+2y)+rx=E \quad \text{.......................(1).}$$

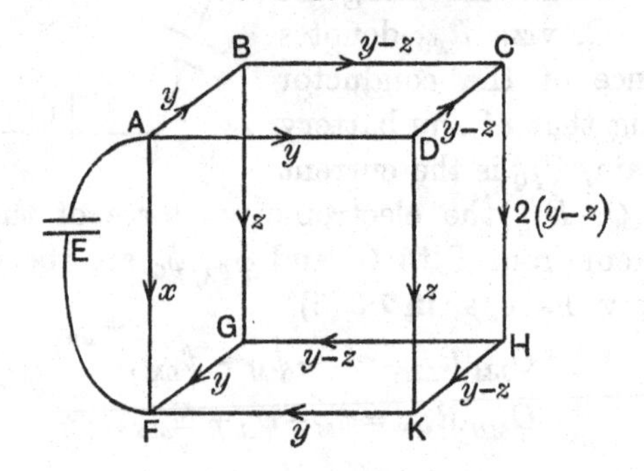

From the square $ABGF$, the resistances being equal and there being no battery in the circuit,

$$x-y-z-y=0 \text{ or } x-2y-z=0 \quad \text{...........(2).}$$

Similarly, from the square $BCHG$

$$z-4(y-z)=0 \text{ or } 4y=5z \quad \text{..................(3).}$$

Hence

$$\frac{y}{5}=\frac{z}{4}=\frac{x}{14}=\frac{E}{24R+14r} \quad \text{......................(4).}$$

And thence the currents in each of the twelve wires can be written down.

(ii) *Two long straight parallel wires are joined by cross wires of the same material at equal distances, forming an infinite ladder of equal squares, the resistance of a side of a square being r. A current enters and leaves the network at the ends A, B of one of the cross wires. If the currents in successive segments of one of the long wires, measured from A, are $u_1, u_2, u_3, \ldots$, shew that*

$$u_n-4u_{n+1}+u_{n+2}=0 \quad (n \geqslant 1).$$

Shew also that $u_n=u_1(2-\sqrt{3})^{n-1}$, and that the equivalent resistance of the network is $r/\sqrt{3}$. [M. T. 1932]

Let C be the current which enters at A and leaves at B, x the current in AB, $v_1, v_2, v_3, \ldots$ the currents in successive cross wires. We notice that there is symmetry about the line AB, and also a symmetry made evident by making the current enter at B and leave at A.

From Kirchhoff's first law

$$u_n = u_{n+1} + v_n, \text{ and } u_{n+1} = u_{n+2} + v_{n+1};$$

and from the second law

$$u_{n+1} + v_{n+1} + u_{n+1} - v_n = 0.$$

Therefore, by eliminating v_n and v_{n+1}, we get

$$u_n - 4u_{n+1} + u_{n+2} = 0 \quad\text{.......................(1).}$$

To solve this linear difference equation, we note that the roots of the quadratic $z^2 - 4z + 1 = 0$ are $2 \pm \sqrt{3}$ and find that (1) has a solution of the form

$$u_n = A(2+\sqrt{3})^n + B(2-\sqrt{3})^n \quad\text{..................(2),}$$

where A and B are constants to be determined.

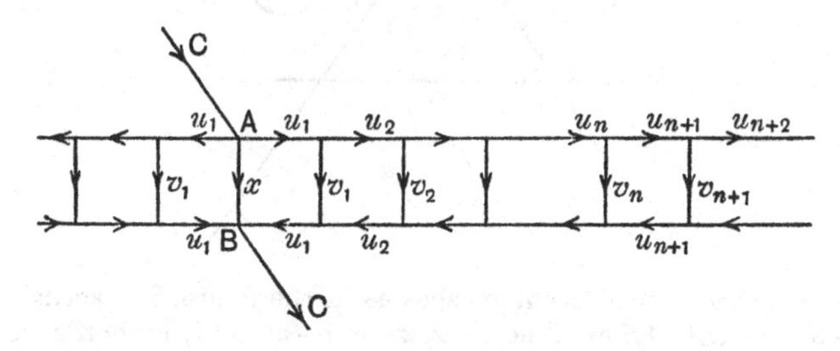

But as $n \to \infty$ it is evident that $u_n \to 0$, so that A must be zero, and

$$u_n = B(2-\sqrt{3})^n.$$

Therefore $$u_1 = B(2-\sqrt{3})$$

and $$u_n = u_1(2-\sqrt{3})^{n-1} \quad\text{.....................(3).}$$

Again, from the central squares, we have

$$C = x + 2u_1 \quad\text{..............................(4),}$$

$$u_1 = u_2 + v_1$$

and $$u_1 + v_1 + u_1 - x = 0.$$

By eliminating v_1 we get

$$x = 3u_1 - u_2,$$

and from (3) this gives $x = 3u_1 - u_1(2-\sqrt{3})$

$$= u_1(1+\sqrt{3}).$$

Therefore from (4), by eliminating u_1, we get

$$C = \sqrt{3}x.$$

But if R is the total resistance of the network

$$CR = \phi_A - \phi_B = rx,$$

so that, since $C = \sqrt{3}x$, therefore $R = r/\sqrt{3}$.

7·52. When a network consists of a number of meshes it is often convenient to use symbols for the currents round the meshes instead of currents in the separate wires. Then provided the symbols denote currents travelling round the meshes in the same sense, the actual current in any wire which is common to two meshes is the difference between the currents in these meshes.

Thus let ABC be a triangular mesh, the resistances of whose sides BC, CA, AB are a, b, c, and let x denote the current in the mesh ABC

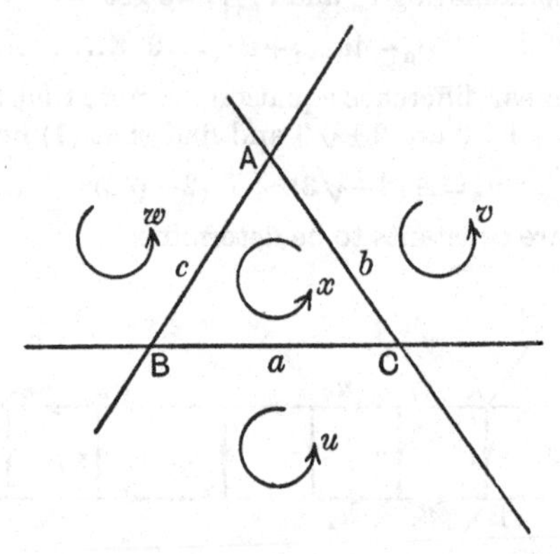

and u, v, w those in adjacent meshes as in the figure. The actual currents in BC, CA, AB are then $x-u$, $x-v$, $x-w$ and Kirchhoff's second law gives

$$a(x-u)+b(x-v)+c(x-w)=E,$$

where E is the total electromotive force in the wires of the mesh; and this may be written

$$(a+b+c)x-au-bv-cw=E;$$

i.e. the mesh current is multiplied by the whole resistance and the products of adjacent mesh currents and corresponding resistances are subtracted. The only advantage of this method is that it enables us to write down equations in a form immediately ready for solution.

We shall make use of the method in the following article.

7·6. Wheatstone's bridge.* This is an apparatus used for comparing resistances. It consists of four conductors BA, BD, CA, CD connected as in the figure, the points B, C being connected to the terminals of a battery and the points A, D to the terminals of a galvanometer—an instrument which indicates and measures a current passing through it.

Let the resistances of BC, CA, AB, DA, DB, DC, including those of the battery and galvanometer in BC and DA, be

* Sir Charles Wheatstone (1802–1875).

$a, b, c, \alpha, \beta, \gamma$. Let x, y, z denote the currents round the meshes DBC, DCA, DAB, and E the electromotive force of the battery.

Then by applying Kirchhoff's second law to the three meshes in turn, we have

$$(a+\beta+\gamma)x-\gamma y-\beta z=E,$$
$$-\gamma x+(b+\gamma+\alpha)y-\alpha z=0$$

and

$$-\beta x-\alpha y+(c+\alpha+\beta)z=0.$$

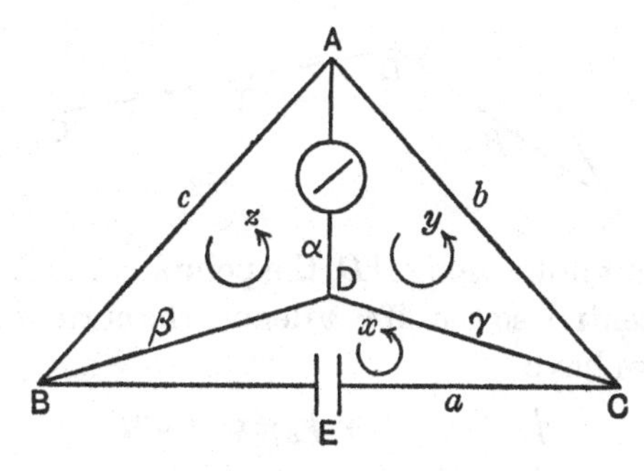

These equations may then be solved to find the currents. The current through the galvanometer is $y-z$; where

$$\begin{vmatrix} a+\beta+\gamma & -\gamma & -\beta \\ -\gamma & b+\gamma+\alpha & -\alpha \\ -\beta & -\alpha & c+\alpha+\beta \end{vmatrix} y = \begin{vmatrix} a+\beta+\gamma & E & -\beta \\ -\gamma & 0 & -\alpha \\ -\beta & 0 & c+\alpha+\beta \end{vmatrix}$$

or

$$\Delta y=E(\alpha\beta+\beta\gamma+\gamma\alpha+c\gamma) \quad \text{..............(1)},$$

where Δ stands for the determinant on the left. Similarly

$$\Delta z=E(\alpha\beta+\beta\gamma+\gamma\alpha+b\beta) \quad \text{..............(2)}.$$

Hence $y=z$, or there will be no current through the galvanometer if

$$b\beta=c\gamma \quad \text{..........................(3)}.$$

Further, we notice that Δ is symmetrical in the resistances, and the expression on the right of (1) for y, the current in CA, does not contain the resistance of BC or CA, so that if the battery were transferred to CA, this same equation would then determine the current in BC. This is a particular case of a general theorem which will be proved later.

When a battery in BC produces no current in DA, these are called *conjugate conductors* and the bridge is said to be balanced.

The condition (3) for the balancing of the bridge may be obtained independently as follows:

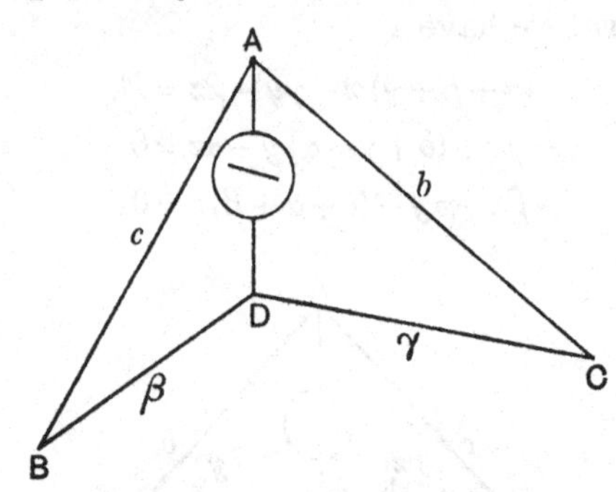

If no current flows along AD, the points A and D must be at the same potential, so that if u, v denote the currents long BAC and BDC, we have

$$\phi_B - \phi_A = cu, \quad \phi_A - \phi_C = bu,$$
$$\phi_B - \phi_D = \beta v, \quad \phi_D - \phi_C = \gamma v,$$

where $\phi_A = \phi_D$. Hence

$$cu = \beta v \text{ and } bu = \gamma v,$$

so that

$$b\beta = c\gamma.$$

7·61. A Wheatstone's bridge may be used for obtaining a potential difference which is a small fraction of that of the battery. For, reverting to the general case

$$\phi_A - \phi_D = \alpha(y - z)$$
$$= \alpha E(c\gamma - b\beta)/\Delta, \text{ from } \mathbf{7{\cdot}6}\ (1), (2),$$

and by adjusting the resistances of the arms so as to make $c\gamma - b\beta$ small, the ratio of $\phi_A - \phi_D$ to E can be made as small as we please.

7·62. By using standard coils of known resistance, and adjustable arrangements for varying the resistances of three of the four conductors BA, BD, CA, CD until a balance is attained, it is clear that the relation $b\beta = c\gamma$ enables a fourth unknown resistance to be determined.

7·7. Currents in a network. Reciprocal Theorems. Let $A_1, A_2, \ldots A_n$ be the n junctions of a network of wires. We can include the possibility of there being two or more wires directly joining A_p, A_q by supposing another and different A to be inserted between A_p and A_q on each of these wires.

Let R_{pq} = the resistance of A_pA_q,
E_{pq} = the electromotive force of the battery in the wire A_pA_q driving current from A_p to A_q,
C_{pq} = the current along A_pA_q,
ϕ_p = the potential at A_p,
Q_p = the current, if any, entering the network at A_p;

and let dashes be used to denote any other set of values of the last four quantities. The resistances of course are invariable.

Then as in **7·3** (5)

$$C_{pq}R_{pq} = E_{pq} + \phi_p - \phi_q$$

and

$$E'_{pq} + \phi_p' - \phi_q' = C'_{pq}R_{pq}.$$

Multiply these equations together, cancel out the factor R_{pq} and add similar equations for all the wires, giving

$$\Sigma E'_{pq}C_{pq} + \Sigma C_{pq}(\phi_p' - \phi_q') = \Sigma E_{pq}C'_{pq} + \Sigma C'_{pq}(\phi_p - \phi_q) \quad \ldots\ldots(1),$$

where Σ means a summation over all different pairs p, q, $(p \neq q)$ such that A_p, A_q are directly joined by a wire; i.e. there is no further junction on the wire between A_p and A_q. And we may suppose the summation extended to all different pairs p, q, $(p \neq q)$ if we use the conventions $C_{pq} = 0$, $C'_{pq} = 0$ when A_p, A_q are not joined by a wire.

Now $\Sigma C_{pq}(\phi_p' - \phi_q')$

$$\begin{aligned} &= C_{12}(\phi_1' - \phi_2') + C_{13}(\phi_1' - \phi_3') + C_{14}(\phi_1' - \phi_4') + \ldots \\ &\qquad\qquad + C_{1n}(\phi_1' - \phi_n') \\ &+ C_{23}(\phi_2' - \phi_3') + C_{24}(\phi_2' - \phi_4') + \ldots + C_{2n}(\phi_2' - \phi_n') \\ &+ \qquad\qquad + \qquad\qquad + \\ &\qquad\qquad + C_{n-1,n}(\phi'_{n-1} - \phi_n'). \end{aligned}$$

But

$$Q_p = C_{p1} + C_{p2} + \ldots + C_{pn}, \quad C_{pp} = 0 \text{ and } C_{qp} = -C_{pq}.$$

Therefore

$$\Sigma C_{pq}(\phi_p' - \phi_q') = \phi_1'Q_1 + \phi_2'Q_2 + \ldots + \phi_n'Q_n = \Sigma \phi_p'Q_p.$$

Similarly $$\Sigma C'_{pq}(\phi_p - \phi_q) = \Sigma \phi_p Q_p',$$
and therefore, from (1),

$$\Sigma E'_{pq} C_{pq} + \Sigma \phi_p' Q_p = \Sigma E_{pq} C'_{pq} + \Sigma \phi_p Q_p' \quad \text{...(2).}$$

We now consider two cases:

(i) *When no currents enter the network at the junctions, the Q's are all zero and the theorem* (2) *becomes*

$$\Sigma E'_{pq} C_{pq} = \Sigma E_{pq} C'_{pq} \quad \text{.................(3).}$$

Now consider two states:

(α) when all the E's are zero except E_{pq} which is equal to E; and (β) when all the E''s are zero except E'_{rs} which is equal to E; then (3) gives

$$EC_{rs} = EC'_{pq} \text{ or } C_{rs} = C'_{pq},$$

which means that the current in $A_r A_s$ due to an electromotive force in $A_p A_q$ is equal to the current in $A_p A_q$ due to an equal electromotive force in $A_r A_s$. We had a special case of this theorem in 7·6.

(ii) *When no batteries are present, i.e. when all E's are zero, the theorem* (2) *becomes*

$$\Sigma \phi_p' Q_p = \Sigma \phi_p Q_p' \quad \text{..................(4).}$$

Now consider two states:

(α) a current Q enters the network at A_p and leaves it at A_q, i.e. $Q_p = Q$, $Q_q = -Q$ and all other Q's are zero; (β) a current Q enters the network at A_r and leaves it at A_s, i.e. $Q_r' = Q$, $Q_s' = -Q$ and all other Q''s are zero; then (4) gives

$$(\phi_p' - \phi_q') Q = (\phi_r - \phi_s) Q \text{ or } \phi_p' - \phi_q' = \phi_r - \phi_s,$$

which means that the potential difference between A_p, A_q when a current enters at A_r and leaves at A_s is equal to the potential difference between A_r, A_s when the same current enters at A_p and leaves at A_q.

7·71. Minimum rate of heat production. (i) *To shew that when there is a steady flow in a network of wires containing no batteries, of all possible distributions of current in accordance with Kirchhoff's first law that one which is also in accordance with the second law gives the minimum rate of heat production.*

With the notation of **7·7** let C_{pq} denote the true current in the conductor of resistance R_{pq} connecting the junctions A_p, A_q; then by **7·4** the rate of heat production is

$$H = \Sigma R_{pq} C^2_{pq} \quad \text{.....................(1)},$$

where the summation may be taken over every pair p, q, $(p \neq q)$, if we make a convention that $C_{pq} = 0$ when A_p, A_q are not directly joined by a wire.

In any other distribution of current in steady flow let the current in A_pA_q be X_{pq}, and let $X_{pq} = C_{pq} + \epsilon_{pq}$ with a like convention. Then the corresponding rate of heat production is

$$H' = \Sigma R_{pq} (C_{pq} + \epsilon_{pq})^2,$$

or

$$H' = H + 2\Sigma R_{pq} C_{pq} \epsilon_{pq} + \Sigma R_{pq} \epsilon^2_{pq} \quad \text{.........(2)}.$$

Now $\Sigma R_{pq} C_{pq} \epsilon_{pq} = \Sigma (\phi_p - \phi_q) \epsilon_{pq}$

$$= \epsilon_{12} (\phi_1 - \phi_2) + \epsilon_{13} (\phi_1 - \phi_3) + \epsilon_{14} (\phi_1 - \phi_4) + \ldots + \epsilon_{1n} (\phi_1 - \phi_n)$$
$$+ \epsilon_{23} (\phi_2 - \phi_3) + \epsilon_{24} (\phi_2 - \phi_4) + \ldots + \epsilon_{2n} (\phi_2 - \phi_n)$$
$$+ \ldots \qquad + \ldots \qquad + \epsilon_{n-1,n} (\phi_{n-1} - \phi_n)$$
$$= \phi_1 (\epsilon_{12} + \epsilon_{13} + \ldots + \epsilon_{1n}) + \phi_2 (\epsilon_{21} + \epsilon_{23} + \ldots + \epsilon_{2n}) + \ldots$$
$$+ \phi_n (\epsilon_{n1} + \epsilon_{n2} + \ldots + \epsilon_{n,n-1})$$
$$= 0,$$

from Kirchhoff's first law, since the total additional flow away from each junction must vanish. Therefore (2) reduces to

$$H' = H + \Sigma R_{pq} \epsilon^2_{pq},$$

and as the terms in the expression $\Sigma R_{pq} \epsilon^2_{pq}$ are positive, therefore

$$H < H'.$$

(ii) *To shew that when there are batteries in the wires and the circumstances are in other respects the same as above, then the actual distribution of currents is such as to make* $H - 2\Sigma E_{pq} C_{pq}$ *a minimum, where* H *is the rate of heat production.*

Let

$$F = H - 2\Sigma E_{pq} C_{pq},$$

or

$$F = \Sigma R_{pq} C^2_{pq} - 2\Sigma E_{pq} C_{pq} \quad \text{............(3)}.$$

Then if $C_{pq} + \epsilon_{pq}$ denote the current in A_pA_q in any other state of continuous flow, the corresponding function is

$$F' = \Sigma R_{pq} (C_{pq} + \epsilon_{pq})^2 - 2\Sigma E_{pq} (C_{pq} + \epsilon_{pq})$$
$$= F + 2\Sigma R_{pq} C_{pq} \epsilon_{pq} + \Sigma R_{pq} \epsilon^2_{pq} - 2\Sigma E_{pq} \epsilon_{pq} \quad \text{...(4)}.$$

But $$\sum (R_{pq}C_{pq} - E_{pq})\,\epsilon_{pq} = \sum (\phi_p - \phi_q)\,\epsilon_{pq}$$
$$= 0, \text{ as above.}$$

Therefore $$F' = F + \sum R_{pq}\epsilon^2_{pq},$$

and $$F < F'.$$

7·72. Example. We will apply the minimum method of **7·71** to **7·51** (i). The distribution of currents depicted in the figure is in accordance with Kirchhoff's first law, so the function to be made a minimum is

$$F = R\,(x+2y)^2 + r\,\{x^2 + 4y^2 + 2z^2 + 8\,(y-z)^2\} - 2E\,(x+2y).$$

We have therefore

$$\frac{\partial F}{\partial x} \equiv 2R\,(x+2y) + 2rx - 2E = 0,$$

$$\frac{\partial F}{\partial y} \equiv 4R\,(x+2y) + 8r\,(3y-2z) - 4E = 0$$

and $$\frac{\partial F}{\partial z} \equiv 4r\,(5z - 4y) = 0.$$

These equations are equivalent to **7·51** (i) (1), (2) and (3) and therefore lead to the same solution.

7·8. Telegraph wire with a 'fault'. A current is sent along a telegraph wire by a battery, one terminal of which is connected to the earth; and in the same way one terminal of the galvanometer at the receiving end is also put to earth, so that the earth forms the return half of the circuit and its resistance is negligible. A leakage from the wire is called a fault. Examples of wires with specified faults are solved by simple application of Kirchhoff's laws, others may be solved by expressing the currents as linear functions of the potentials.

(i) *A fault of given resistance develops in a telegraph wire. Shew that, with a given battery transmitting signals, the current received is least when the fault is at the middle of the wire.*

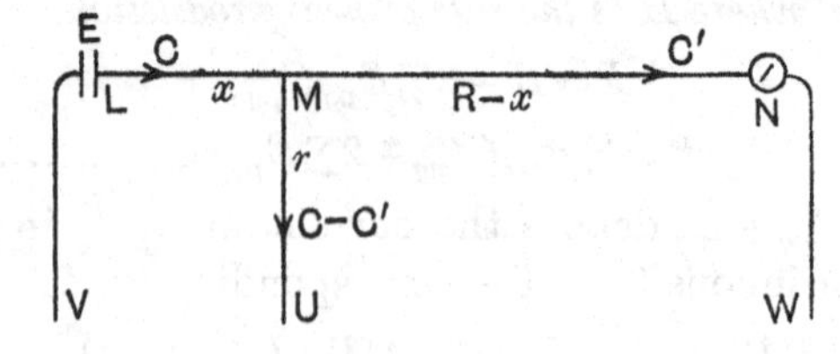

Let R be the resistance of the wire LN between the battery and the galvanometer, r the resistance of the fault at M, x that of the length

LM of the wire, so that $R-x$ is the resistance of the remainder MN. Let E be the electromotive force of the battery, C the current transmitted and C' that received, a current $C-C'$ leaking through the fault.

From the circuit $LMUV$ in the figure, we have

$$Cx+(C-C')r=E;$$

and from the circuit $MNWU$

$$C'(R-x)-(C-C')r=0.$$

By eliminating C we get

$$\{R(r+x)-x^2\}\,C'=Er,$$

so that C' is least when $Rr+Rx-x^2$ is greatest, i.e. when

$$Rr+\tfrac{1}{4}R^2-(\tfrac{1}{2}R-x)^2 \text{ is greatest,}$$

and that is when $x=\frac{1}{2}R$, or when the fault is in the middle of the wire.

(ii) *A, B are the ends of a telegraph wire with a number of faults. R and S are the resistances to a current from A when B is to earth and insulated respectively; and R', S' are the corresponding resistances to a current sent from B. Prove that $RS'=R'S$.* [I. 1894]

Let C be the current emitted from A and C' the current received at B; and let ϕ_A, ϕ_B denote the potentials of A and B. Then the currents must be linear functions of the potentials, say

$$C=k\phi_A+l\phi_B \quad \text{..........................(1)}$$

and

$$C'=m\phi_A+n\phi_B \quad \text{..........................(2),}$$

where k, l, m, n are constants which depend only on resistances, so that the same relations hold whichever way the current passes.

Then we have the following cases:

(*a*) B to earth; resistance is R; so that $\phi_B=0$, $\phi_A=RC$ and (1) gives $C=kRC$, so that $k=1/R$.

(*b*) B insulated, so that $C'=0$; and resistance is S, so that $\phi_A=SC$, the whole current going to earth through the faults.

Hence (1) gives

$$C=\frac{S}{R}C+l\phi_B$$

and (2) gives

$$0=mSC+n\phi_B;$$

therefore, by eliminating the ratio $C:\phi_B$, we get

$$(S-R)\,n=lmRS \quad \text{......................(3).}$$

Next let C' denote the current emitted from B and C the current received at A; then

(*c*) A to earth; resistance is R'; so that $\phi_A=0$, $\phi_B=R'C'$ and (2) gives $C'=nR'C'$, so that $n=1/R'$.

(*d*) A insulated, so that $C=0$; and resistance is S', so that $\phi_B=S'C'$.

Hence (1) gives $$0=\frac{\phi_A}{R}+lS'C'$$

and (2) gives $$C'=m\phi_A+\frac{S'C'}{R'};$$

therefore by eliminating the ratio $C':\phi_A$, we get

$$S'-R'=lmRR'S' \quad \text{.......................(4).}$$

Then from (3) and (4), and putting $n=\frac{1}{R'}$, we get

$$\frac{S-R}{S'-R'}=\frac{S}{S'} \text{ or } RS'=R'S.$$

7·9. Examples. (i) *A current is led to and from an electric lamp at a distance of* 40 *feet by a pair of wires, each of resistance* 13 *ohms per* 1000 *yards. If the difference of potential across the lamp is* 25 *volts and the current supplies energy to the lamp at the rate of* 30 *watts, at what rate in watts is energy lost in the leads? (A watt is the rate at which work is done by a current of* 1 *ampère working through* 1 *volt.)* [M. T. 1909]

The potential difference across the lamp is 25 volts, and energy is being supplied to the lamp at the rate of 30 watts, therefore the current is $\frac{30}{25}$ or $\frac{6}{5}$ ampères; and this is the current in the leads.

But the leads consist of 80 feet of wire of resistance 13 ohms per 1000 yards, i.e. of resistance $\frac{26}{75}$ ohm.

The rate of heat production RC^2 in the leads is therefore $\frac{26}{75}\times\frac{36}{25}=0·5$ watt, nearly.

(ii) *A galvanometer has a resistance of* 1 *ohm and each graduation represents* 1 *milliampère. Find the resistance with which it must be shunted to make it read ampères on the same scale. It is required to use the same instrument as a voltmeter. Find the least resistance which must be connected in series with it, to make its readings accurate within* 1 *per cent. if the resistance of the batteries whose voltage it is required to measure never exceeds* 2 *ohms but is otherwise unknown. On what scale will it then read volts?* [M. T. 1921]

The theory of galvanometers will be considered in a later chapter. For present purposes it is sufficient to know that a galvanometer is an instrument which measures the strength of a current passing through it: it is often called an *ammeter* in the sense that it measures the number of ampères of current; but if the resistance between the terminals is known, then the instrument can clearly be used to measure the difference of potential of the terminals and in this sense it constitutes a *voltmeter*. In order to be able to use the same instrument for measuring wide ranges of current or of electromotive force other resistances are coupled with it, in the former case in parallel and in the

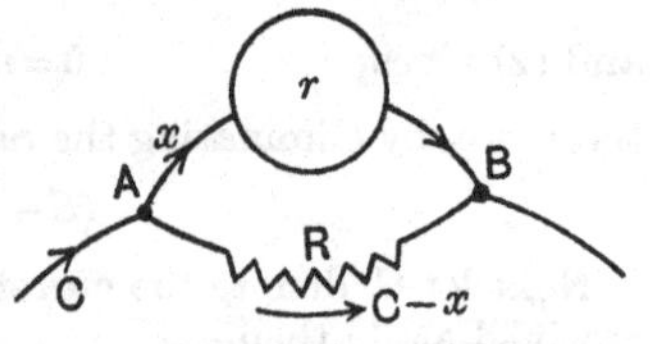

latter case in series. A resistance fixed across the terminals in parallel with the instrument is called a *shunt*. This example illustrates the use of a galvanometer both as an ammeter and as a voltmeter.

Let A, B be the terminals of the galvanometer and r its resistance. Let A, B be joined by a 'shunt' of resistance R. Let C be the current to be measured and suppose that a current x passes through the galvanometer and $C-x$ through the shunt. Then, by Kirchhoff's second law,

$$R(C-x)=rx \quad\text{.........................(1).}$$

In the example above $r=1$ ohm, and the passage of a milliampère through the instrument is to measure the passage of an ampère in the circuit as a whole, so that if $C=1$ then $x=\frac{1}{1000}$. By substituting these data in (1), we find that $R=\frac{1}{999}$ ohm, a shunt of small resistance taking the greater part of the current.

When used as a voltmeter the insertion of a resistance in series with the galvanometer serves to reduce the potential difference of the terminals, so that by the substitution of different resistances a wider range of voltages can be measured by the same instrument.

Let E be the electromotive force of a battery in volts, r' its resistance in ohms, and S ohms a resistance joined in series with the battery and galvanometer. Then

$$E=C(r+r'+S) \quad\text{.........................(2),}$$

where C is the current through the whole circuit.

But $r=1$ and $0<r'<2$, so to reduce the possible variation in $r+r'+S$ to 1 per cent. we must have $\dfrac{r'}{1+r'+S}<\dfrac{1}{100}$, or $S>99r'-1$, where r' may be as great as 2, so that the least value of S is 197.

Then from (2) $E=200C$ approximately, and one graduation means that $C=\frac{1}{1000}$ amp. and therefore $E=\frac{1}{5}$ volt.

(iii) *Current is supplied by a motor working at rate W to two circuits in parallel, one of which contains accumulators of voltage E_0 and resistance r on charge, and the other lamps of total resistance R. Prove that the current through the lamps is*

$$\frac{E_0+\left\{E_0^2+4rW\left(1+\dfrac{r}{R}\right)\right\}^{\frac{1}{2}}}{2(R+r)}.$$

If r/R is very small explain why disconnecting the accumulators may result in damage to the lamp circuit. [M. T. 1929]

The work done by the motor is partly converted into heat in the circuits and the lamps, and partly used in driving current through the accumulators which are absorbing energy at the rate E_0 for every unit of current which passes.

If A, B are the common terminals of the circuits, ϕ_A, ϕ_B their potentials and x, y the currents through the lamps and the accumulators, we have

$$\phi_A-\phi_B=Rx=ry+E_0 \quad\text{.......................(1)}$$

for the voltage of the accumulators must be regarded as opposed to the passage of the current y; and

$$W = Rx^2 + ry^2 + E_0 y \quad \text{.......................(2).}$$

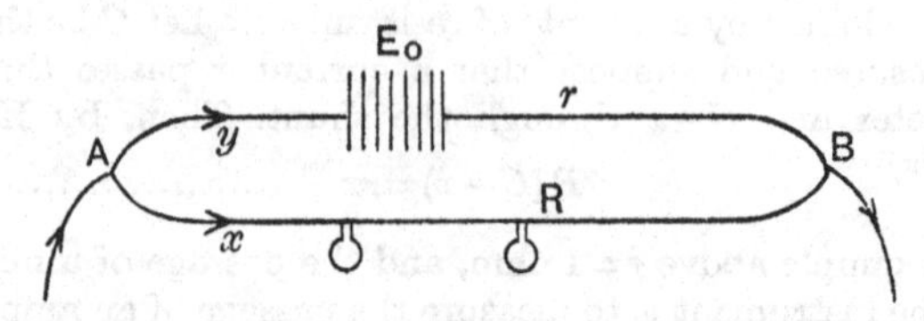

By substituting for y in terms of x from (1) in (2), we get a quadratic for x

$$(R+r)x^2 - E_0 x - rW/R = 0,$$

giving

$$x = \frac{E_0 \pm \sqrt{\left\{E_0^2 + 4rW\left(1+\frac{r}{R}\right)\right\}}}{2(R+r)},$$

and since x is positive the positive sign must be taken.

If r/R is very small, it is evident from (1) that the lamps only receive a small portion of the current from the motor; and if the accumulators are disconnected, the whole current will pass through the lamps and they may break as the result.

(iv) *Two condensers having capacities C_1 and C_2 are arranged as shewn, and G is a galvanometer. The resistances R_3 and R_4 are adjusted so that on closing the battery circuit not even a transient current flows through G. Treating the transient charging currents as steady, shew that*

$$R_3 C_1 = R_4 C_2. \quad \text{[M. T. 1928]}$$

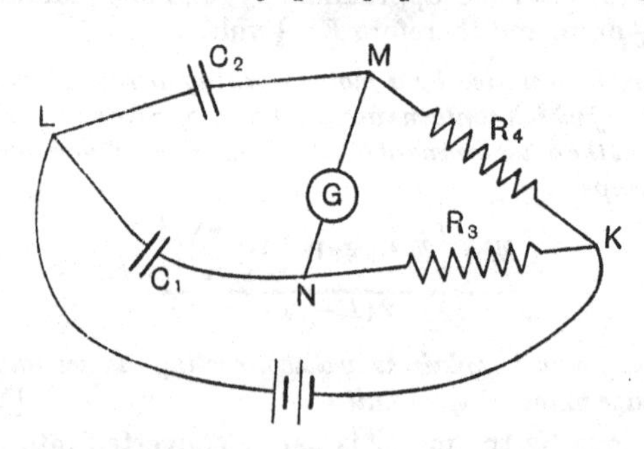

Let us consider the effect of inserting a condenser of capacity C in a wire LN in which a current x is flowing from L to N. If the wire meets

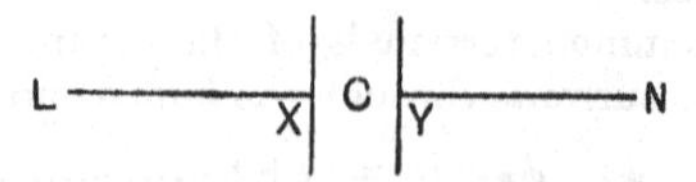

the condenser plates at X and Y and R is its total resistance, by Ohm's Law

$$\phi_L - \phi_X + \phi_Y - \phi_N = Rx.$$

But if Q is the charge on the positive plate of the condenser,

$$Q = C(\phi_X - \phi_Y);$$

so that

$$Rx = \phi_L - \phi_N - Q/C.$$

Hence the presence of the condenser diminishes the electromotive force in the wire by Q/C. And the current x is the rate at which Q is increasing, or $x = \dot{Q}$.

In the particular problem there is no current in MN so that $\phi_M = \phi_N$. Let x, y denote the currents along LNK and LMK, and let R_1, R_2 be the resistances of LN and LM.

Then, as shewn above,

$$R_1 x = \phi_L - \phi_N - Q_1/C_1 \text{ and } R_2 y = \phi_L - \phi_M - Q_2/C_2,$$

where Q_1, Q_2 are the positive charges on the condenser plates. Also

$$\phi_N - \phi_K = R_3 x \text{ and } \phi_M - \phi_K = R_4 y.$$

Hence, since $\phi_M = \phi_N$, we have

$$R_1 x + \frac{Q_1}{C_1} = R_2 y + \frac{Q_2}{C_2} \text{ and } R_3 x = R_4 y.$$

But in this problem the currents are regarded as steady so that the charges which accumulate are proportional to the currents, and we may write

$$Q_1 = tx, \; Q_2 = ty.$$

Then by eliminating x and y, we get

$$\left(R_1 + \frac{t}{C_1}\right) R_4 = \left(R_2 + \frac{t}{C_2}\right) R_3;$$

and this relation must hold throughout the whole process of charging and is therefore true for all values of t. Consequently we must have

$$R_1 R_4 = R_2 R_3 \text{ and } R_3 C_1 = R_4 C_2.$$

EXAMPLES

1. If the resistance of a copper wire 1 metre long and weighing 1 gram be 0·15 ohm, find the length of a wire of the same material which weighs a million grams and whose resistance is 6000 ohms. [I. 1891]

2. The resistance of 1 mile of copper wire whose diameter was 0·065 inch was found to be 15·6 ohms. The resistance of a wire of pure copper 1 foot long and 0·001 inch in diameter is 9·94 ohms. Compare the specific resistance of the copper used in the first wire with that of pure copper. [I. 1905]

3. Two cells whose E.M.F.'s are 2 and 3 volts and internal resistances 2 ohms are connected in parallel with a resistance of 1 ohm. Find the currents through the external resistance and each cell. [M. T. 1908]

4. Three cells, each of E.M.F. 1·8 volts and internal resistance 1·1 ohms, are used to send a current through a resistance of 1 ohm. Find whether the current is greater when the cells are joined in series or in parallel. [M. T. 1915]

5. Find the least number of Grove's cells, each having an E.M.F. of 1·87 volts and a resistance of 0·17 ohm, which will send a current of 16 ampères through a resistance of 1·5 ohms, and shew how the cells should be arranged. [M. T. 1919]

6. Shew that the least number of Leclanché cells (E.M.F. 1·55 volts, internal resistance 0·7 ohm) by which an incandescent lamp requiring a current of 2 ampères and a potential difference of 10 volts can be worked is 24. [M. T. 1893]

7. A battery of mn equal cells is such that when it is arranged in m parallel sets of n cells in series the maximum current C is produced for a given external circuit. Shew that when the cells are arranged in n parallel sets of m cells the current is $2mnC/(m^2+n^2)$. [I. 1902]

8. A current of not less than 1 ampère has to be sent through a wire whose resistance is 100 ohms by a number of cells (E.M.F. 1·5 volts, internal resistance 4 ohms). Shew that it can be done by using 715 cells, and find the current which passes. [I. 1894]

9. A certain kind of cell has a resistance of 10 ohms, and an electromotive force of 0·85 of a volt. Shew that the greatest current which can be produced in a wire whose resistance is 22·5 ohms, by a battery of five such cells, arranged in a single series of which any element is either one cell or a set of cells in parallel, is exactly 0·06 of an ampère. [M. T. 1901]

10. A certain kind of cell has a resistance of 8 ohms and an E.M.F. of 1·2 volts. Twelve cells are arranged in three elements, each element consisting of four cells in parallel, and the elements being arranged in series. What current will such a battery produce in a wire whose resistance is 20 ohms? [St John's Coll. 1906]

11. A battery whose internal resistance is 5 ohms is connected by a wire of resistance 20 ohms with a galvanometer whose resistance is 80 ohms; the galvanometer is shunted with 20 ohms and the current through it is 0·1 ampère. What is the E.M.F. of the battery? [I. 1910]

12. A battery, whose E.M.F. is 3 volts and internal resistance 8 ohms, is connected through a resistance of 92 ohms with a galvanometer of resistance 80 ohms. What is the current through the galvanometer, and what would it be if the galvanometer were shunted with 20 ohms? [St John's Coll. 1907]

13. In an overhead-wire tramway the resistance of the wire is 0·3 of an ohm per mile. A battery at one end gives 500 volts and the

current passes along the wire and through three tramcars, returning to the battery along a rail of negligible resistance. The cars are at distances 1, 2, and 3 miles from the battery and each take a current of 60 ampères from the circuit. Find the difference of potential between the wire and the rail at the farthest car. [I. 1907]

14. ABC is the trolley wire of an electric tramway 2000 yards long. B is its middle point and O is the position of the generating station. Current is supplied along three feeders, OA, OB, OC, the resistances of these being 0·35, 0·25, 0·10 ohm respectively. The resistance of the trolley wire is 0·80 ohm. A car situated at a point 500 yards distant from A is taking 100 ampères from the trolley wire. Find the current in each of the three feeders.

If the resistance of the return circuit is negligible, and the potential difference at the generating station is 500 volts, determine the potential difference at the car. [M. T. 1912]

15. It is required to light 40 electric lamps arranged in parallel. Each lamp needs a potential difference of 100 volts between its terminals and uses 0·16 ampère. If the resistance of the leads to the dynamo is 2 ohms, calculate the E.M.F. required for the dynamo, the energy lost per second in the leads, and that used in the lamps. [M. T. 1922]

16. The points A, B, C, D are joined by five wires AB, BC, CD, DA and BD. The resistances in these wires are respectively 5, 5, 5, 3 and 8 ohms. Find the equivalent resistance of the network for a current entering at A and leaving at C. [M. T. 1915]

17. The arms AB, AD, AC of a Wheatstone's bridge are formed of wires of resistance 1 ohm, and BC, BD by wires of resistances 10 ohms and 9 ohms respectively. The points A, D are also connected by a wire including a galvanometer of total resistance 9 ohms. If CD contains a battery of E.M.F. 1 volt and has a total resistance of 3 ohms, shew that no current will pass along AB, and determine that through the galvanometer. [Trinity Coll. 1898]

18. A cable AB, 50 miles in length, is known to have one fault and it is necessary to localize it. If the end A is attached to a battery and has its potential maintained at 200 volts, while the other end B is insulated, it is found that the potential of B when steady is 40 volts. Similarly when A is insulated the potential to which B must be raised to give A a steady potential of 40 volts is 300 volts. Shew that the distance of the fault from A is 19·05 miles. [Trinity Coll. 1896]

19. An electric current passes along a uniform cable from A to B. The resistance from A to B is R, the potential at A is V, and the cable is earthed at B. If, owing to a leak at some intermediate point X,

current of amount C' reaches B when current of amount C is sent out from A, find the ratio of the length AX to the length AB.

Shew that the resistance of the leak is

$$\frac{C'}{(C-C')^2}(CR-V). \qquad \text{[M. T. 1931]}$$

20. A cable AB of length l of uniform wire develops a leak at a certain point. To locate the fault two observations are made. The resistance between A and the earth through the cable when B is earthed is found to be that of a length a of the wire, and that between B and the earth when A is earthed that of a length b. Shew that the point at which the fault exists divides AB in the ratio $\{a(l-b)/b(l-a)\}^{\frac{1}{2}}$, and find the resistance of the leak. [M. T. 1934]

21. A telegraph wire joining two places A, B drops from one of its supports at a place C and rests on another wire which is earthed at both ends. If λ is the ratio of the current strength at A to that at B when the current in AB is sent from A, and μ is the ratio when the current is sent from B, shew that C divides AB in the ratio $(\mu^{-1}-1):\lambda-1$. [M. T. 1908]

22. AB is a telegraph wire. Faults whose resistances measured in lengths of wire are r and s develop at distances $y+z$ and z from the end B. Currents are sent from A. Shew that the difference of the resistances to the current from A according as B is insulated and put to earth is

$$r^2s^2/(y+r+s)(ys+yz+rz+sz+rs). \qquad \text{[I. 1922]}$$

23. A, B, C are three stations on the same telegraph wire. An operator at A knows that there is a fault between A and B, and observes that the current at A when he uses a given battery is i, i' or i'' according as B is insulated and C to earth, B to earth, or B and C both insulated. Shew that the distance of the fault from A may be obtained in the form

$$\{\kappa a-\kappa' b+(b-a)^{\frac{1}{2}}(\kappa a-\kappa' b)^{\frac{1}{2}}\}/(\kappa-\kappa'),$$

where $AB=a$, $BC=b-a$, $\kappa=\dfrac{i''}{i-i''}$, and $\kappa'=\dfrac{i''}{i'-i''}$. [M. T. 1904]

24. A 'fault' occurs in a telephone wire, AB, due to accidental connection to earth at an unknown point C, by a conductor of appreciable, but unknown, resistance r. The wire AB is uniform and has a known resistance R. Measurements of the resistance between the end A and earth are made by a Wheatstone bridge, (1) when the end B is insulated, (2) when the end B is earthed, the results being R_1 and R_2 respectively. Shew that, if the resistance of the earth is negligible,

$$\frac{AC}{AB}=\frac{R_2-\sqrt{(R_1-R_2)(R-R_2)}}{R},$$

and find the value of r. [M. T. 1925]

25. The 'universal' shunt galvanometer is arranged as follows: between two points A and B there are two wires, one of which contains the coil of the galvanometer, and the other admits of a connection being made at a variable point C. If a current enters at A and leaves at C, shew that the fraction of this current measured by the galvanometer is proportional to the resistance of AC. [M. T. 1909]

26. A cell of electromotive force E and internal resistance r is connected by wires of resistance r_1, r_2 to the terminals of a galvanometer of resistance G, there being a shunt of resistance R across the terminals. Find the current through the galvanometer. [M. T. 1933]

27. Two arms AB, AC of a Wheatstone's bridge have resistances 10 ohms each, BD is a standard coil of 5 ohms, and CD a coil of approximately 5 ohms, whose exact resistance is required. The galvanometer (in BC) has a resistance of 50 ohms and can detect 10^{-5} ampère. If the maximum current that may be passed through the resistance BD is $\frac{1}{10}$ ampère, find to two significant figures the smallest difference between the unknown coil and the standard which can be detected.
[M. T. 1922]

28. Two conducting circuits OPQ, $O'P'Q'$ are connected from P to P' and Q to Q' by wires of resistances r and r' respectively. A current enters the circuit at O and leaves at O'. Shew that, if the resistances of the lengths OP, PQ, QO are A, B, C and of the lengths $O'P'$, $P'Q'$, $Q'O'$ are a, b, c respectively, the currents in PP' and QQ' are in the ratio

$$\frac{BC}{A+B+C}+\frac{bc}{a+b+c}+r' : \frac{AB}{A+B+C}+\frac{ab}{a+b+c}+r.$$

[M. T. 1916]

29. A network is made up of the sides of a regular tetrahedron in which the mid-points A and B of a pair of opposite sides are connected by a wire whose resistance is the same as that of any one of the sides of the tetrahedron. A current is led in at A and out at B. Shew that the equivalent resistance of the network between A and B is $\frac{3}{7}r$, where r is the resistance of one of the sides of the tetrahedron. [M. T. 1934]

30. $ABCD$ is a uniform circular wire of resistance 4 ohms, and AOC, BOD are two wires forming diameters at right angles, each of resistance 2 ohms; prove that if a battery be placed in AD, the resistance of the network is $\frac{15}{7}$ ohms, and if in AO, $\frac{15}{8}$ ohms. [I. 1909]

31. A, B, C, D are four points in succession at equal distances along a wire; and A, C, and B, D are also joined by two other wires of the same length as the distances between those pairs of points measured along the original wire. A current enters the network thus formed at A and leaves at D; shew that $\frac{1}{5}$ of it passes along BC. [M. T. 1909]

32. A square $ABCD$ is formed of a uniform piece of wire, and the centre is joined to the middle points of the sides by straight wires of the same material and cross-section. A current is taken in at A and s

drawn off at the middle point of BC. Prove that the equivalent resistance of the network between these points is $\frac{29}{48}$ of that of a side of the square. [M. T. 1906]

33. The sides of a hexagon $ABCDEF$ are formed of wires each of resistance R, and wires each of the same resistance R join BF and BE; shew that the equivalent resistance of the conductors between A and D is $\frac{40}{29}R$. [I. 1901]

34. A network is made up of 13 feet of uniform wire, placed so as to form four equal squares side by side. A unit current enters at one extreme corner and leaves by the diagonally opposite corner. Shew that the total resistance is equal to that of $2\frac{7}{19}$ feet of the wire, and find the current in each of the cross pieces. [I. 1902]

35. A network is formed of uniform wire in the shape of a rectangle of sides $2a$, $3a$ with parallel wires arranged so as to divide the internal space into six squares of side a, the contact at points of intersection being perfect. Shew that if a current enter the framework by one corner and leave it by the opposite, the resistance is equivalent to that of length $\frac{121}{69}a$ of the wire. [Trinity Coll. 1895]

36. An octahedron is formed of twelve bars of equal length and thickness and of the same material; a current enters the system at one end of a bar and leaves at the other end of the same bar; shew that the resistance of the octahedron is $\frac{5}{12}$ of that of a single bar. [I. 1902]

37. Five points are connected by ten wires, each pair being joined by a wire of the same resistance R. Shew that the resistance to a current entering at one of the points and leaving at any other point is $\frac{2}{5}R$. [I. 1926]

38. Six equal and uniform wires are connected so as to form a regular tetrahedron $ABCD$. A current enters at the middle point of AB, and leaves from the middle point of CD. Shew that the resistance is $\frac{3}{4}r$, where r is the resistance of one of the wires. [I. 1907]

39. A cubical framework consists of 12 equal wires, each of resistance R. Shew that, if a current enters at any corner of the cube, and leaves at the opposite corner, the equivalent resistance of the cube is $\frac{5}{6}R$. [I. 1908]

40. A cube is formed of twelve uniform wires of the same resistance r, the opposite corners are connected by wires of resistance r' which are otherwise insulated. Prove that the resistance to a current which enters at one corner of the cube and leaves at the opposite corner is

$$\frac{rr'(2r+5r')}{2(r^2+4rr'+3r'^2)}. \qquad \text{[I. 1899]}$$

41. A tetrahedron frame $ABCD$ is formed by six wires, the resistances of opposite edges being equal. Prove that the whole resistance of the frame for a current entering at A and leaving at D is

$$\frac{(r_1r_3+2r_1r_2+r_2r_3)r_3}{2(r_1+r_3)(r_2+r_3)},$$

where r_1 is the resistance of AB or CD, r_2 that of AC or BD and r_3 that of AD or BC. [I. 1902]

42. If every pair of n electrodes are connected by a conductor of resistance R, shew that the equivalent resistance of the network between any pair of electrodes is $2R/n$. [M. T. 1894]

43. In Wheatstone's bridge the resistances of the conductors AB, BC, AD, DC are R_1, R_2, R_3, R_4 respectively. There is a galvanometer in the conductor BD and a battery in AC. The resistances R_1 and R_4 being unknown, and the bridge being not quite in adjustment, it is found that it can be brought into adjustment by shunting R_1 with a large resistance X, or by shunting R_4 with a large resistance Y. Shew that approximately

$$R_1 = \sqrt{\frac{XR_2R_3}{Y}} + \frac{1}{2}\frac{R_2R_3}{Y},$$

$$R_4 = \sqrt{\frac{YR_2R_3}{X}} + \frac{1}{2}\frac{R_2R_3}{X},$$

where it is assumed that X and Y are large compared with R_1, R_2, R_3, R_4. [M. T. 1924]

44. A cell of resistance r is connected to the ends of a wire AB. The cell is then replaced by two different cells, of resistances R, R', arranged in parallel, producing the same current in AB and having the combined resistance r when in parallel. Shew that the total heat production is greater in the second case than in the first, by the amount which would be produced in the circuit of the two cells if the wire AB were broken. [M. T. 1914]

45. A network consists of a rectangle $ABCD$ of wire, in which $AB = CD = 2a$, $BC = DA = a$, and a wire XY of the same material of length a, containing a battery of E.M.F. E and negligible resistance; X can slide along AB and Y along DC so that XY remains parallel to BC. Shew that the heat evolved is $6aE^2/r\{15a^2 - 4\xi^2\}$, where r is the resistance of unit length of wire and ξ is the distance of X from the mid-point of AB. [M. T. 1935]

46. Three wires APB, AQB, ARB are arranged in parallel between the points A, B; their resistances are p, q, r ohms respectively. Batteries of negligible resistances and of electromotive forces E, F volts are now inserted in the branches APB, AQB; the negative pole of each battery is connected to A. If the rate at which electrical energy is expended by the two batteries together is W watts, shew that

$$W = \frac{r(E-F)^2 + pF^2 + qE^2}{pq + rp + rq}. \qquad \text{[M. T. 1914]}$$

47. Coils of resistances R_1 and R_2 are connected, in parallel, to a battery of internal resistance r. Shew that the rate of expenditure of energy in the two coils will exceed that in either of the coils if connected separately to the battery, provided

$$\frac{1}{r^2} > \frac{1}{R}\left(\frac{1}{R_1} + \frac{1}{R_2}\right),$$

where R is the smaller of R_1 and R_2. [M. T. 1923]

48. A Wheatstone's bridge is represented diagrammatically by a quadrilateral $ABCD$, in which A, C are the battery and B, D the galvanometer terminals, and the resistances of AB, BC, CD, DA are P, Q, R, S. A balance is obtained by putting resistances α, β in parallel with R, S. The latter pair are then interchanged and a new balance obtained by putting resistances α' in parallel with R and β' in parallel with S. Prove that

$$(P+Q)\left(\frac{1}{R} - \frac{1}{S}\right) = P\left(\frac{1}{\beta} - \frac{1}{\alpha'}\right) + Q\left(\frac{1}{\beta'} - \frac{1}{\alpha}\right).$$ [I. 1914]

49. A quadrilateral $ABCD$ is formed of four wires and the diagonal AC is formed of another wire. In the wire AB is a battery of electromotive force E in the direction BA, in AD another of E.M.F. E' in direction AD and in AC a galvanometer; if the total resistances in AB, AD, BC, CD are c, b, β, γ and $b\beta = c\gamma$, prove that there is no current through the galvanometer provided $Eb = E'c$. [I. 1912]

50. An electric circuit contains a galvanometer and a battery of constant electromotive force V. The resistance of the galvanometer is G and that of the rest of the circuit including the battery R. Shew that on shunting the galvanometer with resistance S the current through the galvanometer is decreased by

$$\frac{VRG}{(R+G)(RS+RG+GS)}.$$

A circuit contains two lamps, each of resistance R, in parallel on leads each of resistance S. The resistance of the rest of the circuit, including the battery of constant voltage V, is r. Shew that if one lamp is broken, the heat emitted in unit time by the other is increased by

$$\frac{V^2R\{3r^2 + 2r(R+S)\}}{(r+R+S)^2(2r+R+S)^2}.$$ [M. T. 1913]

51. The circuit of an electric battery is completed through a galvanometer of high resistance G, and its observed deflection is D. The poles of the battery are then connected by a shunt of resistance S, and the deflection falls to D'. Prove that the resistance of the battery is

$$S\frac{(D-D')G}{D'(G+S)-DS}.$$ [M. T. 1911]

52. A battery is sending current through an external resistance R. The terminals of the battery are connected to a condenser and the charge communicated is found to be Q_1. On repeating the experiment

with an infinite external resistance, the charge given to the condenser is Q_2. Prove that the internal battery resistance is

$$R(Q_2-Q_1)/Q_1.$$ [M. T. 1912]

53. A tramcar takes its current from a trolley wire of total resistance r whose terminals are both kept at potential V. If the power taken by the car is assumed constant and equal to H, shew that the minimum potential difference between the trolley and the earth is

$$V'=\tfrac{1}{2}\{V+(V^2-rH)^{\frac{1}{2}}\},$$

and that this occurs when the car is midway between the terminals. Shew that the waste of power in the wire is then a maximum, and is equal to $H(V/V'-1)$. [St John's Coll. 1915]

54. A, B are the ends of a long telegraph wire with a number of faults and C is an intermediate point on the wire; the resistance to a current sent from A is R when C is earth connected, but if C is not earth connected the resistance is S or T according as the end B is to earth or insulated. If R', S', T' denote the resistances under similar circumstances when a current is sent from B towards A, shew that

$$T'(R-S)=R'(R-T).$$ [M. T. 1903]

ANSWERS

1. 200 kilom.

2. As 5 : 4 approx.

3. 0·375, 0·875, 1·25 amps.

4. Less in series in the ratio 41 : 43.

5. 75 arranged in three parallel sets each containing 25 cells in series.

8. 1·00046 amps.

10. 0·138 amp.

11. 20·5 volts.

12. $\frac{1}{60}$ amp., $\frac{3}{580}$ amp.

13. 392 volts.

14. 40, 40, 20 amps.; 478 volts.

15. 112·8 volts; the leads consume at the rate of 81·92 watts and the lamps at 640 watts.

16. $4\frac{47}{112}$ ohms.

17. $\frac{1}{52}$ amp.

20. That of a length $\dfrac{a\sqrt{\{b(l-a)\}}-b\sqrt{\{a(l-b)\}}}{\sqrt{\{b(l-a)\}}+\sqrt{\{a(l-b)\}}}\,\dfrac{l}{b-a}$ of wire.

26. $RE/\{RG+(R+G)(r+r_1+r_2)\}$.

27. 0·0085 ohm.

34. $\frac{7}{19}, \frac{2}{19}, \frac{1}{19}, \frac{2}{19}, \frac{7}{19}$.

Chapter VIII

MAGNETISM

8·1. The peculiar properties of the oxide of iron Fe_3O_4 were known to the ancient world. It was called lodestone or leading stone because of its power of attracting iron, and its ancient name λίθος Μαγνῆτις was attributed by Lucretius to Magnesia in Thessaly, where the ore was found in large quantities. The modern science of Magnetism may be said to date from William Gilbert of Colchester who published his great work *De Magnete* in 1600.

If a body such as a steel needle has been rubbed with a lodestone, and so acquired magnetic properties, and is then suspended so that it can turn freely about its centre of gravity, it sets itself in equilibrium so that a definite line in the body—called *the axis of the magnet*—is parallel to a definite direction in space—called *the direction of the earth's magnetic field*, roughly north and south. If the ends of the axis are marked on the magnet, it is found that, when the magnet is free to move, one end always points in a northerly direction and the other southerly. The azimuth, or direction of the magnetic field west of north, is called *the variation* or *the magnetic declination*, and the angle between the direction of the earth's magnetic field and the horizontal plane is called *the magnetic dip*. The earth's magnetic field is completely specified by its intensity, declination and dip.

8·11. Magnetic poles. It is found that if the axes of several magnets are marked and the magnets are freely suspended, ends which both point to the north or both point to the south repel one another and an end which points to the north attracts an end which points to the south.

The ends of a long thin magnet are called its poles. Like poles repel and unlike attract. The two kinds of poles may be called positive and negative. They act as centres of force. In a long thin bar magnet the rest of the bar appears to be almost devoid of magnetic properties; but, if the bar be broken in two, each portion appears to have poles similar to those of the original magnet. This and repeated experiments of the same kind lead to the conclusion that a uniformly magnetized body is composed of a great number of small magnets with their axes in the same direction and like poles all pointing the same way.

It is impossible however to isolate a magnetic pole or obtain a body in which the total sum of magnetism is not zero. This fact may be verified by a simple experiment, viz. that if a magnet be placed floating on water supported on a cork, though the earth's magnetic field exerts on the magnet a couple which causes it to rotate until it settles in the plane of the magnetic meridian, yet the magnet then remains at rest. Whereas if it contained an excess of either positive or negative poles, the earth's field would exert on the magnet a resultant force which would give it an acceleration.

8·12. The fundamental vectors. Magnetic fields can be explored by small magnets such as compasses, and the couples exerted on the same magnet in different positions in the field, or on different magnets in the same position, can be compared by observing their periods of oscillation. Such experiments lead to the conclusion that for a given small magnet at a fixed point in a given field there are two vectors **M** and **H**, the former in the direction of the axis of the magnet and the latter in a fixed direction, such that the couple on the magnet is the vector product* [**MH**], i.e. $MH \sin\theta$, where M, H denote the magnitudes of the vectors **M**, **H**, and θ is the angle between them, and the vector **M** is invariable for the same magnet. **M** is called **the moment of the magnet** and **H the intensity of the magnetic field or the magnetic force.**

Hence a small magnet of moment M in a field of intensity H is acted upon by a couple of moment $MH \sin\theta$, where θ is the angle between the positive directions of the axis of the magnet

* The vector product of two vectors **A**, **B** inclined at an angle θ is a vector **C** at right angles to the plane of **A** and **B**, of magnitude $AB \sin\theta$ and in such sense that **C** is along the axis of a right-handed screw which would turn from **A** to **B** through an angle θ less than two right angles. The reader will remember that a couple can be represented by a vector perpendicular to its plane. The vector product of **A** and **B** is variously denoted by [**AB**], $\mathbf{A} \wedge \mathbf{B}$ and $\mathbf{A} \times \mathbf{B}$.

and the line of force, and the couple tends to reduce the angle θ, i.e. to set the axis of the magnet along the line of force.

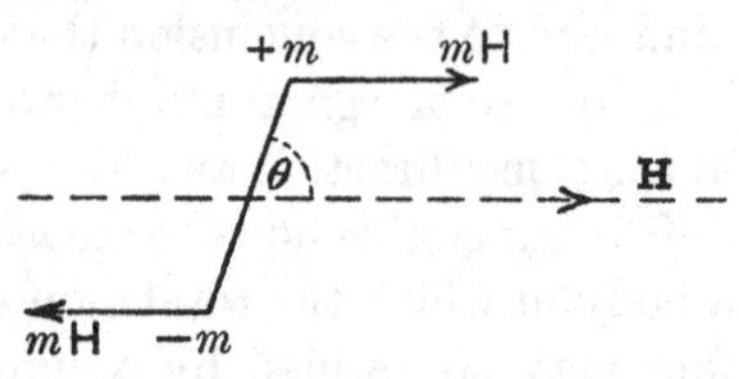

If we regard a thin magnet as composed of poles of strength $\pm m$ at a distance l apart and set its axis at an inclination θ to the direction of the field **H**, we see that the hypothesis that a field of intensity **H** exerts a force $m\mathbf{H}$ on a pole of strength m implies that the field exerts on the small magnet a couple $mHl \sin\theta$, or $[ml\mathbf{H}]$, so that **M** must be identical with $m\mathbf{l}$.

8·2. Field of a magnetic bipole. Experiment confirms the fact that the forces exerted in a magnetic field are consistent with a law of force between magnetic poles of the same form as that between electric charges, i.e. that like poles of strengths m, m' at a distance r apart repel one another with a force mm'/r^2. We therefore assume that each pole which goes to compose a magnet produces a field of potential like that of a point charge of electricity.

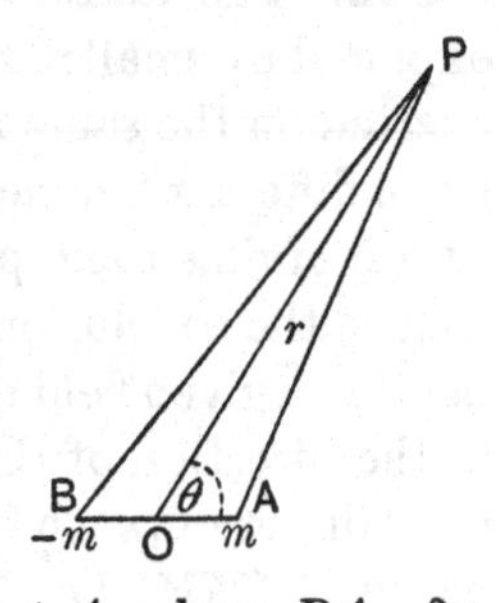

To find the potential of the field produced by a small magnet or magnetic bipole. Regard the magnet as composed of two poles of strengths $-m$ at B and m at A, where $BA = \delta s$, and $m\,\delta s = M$, the moment of the magnet.

Take an origin O at the middle point of BA, as in **6·6**, and let P be the point (r, θ) referred to OA as axis. Then the potential at P is given by

$$\phi = -\frac{m}{BP} + \frac{m}{AP} = -\frac{m}{r + \frac{1}{2}\delta s \cos\theta} + \frac{m}{r - \frac{1}{2}\delta s \cos\theta}$$

$$= -\frac{m}{r}\left(1 - \frac{\delta s \cos\theta}{2r} - \dots\right) + \frac{m}{r}\left(1 + \frac{\delta s \cos\theta}{2r} + \dots\right)$$

$$= \frac{m\,\delta s \cos\theta}{r^2} \text{ to the first power of } \delta s,$$

or $$\phi = \frac{M\cos\theta}{r^2} \qquad \dots\dots(1).$$

This result may also be written

$$\phi = \frac{(\mathbf{Mr})}{r^3} \quad \text{.....................(2)},$$

where $(\mathbf{Mr})$ or $\mathbf{Mr}$ denotes the scalar product* of the vector $\mathbf{M}$ and the position vector $\mathbf{r}$ of the point P relative to the centre of the magnet.

Again, since in the figure $\cos\theta = -dr/ds$, another form for the same result is

$$\phi = M \frac{\partial}{\partial s}\left(\frac{1}{r}\right) \quad \text{.....................(3)}.$$

8·21. We may now establish the vectorial property of the moment of a magnetic bipole thus:

Since $$\frac{(\mathbf{M}_1 + \mathbf{M}_2)\,\mathbf{r}}{r^3} = \frac{\mathbf{M}_1\mathbf{r}}{r^3} + \frac{\mathbf{M}_2\mathbf{r}}{r^3},$$

therefore the potential of a magnet of moment $\mathbf{M}_1 + \mathbf{M}_2$ is the sum of the potential due to $\mathbf{M}_1$ and $\mathbf{M}_2$ separately. Hence the 'moment of a magnet' obeys the vector law of composition and resolution.

8·22. Components of magnetic force due to a magnetic bipole. The magnetic force $\mathbf{H}$ is the negative gradient of the potential so that its components along and perpendicular to the radius vector are

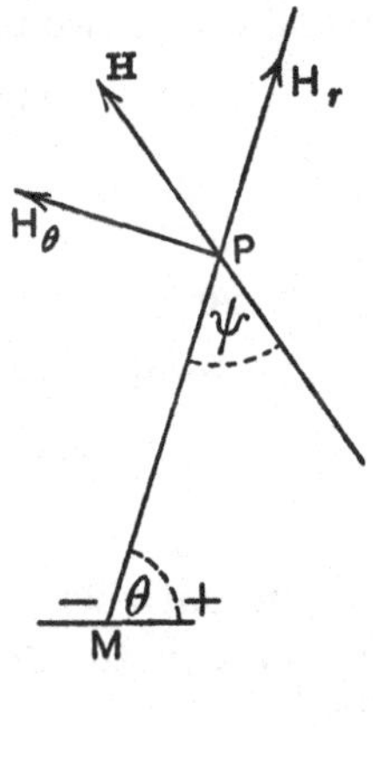

$$\left.\begin{aligned} H_r &= -\frac{\partial\phi}{\partial r} = \frac{2M\cos\theta}{r^3} \\ \text{and} \qquad H_\theta &= -\frac{1}{r}\frac{\partial\phi}{\partial\theta} = \frac{M\sin\theta}{r^3} \end{aligned}\right\} \quad \text{......(1)}.$$

The resultant $\mathbf{H}$ of these two components is a tangent to the line of force at P. Let it make an angle ψ with the radius vector. Then we have, along the line of force through P,

$$r\frac{d\theta}{dr} = \tan\psi = \frac{H_\theta}{H_r} = \tfrac{1}{2}\tan\theta,$$

so that $$\frac{dr}{r} = \frac{2\cos\theta}{\sin\theta}\,d\theta,$$

or $$r = C\sin^2\theta \quad \text{.....................(2)},$$

* The *scalar product* of two vectors $\mathbf{A}$, $\mathbf{B}$ is defined to be $AB\cos\theta$, where θ is the angle between their positive directions.

where C is an arbitrary constant, is the equation of the lines of force.

The lines of force in any section of a magnetic field may be exhibited by the help of iron filings. Thus in any such case as that just considered, if iron filings are scattered on a piece of cardboard laid on the magnet, and the cardboard is gently tapped, the filings become magnetized by induction and set themselves along the lines of force.

8·23. Other expressions for the field due to a bipole. Taking the expressions H_r and H_θ of **8·22** for the components of $\mathbf{H}$, we observe that H_θ can be resolved obliquely into $H_\theta \cot\theta$ or $\dfrac{M\cos\theta}{r^3}$ in the direction of $\mathbf{r}$, and $-H_\theta \operatorname{cosec}\theta$ or $-\dfrac{M}{r^3}$ in the direction of $\mathbf{M}$; and H_r being $\dfrac{2M\cos\theta}{r^3}$ in the direction of $\mathbf{r}$, it follows that the field $\mathbf{H}$ is the resultant of two oblique components, viz.

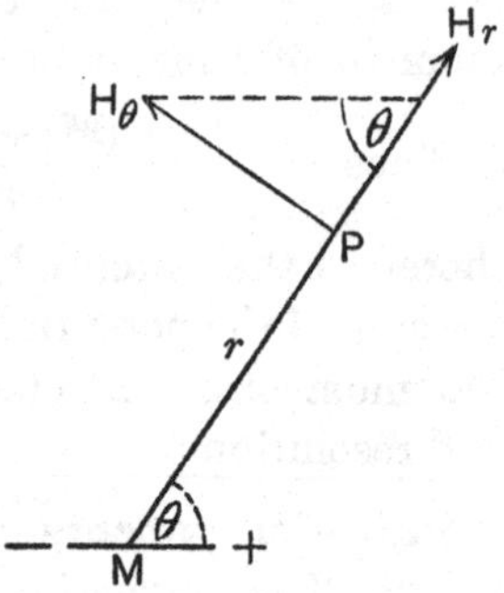

H_r H_θ θ P r θ $-$ M $+$

$$\left.\begin{array}{ll} & \dfrac{3M\cos\theta}{r^3} \text{ in the direction of } \mathbf{r} \\ \text{and} & -\dfrac{M}{r^3} \text{ in the direction of } \mathbf{M} \end{array}\right\} \quad \text{.........(1).}$$

This result may be written in vector form thus:

$$\mathbf{H} = -\frac{\mathbf{M}}{r^3} + \frac{3\,(\mathbf{Mr})\,\mathbf{r}}{r^5} \quad \text{..................(2).}$$

Alternatively, we may take a small arbitrary displacement $\delta\mathbf{r}$ and equate the work done by $\mathbf{H}$ to the loss of potential. Thus

$$\mathbf{H}\delta\mathbf{r} = -\delta\phi = -\delta\left\{\frac{(\mathbf{Mr})}{r^3}\right\} \qquad \text{from } \mathbf{8{\cdot}2}\,(2)$$

$$= -\frac{\mathbf{M}\delta\mathbf{r}}{r^3} + \frac{3\,(\mathbf{Mr})\,\delta r}{r^4}.$$

But, if α is the angle between $\mathbf{r}$ and $\delta\mathbf{r}$, $\delta r = |\delta\mathbf{r}| \cos\alpha = \mathbf{r}\,\delta\mathbf{r}/r$, so that

$$\mathbf{H}\,\delta\mathbf{r} = \delta\mathbf{r}\left\{-\frac{\mathbf{M}}{r^3} + \frac{3\,(\mathbf{M}\mathbf{r})\,\mathbf{r}}{r^5}\right\},$$

which gives the result (2).

8·24. Oscillations of a small magnet. Let a small magnet of moment $\mathbf{M}$ be free to turn about its centre in a magnetic field whose intensity is $\mathbf{H}$ at the centre of the magnet. In equilibrium the axis of the magnet coincides in direction with $\mathbf{H}$. Let the magnet be turned through an angle θ in *any* plane through the direction of $\mathbf{H}$. The couple tending to restore the magnet to its equilibrium position is $MH \sin\theta$ (**8·12**), therefore if K is its moment of inertia about its centre, its equation of motion is

$$K\ddot{\theta} = -MH \sin\theta,$$

and for small values of θ this represents harmonic oscillations of period

$$2\pi\sqrt{(K/MH)}.$$

We note that though the magnet has two degrees of freedom and a body with two degrees of freedom has in general two independent principal oscillations, yet in this case its period of oscillation in every plane through the direction of the resultant force at its centre is the same, so that oscillation in any such plane may be regarded as a principal oscillation, and the two periods of principal oscillations are the same.

8·25. Examples. (i) *Two small magnets of moments m, m' are fixed at two corners of an equilateral triangle with their axes bisecting the angles. A third small magnet is free to turn about the other angular point. Shew that its axis makes with the bisector of the third angle an angle*

$$\tan^{-1}\left(\frac{\sqrt{3}}{7}\frac{m-m'}{m+m'}\right).$$

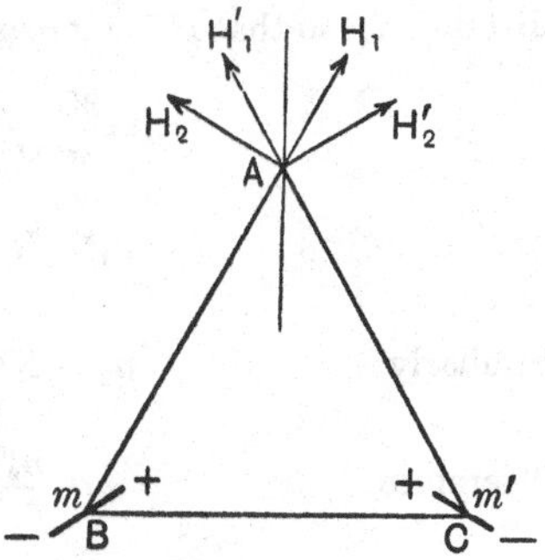

The magnets m, m' at B, C produce at A fields with components

$$H_1 = \frac{2m\cos 30^\circ}{r^3} = \frac{m\sqrt{3}}{r^3}, \quad H_2 = \frac{m \sin 30^\circ}{r^3} = \frac{m}{2r^3};$$

$$H_1' = \frac{2m'\cos 30^\circ}{r^3} = \frac{m'\sqrt{3}}{r^3}, \quad H_2' = \frac{m' \sin 30^\circ}{r^3} = \frac{m'}{2r^3};$$

in the directions indicated in the figure, where r is the length of a side of the triangle.

Resolving along and perpendicular to the bisector of the angle BAC, the components are

$$(H_1+H_1')\cos 30°+(H_2+H_2')\cos 60°, \text{ or } 7(m+m')/4r^3;$$

and $$(H_1-H_1')\sin 30°+(H_2'-H_2)\sin 60°, \text{ or } (m-m')\sqrt{3}/4r^3.$$

Hence the line of force at A, which is the direction that the third magnet takes up in equilibrium, makes with the bisector of the angle BAC an angle $\tan^{-1}\left(\frac{\sqrt{3}}{7}\frac{m-m'}{m+m'}\right)$.

(ii) *A small magnet is suspended by a torsionless fibre so that it can oscillate about the magnetic meridian in a horizontal plane, when it is observed to execute N vibrations per minute. A bar magnet is placed with its axis horizontal and parallel to the meridian and passing through the centre of the small magnet, at a distance d_1 measured from its mid-point and large compared to its own length. The small magnet now executes n_1 oscillations per minute, and when d_1 is changed to d_2, n_1 becomes n_2. Shew that*

$$\frac{n_2^2-N^2}{n_1^2-N^2}=\frac{d_1^3}{d_2^3}.$$

The oscillations may be supposed of small amplitude. [M. T. 1926]

As in **8·24**, if K is the moment of inertia of the small magnet, and M its moment, its period of oscillation in a horizontal plane in the earth's field is $2\pi\sqrt{(K/MH)}$, where H is the horizontal component of the earth's magnetic force; so that $N=\frac{30}{\pi}\sqrt{\left(\frac{MH}{K}\right)}$.

The effect of a bar magnet of moment M' with its axis along the magnetic meridian is to produce a force $2M'/d_1^3$ along the meridian at distance d_1; so that H is increased by $2M'/d_1^3$, and therefore

$$n_1=\frac{30}{\pi}\sqrt{\left\{\frac{M}{K}\left(H+\frac{2M'}{d_1^3}\right)\right\}},$$

or $$n_1^2-N^2=\frac{900}{\pi^2}\cdot\frac{2MM'}{Kd_1^3}.$$

Similarly $$n_2^2-N^2=\frac{900}{\pi^2}\cdot\frac{2MM'}{Kd_2^3}.$$

Therefore $$\frac{n_2^2-N^2}{n_1^2-N^2}=\frac{d_1^3}{d_2^3}.$$

8·3. Potential energy of a magnetic bipole in a given field. Let the small magnet be represented by a pole $-m$ at B and a pole m at A, where $BA=\delta s$ is drawn in a definite direction.

Let ϕ be the potential of the given field at B, then $\phi + \frac{\partial \phi}{\partial s}\delta s$ is the potential at A, and the potential energy of the two poles or the work which an external agent would have to do in order to place them in their assigned positions in the given field is given by

$$W = -m\phi + m\left(\phi + \frac{\partial \phi}{\partial s}\delta s\right)$$

$$= m\delta s \, . \frac{\partial \phi}{\partial s} = -M \, . H_s \quad \text{..................(1)},$$

where H_s is the component of the given field in the direction of the axis of the magnet; or

$$W = -MH\cos\theta \quad \text{.....................(2)},$$

where θ is the angle between the positive directions of **M** and **H**; or

$$W = -(\mathbf{MH}) \quad \text{........................(3)},$$

where the formula on the right denotes minus the 'scalar product' of the vectors **M** and **H**.

8·31. Mutual potential energy of two small magnets; i.e. the potential energy of one in the field of the other. The mutual relations of two small magnets lend themselves to very simple treatment by vectorial methods. But we shall first set out the arguments in such a way as to be intelligible without vector algebra.

Let M, M' be the moments of the magnets, r the distance between their centres, θ, θ' the angles which the line joining the centres makes with the positive directions of the axes and ϵ the angle between the axes (the magnets are not in general coplanar).

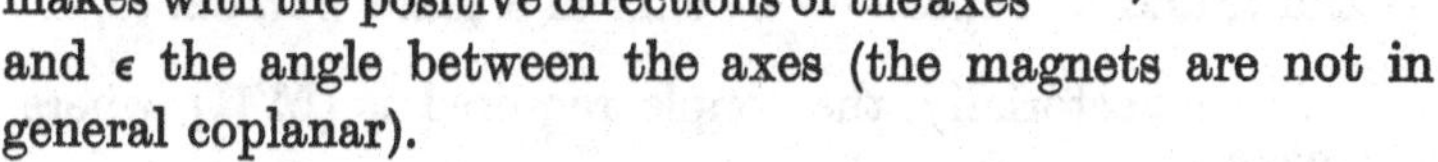

Then, from **8·3**, the potential energy of M' in the field of M is given by

$$W = -M'H_{M'},$$

where $H_{M'}$ denotes the component along M' of the field due to M. But by **8·23** the field **H** due to M has components $-\frac{M}{r^3}$

in the direction of M and $\frac{3M\cos\theta}{r^3}$ in the direction of r. So by resolving these in the direction of M', we get

$$W = \frac{MM'}{r^3}(\cos\epsilon - 3\cos\theta\cos\theta') \quad \text{........(1)}.$$

Using vectors, we write, from 8·3,

$$W = -(\mathbf{M'H}),$$

where from 8·23 (2)

$$\mathbf{H} = -\frac{\mathbf{M}}{r^3} + \frac{3(\mathbf{Mr})\,\mathbf{r}}{r^5},$$

so that
$$W = \frac{(\mathbf{MM'})}{r^3} - \frac{3(\mathbf{Mr})(\mathbf{M'r})}{r^5} \quad \text{...........(2)};$$

or if we put $\mathbf{r} = r\mathbf{r}_1$, so that $\mathbf{r}_1$ is a unit vector in the direction of $\mathbf{r}$, the result may be written

$$W = \{(\mathbf{MM'}) - 3(\mathbf{Mr}_1)(\mathbf{M'r}_1)\}/r^3 \quad \text{........(3)},$$

and this is clearly equivalent to (1).

8·32. Couple exerted by one small magnet on another. The couple exerted on the magnet M' by the magnet M may be represented, as in 8·12, by the vector product $[\mathbf{M'H}]$, where $\mathbf{H}$ is the field due to M; i.e. the couple is $M'H\sin\psi$, if ψ is the angle between M' and $\mathbf{H}$. But by 8·23 $\mathbf{H}$ is the resultant of components $\frac{3M\cos\theta}{r^3}$ in the direction of r and $-\frac{M}{r^3}$ in the direction of $\mathbf{M}$; therefore the couple exerted by M on M' is compounded of a couple $\frac{3M'M}{r^3}\cos\theta\sin\theta'$ in the plane of M' and r, and a couple $-\frac{M'M}{r^3}\sin\epsilon$ in a plane parallel to M' and M. Or vectorially, the couple required is $[\mathbf{M'H}]$, where by 8·23 (2)

$$\mathbf{H} = -\frac{\mathbf{M}}{r^3} + \frac{3(\mathbf{Mr})\,\mathbf{r}}{r^5},$$

so that the couple is

$$-\frac{[\mathbf{M'M}]}{r^3} + \frac{3[\mathbf{M'r}](\mathbf{Mr})}{r^5};$$

or, if $\mathbf{m}_1$, $\mathbf{m}_1'$, $\mathbf{r}_1$ denote unit vectors in the directions of $\mathbf{M}$, $\mathbf{M}'$ and $\mathbf{r}$, the result may be written

$$\frac{MM'}{r^3}\{[\mathbf{m}_1\mathbf{m}_1'] + 3[\mathbf{m}_1'\mathbf{r}_1](\mathbf{m}_1\mathbf{r}_1)\} \quad \text{.........(1).}$$

8·321. We observe that the components of the couple in **8·32** can also be deduced from the expression for the potential energy of $\mathbf{M}'$ in the field of $\mathbf{M}$, viz.

$$W = \frac{MM'}{r^3}(\cos\epsilon - 3\cos\theta\cos\theta').$$

We suppose that M is fixed in position and that M' is free to turn about its centre. Then the direction of M' is determined by the angles ϵ and θ', for θ is a fixed angle.

If for example we draw radii OP, OP', OR of a unit sphere parallel to $\mathbf{M}$, $\mathbf{M}'$ and $\mathbf{r}$, then P and R are fixed points and the arcs PP', RP' equal respectively to ϵ, θ', determine P'.

Hence M' is acted upon by a couple tending to increase ϵ of magnitude $-\frac{\partial W}{\partial \epsilon}$ or $\frac{MM'\sin\epsilon}{r^3}$, and this must act in a plane parallel to M and M'; and a couple tending to increase θ' of magnitude $-\frac{\partial W}{\partial \theta'}$ or $-\frac{3MM'}{r^3}\cos\theta\sin\theta'$ in the plane of r and M'. As before, these two couples can be written as in **8·32** (1), paying due regard to signs and remembering that

$$[\mathbf{m}_1'\mathbf{r}_1] = -[\mathbf{r}_1\mathbf{m}_1'].$$

8·33. Force exerted by one small magnet on another. The force exerted by M on M' and tending to move the latter bodily without rotation must be deduced from the potential energy. This may be written

$$W = \frac{MM'}{r^3}\left(\cos\epsilon - \frac{3pp'}{r^2}\right) \quad \text{...............(1),}$$

where p, p' denote the projections of r on the axes of the magnets. In a general displacement in which M remains fixed and M' alters its position but not its direction, ϵ remains constant and r, p, p' may be regarded as independent variables on which the position of the centre of M' depends.

Thus if through the centre O of M we draw lines OP, OP'

of lengths p, p' in the directions of the magnets, and through P, P' take planes at right angles to OP, OP', these planes intersect in a line, and the centre O' of M' is a point on this line at a distance r from O.

The force tending to move M' parallel to itself therefore has components

$$-\frac{\partial W}{\partial r}=\frac{3MM'}{r^4}\left(\cos\epsilon-\frac{5pp'}{r^2}\right)$$

$$=\frac{3MM'}{r^4}(\cos\epsilon-5\cos\theta\cos\theta') \quad \text{.........(2)}$$

in the direction of r;

$$-\frac{\partial W}{\partial p}=\frac{3MM'p'}{r^5}=\frac{3MM'\cos\theta'}{r^4} \quad \text{............(3)}$$

in the direction of p or M; and

$$-\frac{\partial W}{\partial p'}=\frac{3MM'p}{r^5}=\frac{3MM'\cos\theta}{r^4} \quad \text{............(4)}$$

in the direction of p' or M'.

It is easy to see that the force may be written as a vector sum in the form

$$(\rho\mathbf{r}_1+\mu\mathbf{M}+\mu'\mathbf{M}')/r^4,$$

where $\rho=3\{(\mathbf{MM}')-5(\mathbf{Mr}_1)(\mathbf{M}'\mathbf{r}_1)\}$,

$\mu=3(\mathbf{M}'\mathbf{r}_1)$

and $\mu'=3(\mathbf{Mr}_1)$.

Alternatively, as in **8·23**, if $\mathbf{F}$ denotes the force exerted by $\mathbf{M}$ on $\mathbf{M}'$, we have

$$\mathbf{F}\,\delta\mathbf{r}=-\delta W=-\delta\left\{\frac{(\mathbf{MM}')}{r^3}-\frac{3(\mathbf{Mr})(\mathbf{M}'\mathbf{r})}{r^5}\right\} \quad \text{from } \mathbf{8·31}\,(2)$$

$$=-\frac{3(\mathbf{MM}')\,\delta r}{r^4}+\frac{3\{(\mathbf{M}\,\delta\mathbf{r})(\mathbf{M}'\mathbf{r})+(\mathbf{Mr})(\mathbf{M}'\delta\mathbf{r})\}}{r^5}-\frac{15(\mathbf{Mr})(\mathbf{M}'\mathbf{r})\,\delta r}{r^6}$$

$$=\delta\mathbf{r}\left\{\frac{3(\mathbf{MM}')\,\mathbf{r}}{r^5}-\frac{15(\mathbf{Mr})(\mathbf{M}'\mathbf{r})\,\mathbf{r}}{r^7}+\frac{3\mathbf{M}(\mathbf{M}'\mathbf{r})}{r^5}+\frac{3\mathbf{M}'(\mathbf{Mr})}{r^5}\right\};$$

so that $F=(\rho\mathbf{r}_1+\mu\mathbf{M}+\mu'\mathbf{M}')/r^4$,

where ρ, μ, μ' have the meanings above.

8·34. Coplanar magnets. The formulae for coplanar magnets can be obtained independently. Thus, as in **8·3**, the potential energy of a small magnet M' in the field of a magnet M is

$$W = -M'H_{M'},$$

where $H_{M'}$ is the component in the direction of $\mathbf{M}'$ of the field $\mathbf{H}$ due to M. But, as in **8·22**, $\mathbf{H}$ has components $2M\cos\theta/r^3$ along r and $M\sin\theta/r^3$ perpendicular to r, and M' makes an angle θ' with r (see the figure of **8·31**), therefore

$$W = -\frac{MM'}{r^3}(2\cos\theta\cos\theta' - \sin\theta\sin\theta') \quad \text{......(1).}$$

This result follows of course directly from **8·31** (1) by putting $\epsilon = \theta - \theta'$.

Let us now find the force and couple exerted by $\mathbf{M}$ on $\mathbf{M}'$. The couple can be written down at once as $[\mathbf{M}'\mathbf{H}]$, but we shall also find it by a general method which will also serve to determine the force. Let R, T denote the forces tending to move the magnet M' without rotation along and perpendicular to r, and let G denote the couple tending to turn M' about its centre. Keeping M fixed, let M' undergo a small displacement of a general type which alters r, θ and θ'. The work done in this small displacement is equal to the loss of potential energy so that

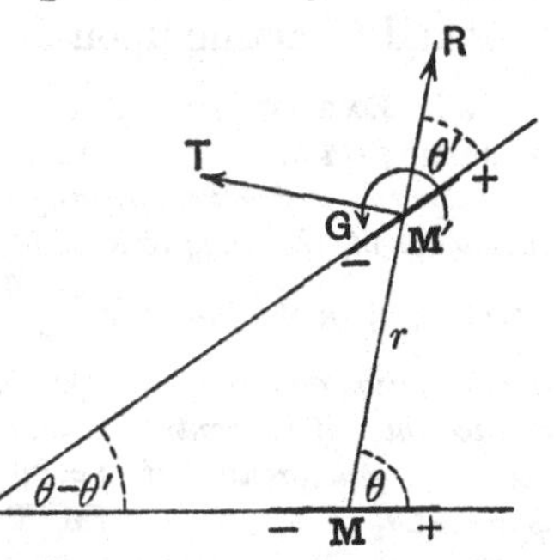

$$Rdr + Trd\theta + Gd(\theta - \theta') = -dW$$

$$= -\frac{\partial W}{\partial r}dr - \frac{\partial W}{\partial \theta}d\theta - \frac{\partial W}{\partial \theta'}d\theta'.$$

Since the displacements are arbitrary, we must have

$$R = -\frac{\partial W}{\partial r} = -\frac{3MM'}{r^4}(2\cos\theta\cos\theta' - \sin\theta\sin\theta') \quad \text{...(2),}$$

$$Tr + G = -\frac{\partial W}{\partial \theta} = -\frac{MM'}{r^3}(2\sin\theta\cos\theta' + \cos\theta\sin\theta') \quad \text{...(3)}$$

and

$$G = \frac{\partial W}{\partial \theta'} = \frac{MM'}{r^3}(2\cos\theta\sin\theta' + \sin\theta\cos\theta') \quad \text{......(4).}$$

From (3) and (4) we get

$$T = -\frac{3MM'}{r^4}\sin(\theta+\theta') \quad \ldots\ldots\ldots\ldots\ldots(5).$$

The forces R, T and the couple G can be compounded into a single force.

In like manner the magnet M is subject to forces exerted by M' equal and opposite in direction to R and T, and to a couple

$$G' = \frac{\partial W}{\partial \theta} = \frac{MM'}{r^3}(2\sin\theta\cos\theta' + \cos\theta\sin\theta') \quad \ldots(6).$$

The couple G and G' are not equal and opposite, but no contradiction is involved, because the transverse forces T are not in the same straight line, but T and G acting on M' compound into a single force which is equal and opposite to $-T$ and G' acting upon M.

8·35. Examples. (i) *A magnetic particle of moment μ lies in a magnetic field at the point (r_1, θ_1) in a plane. The axis of the particle lies in the plane and makes an angle ϕ with the direction given by $\theta=0$. The potential of the magnetic field at the point (r, θ) in the plane is $\frac{\cos\theta}{r}$. Find the potential energy of the particle and deduce that, if its centre is fixed, it can rest in equilibrium, if $\phi = 2\theta_1$ or if $\phi = \pi + 2\theta_1$.* [M. T. 1934]

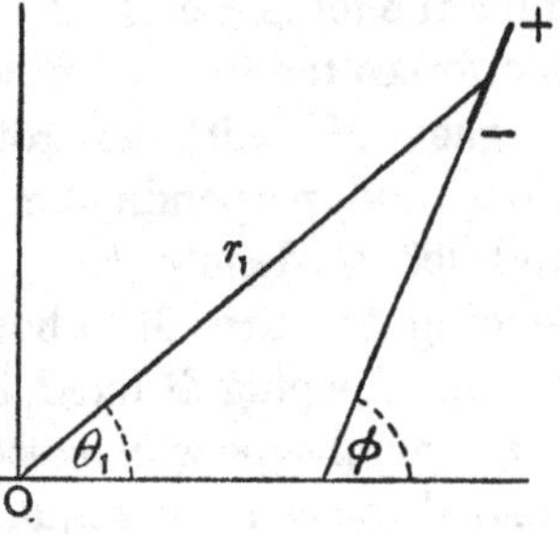

Since the potential of the field is $\frac{\cos\theta}{r}$, therefore at a point (r, θ) the magnetic force has components

$$H_r = \frac{\cos\theta}{r^2}, \quad H_\theta = \frac{\sin\theta}{r^2}.$$

This makes the force at the centre of the magnet μ in direction of its axis

$$\frac{\cos\theta_1\cos(\phi-\theta_1)}{r_1^2} + \frac{\sin\theta_1\sin(\phi-\theta_1)}{r_1^2} \quad \text{or} \quad \frac{\cos(\phi-2\theta_1)}{r_1^2}.$$

Hence the potential energy of the magnet in the given field is

$$W = -\mu\cos(\phi-2\theta_1)/r_1^2.$$

The couple G tending to turn the magnet about its centre is given by

$$G = -\frac{\partial W}{\partial \phi} = \frac{\mu\sin(\phi-2\theta_1)}{r_1^2},$$

and since this vanishes if $\phi = 2\theta_1$, or if $\phi = \pi + 2\theta_1$, these give possible positions of equilibrium.

(ii) *Three small magnets of moments m_1, m_2, m_3 are placed along rectangular axes with their centres at equal distances a from the origin. Shew that if a small magnet be free to turn about its centre at the point $(2a, 2a, 2a)$ it will take up a position of equilibrium in which the direction cosines of its axis are proportional to*

$$-m_1+m_2+m_3, \quad m_1-m_2+m_3$$

and $$m_1+m_2-m_3. \qquad \text{[I. 1927]}$$

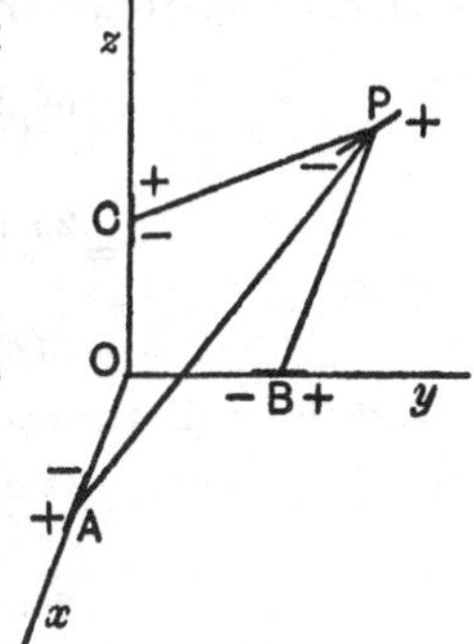

Let A, B, C, P be the positions of the four magnets and λ, μ, ν the direction cosines of the fourth.

The direction cosines of AP, BP, CP are $\frac{1}{3}, \frac{2}{3}, \frac{2}{3}$; $\frac{2}{3}, \frac{1}{3}, \frac{2}{3}$; $\frac{2}{3}, \frac{2}{3}, \frac{1}{3}$ respectively. Therefore if M be the moment of the fourth magnet its potential energy is

$$W=\frac{Mm_1}{27a^3}\{\lambda-\tfrac{1}{3}(\lambda+2\mu+2\nu)\}+\frac{Mm_2}{27a^3}\{\mu-\tfrac{1}{3}(2\lambda+\mu+2\nu)\}$$
$$+\frac{Mm_3}{27a^3}\{\nu-\tfrac{1}{3}(2\lambda+2\mu+\nu)\},$$

or $$W=\frac{2M}{81a^3}\{m_1(\lambda-\mu-\nu)+m_2(-\lambda+\mu-\nu)+m_3(-\lambda-\mu+\nu)\}.$$

For a position of equilibrium W must be stationary in value for all small changes in λ, μ, ν subject to $\lambda^2+\mu^2+\nu^2=1$.

Therefore $$\delta W=0,$$

or

$$\frac{2M}{81a^3}\{(m_1-m_2-m_3)\,\delta\lambda+(-m_1+m_2-m_3)\,\delta\mu+(-m_1-m_2+m_3)\,\delta\nu\}=0,$$

for all small values of $\delta\lambda$, $\delta\mu$, $\delta\nu$ subject to

$$\lambda\delta\lambda+\mu\delta\mu+\nu\delta\nu=0;$$

and this requires that

$$\lambda:\mu:\nu=-m_1+m_2+m_3:m_1-m_2+m_3:m_1+m_2-m_3.$$

8·4. Gauss's verification of the law of inverse squares. Comparison is made of the couples exerted by a small magnet M on a small magnet M' in two particular positions.

We note that in the special cases (i) when M is 'end on' to M', so that in **8·34** $\theta=0$ and $\theta'=\frac{1}{2}\pi$, the couple exerted on M' by M is $2MM'/r^3$; and (ii) when M is broadside to M', so that $\theta=\frac{1}{2}\pi$ and $\theta'=0$, the couple is MM'/r^3; i.e. the couple in fig. (i) is twice the couple in fig. (ii) for magnets at the same distance apart. We shall now shew that if the law of force were a force $\propto 1/r^p$ at distance r, then the couple in fig. (i) would be p times the couple in fig. (ii).

Let us regard the magnet M as composed of two poles $-m$ at B and

m at A, where $BO = OA = a$, and let OO', the distance between the centres of the magnets, be r.

Then in fig. (i) the force at O' due to $M = \frac{m}{(r-a)^p} - \frac{m}{(r+a)^p}$

$$= \frac{m}{r^p}\left(1 + \frac{ap}{r} + \ldots\right) - \frac{m}{r^p}\left(1 - \frac{ap}{r} + \ldots\right)$$

$$= \frac{2map}{r^{p+1}} \text{ to the first power of } a$$

$$= Mp/r^{p+1}, \text{ since } M = 2ma \text{ along } OO'.$$

Also in fig. (ii) the resultant force at O' is clearly parallel to AB and

$$= \frac{m}{r^p}\frac{a}{r} + \frac{m}{r^p}\frac{a}{r} \text{ to the first power of } a$$

$$= \frac{2ma}{r^{p+1}} = M/r^{p+1};$$

so that the couple in fig. (i) is p times the couple in fig. (ii).

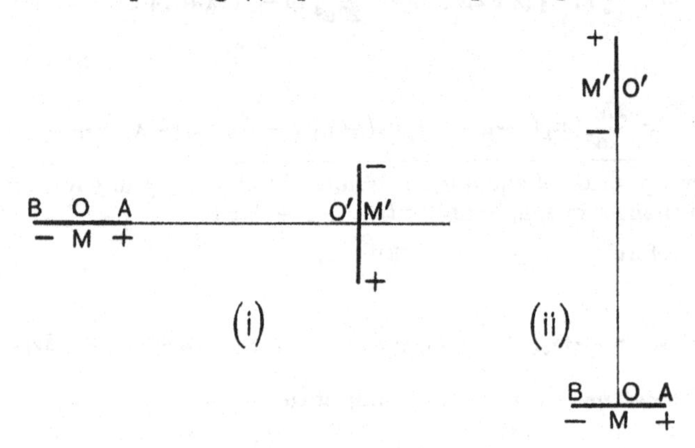

Now let M' be suspended so that it can turn freely about a vertical axis; and let H be the horizontal component of the earth's magnetic field.

Firstly, let M be placed pointing east and west with its centre due west of O' at a distance r; M' will in its equilibrium position make an angle θ with the magnetic meridian, so that the couples produced by the two fields H and Mp/r^{p+1} balance one another; i.e. so that

$$M'H \sin\theta = M' \frac{Mp}{r^{p+1}} \cos\theta \quad \ldots\ldots\ldots\ldots\ldots\ldots\ldots(1).$$

Secondly, let M be placed with its centre due south of O' still pointing east and west and so that $OO' = r$ as in fig. (ii). The deflection of M' will now be θ', where

$$M'H \sin\theta' = M' \frac{M}{r^{p+1}} \cos\theta' \quad \ldots\ldots\ldots\ldots\ldots\ldots(2).$$

Therefore $\tan\theta = p\tan\theta'$; but θ and θ' can be measured and thus it can be confirmed that $p=2$.

8·41. Experimental determination of the magnetic moment of a small magnet. We saw in **8·24** that the period of oscillation of a small magnet free to turn about its centre in a given magnetic field is $2\pi\sqrt{(K/HM)}$, where K is its moment of inertia, M its magnetic moment and H the intensity of the given field.

If therefore the magnet is free to turn about a vertical axis, about which its moment of inertia is known, by observing the time of oscillation we can determine the product HM, where H is now the horizontal component of the earth's magnetic force.

Further, if we place the given magnet pointing east due west of another small magnet M' as in **8·4**, where M' is free to rotate about the vertical, the deflection θ of M' will be given, as in **8·4** (1), by

$$M'H\sin\theta = \frac{2M'M}{r^3}\cos\theta$$

so that

$$\frac{M}{H} = \tfrac{1}{2}r^3\tan\theta.$$

Hence by measuring r and θ the ratio M/H can be determined, and we have seen above how to determine the product MH; so that by this process both M and H can be found.

8·5. To determine the possible positions of relative equilibrium of two small magnets free to turn about their centres in the same plane. Referring to the figure of **8·31**, the mutual potential energy is given by

$$W = -\frac{MM'}{r^3}(2\cos\theta\cos\theta' - \sin\theta\sin\theta'),$$

so that

$$\frac{\partial W}{\partial\theta} = \frac{MM'}{r^3}(2\sin\theta\cos\theta' + \cos\theta\sin\theta')$$

and

$$\frac{\partial W}{\partial\theta'} = \frac{MM'}{r^3}(2\cos\theta\sin\theta' + \sin\theta\cos\theta').$$

In equilibrium W must be stationary so that the first derivatives must vanish. There are four solutions which give different relative positions

(i) $\theta=0,\quad \theta'=0$;
(ii) $\theta=0,\quad \theta'=\pi$;
(iii) $\theta=\tfrac{1}{2}\pi,\quad \theta'=\tfrac{1}{2}\pi$;
(iv) $\theta=\tfrac{1}{2}\pi,\quad \theta'=\tfrac{3}{2}\pi$.

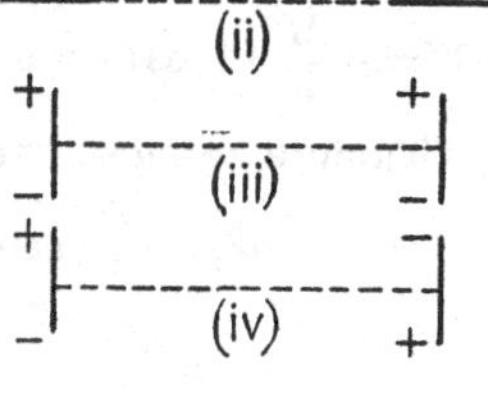

The conditions for a true maximum or minimum of W are that the discriminant $W_{\theta\theta}W_{\theta'\theta'} - W_{\theta\theta'}^2$ must be positive and $W_{\theta\theta}$ negative for a maximum and positive for a minimum, where the suffixes denote partial derivatives.

But $$W_{\theta\theta} = W_{\theta'\theta'} = \frac{MM'}{r^3}(2\cos\theta\cos\theta' - \sin\theta\sin\theta')$$

and $$W_{\theta\theta'} = \frac{MM'}{r^3}(-2\sin\theta\sin\theta' + \cos\theta\cos\theta').$$

Hence the values in the four cases are

$$\begin{array}{lcccc} & \text{(i)} & \text{(ii)} & \text{(iii)} & \text{(iv)} \\ W_{\theta\theta} = W_{\theta'\theta'} = & \dfrac{2MM'}{r^3}, & -\dfrac{2MM'}{r^3}, & -\dfrac{MM'}{r^3}, & \dfrac{MM'}{r^3} \end{array}$$

and $$W_{\theta\theta'} = \frac{MM'}{r^3}, \ -\frac{MM'}{r^3}, \ -\frac{2MM'}{r^3}, \ \frac{2MM'}{r^3};$$

and the signs of the discriminant in the four cases are

(1) +, (ii) +, (iii) −, (iv) −.

Also in (i) $W_{\theta\theta}$ is +, so that W is a minimum and the equilibrium is stable, but in (ii) $W_{\theta\theta}$ is −, so that W is a maximum and the equilibrium is unstable. But in positions (iii) and (iv) W is neither a true maximum nor a true minimum, so the equilibrium must be regarded as unstable as it is certainly not stable for all small displacements.

8·51. Example. *Two small magnetic needles of equal moment M are pivoted on vertical axes which are fixed at a distance c apart, the line of centres being perpendicular to the magnetic meridian. Obtain an expression for the potential energy of the needles in terms of the angles between the line of centres and the axes of the needles, allowing for the earth's horizontal field H.*

Shew that if $H < 3M/c^3$, there is a stable position of equilibrium in which the needles are parallel and make an angle θ with the line of centres, given by

$$3M\sin\theta = Hc^3.$$

Prove also that, if $H > 3M/c^3$, the position with the axes of the needles parallel to H is stable.

[I. 1915]

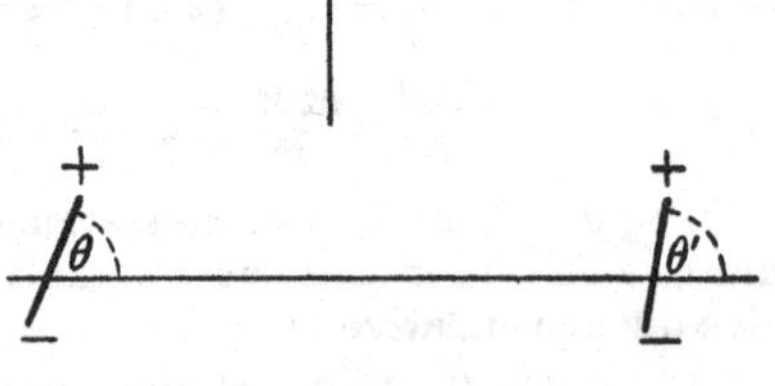

The potential energy is obtained by adding to the mutual potential energy of the magnets that of each magnet in the earth's field, so that, by **8·3** (2) and **8·34** (1), we have

$$W = -\frac{M^2}{c^3}(2\cos\theta\cos\theta' - \sin\theta\sin\theta') - MH\sin\theta - MH\sin\theta' \quad \ldots(1).$$

Therefore, using suffixes to denote partial differentiation,

$$W_\theta = \frac{M^2}{c^3}(2\sin\theta\cos\theta' + \cos\theta\sin\theta') - MH\cos\theta,$$

$$W_{\theta'} = \frac{M^2}{c^3}(2\cos\theta\sin\theta' + \sin\theta\cos\theta') - MH\cos\theta'.$$

Hence W_θ, $W_{\theta'}$ both vanish for a position in which $\theta'=\theta$ and

$$3M\sin\theta = Hc^3 \quad \text{.........................(2)},$$

giving a real angle θ if $H<3M/c^3$.

Further $W_{\theta\theta}=\frac{M^2}{c^3}(2\cos\theta\cos\theta'-\sin\theta\sin\theta')+MH\sin\theta$,

$$W_{\theta'\theta'}=\frac{M^2}{c^3}(2\cos\theta\cos\theta'-\sin\theta\sin\theta')+MH\sin\theta',$$

$$W_{\theta\theta'}=-\frac{M^2}{c^3}(2\sin\theta\sin\theta'-\cos\theta\cos\theta').$$

Hence, in the position given by (2),

$$W_{\theta\theta}W_{\theta'\theta'}-W^2_{\theta\theta'}=\frac{M^4}{c^6}[\{(2\cos^2\theta-\sin^2\theta)+3\sin^2\theta\}^2-(2\sin^2\theta-\cos^2\theta)^2]$$

$$=\frac{M^4}{c^6}3\cos^2\theta(1+3\sin^2\theta).$$

Since this expression is positive and since $W_{\theta\theta}$ is also seen to be positive, therefore W is a minimum and the position is one of stable equilibrium.

If however $H>3M/c^3$, (2) does not give a real value for θ, but W_θ, $W_{\theta'}$ both vanish for $\theta=\theta'=\frac{1}{2}\pi$, and in this case

$$W_{\theta\theta}=W_{\theta'\theta'}=M(H-M/c^3)$$

and

$$W_{\theta\theta'}=-2M^2/c^3,$$

so that

$$W_{\theta\theta}W_{\theta'\theta'}-W^2_{\theta\theta'}=M^2\{(H-M/c^3)^2-4M^2/c^6\}$$

$$=M^2\{(H-3M/c^3)(H+M/c^3)\},$$

and this is positive and so is $W_{\theta\theta}$, so that W is a minimum and the position in which the axes are parallel to H is stable.

8·6. Finite magnets. Intensity of magnetization. A magnet of any form consists of an agglomeration of small magnets or bipoles. A small element of volume contains a number of these, and we assume that the positive poles can be regarded as collected at their mean centre and the negative poles at their mean centre giving a single bipole of a definite moment. Each element of volume of a magnetized body may therefore be regarded as a small magnet and the ratio of the moment of the small magnet to its volume is called the **intensity of magnetization** and usually denoted by **I**. In most magnetic bodies the vector **I** varies from point to point, but a body in which **I** is constant in magnitude and direction at every point is said to be uniformly magnetized.

8·61. Uniformly magnetized sphere. We may regard a bipole as constructed by taking *coincident* equal positive and negative poles m and $-m$ and pulling them apart a short distance δx. In the same way if we consider a small element of volume δv to contain magnetic matter of positive density ρ and at the same time that there is coincident with it a volume δv of magnetic matter of negative density $-\rho$ and we make a relative infinitesimal displacement δx of the two elements, we obtain a resultant bipole of moment $\rho\,\delta v\,.\,\delta x$, wherein $\rho\,\delta x$ is by **8·6** the intensity of magnetization.

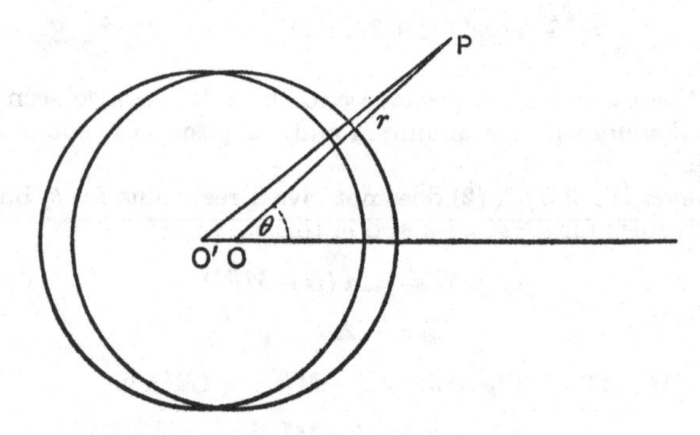

In the case of *uniformly* magnetized bodies of finite size, we may suppose their magnetic condition to have been produced in the same way. Consider, for example, a sphere of radius a uniformly magnetized to intensity **I**. We regard this as produced by taking a sphere of radius a of magnetic matter of uniform density ρ with a like sphere of density $-\rho$, the two coinciding in position, and then giving their centres O, O' a small relative displacement δx.

It was shewn in **3·2** that the potential at an external point due to a uniform spherical charge is the same as if the charge were collected at the centre of the sphere, and if we regard a uniform solid sphere as stratified in concentric shells it is evident that the same result holds good for a uniform solid spherical distribution.

Hence the potential at an external point P, whose coordinates are r, θ, with O as origin and $O'O$ as axis, regarded as due to the positive and negative spheres is

$$\frac{\frac{4}{3}\pi a^3\rho}{r} - \frac{\frac{4}{3}\pi a^3\rho}{r+\delta x\cos\theta} = \frac{\frac{4}{3}\pi a^3\rho\,\delta x\cos\theta}{r^2},$$

to the first power of δx.

Now putting $\rho\,\delta x$ equal to a finite number I, no matter how small δx may be, the potential becomes

$$\phi = \frac{\frac{4}{3}\pi a^3 I\cos\theta}{r^2} \quad \text{.....................(1).}$$

Hence the external potential due to a uniformly magnetized sphere of radius a, magnetized to intensity $\mathbf{I}$, is the same as if a small magnet or bipole of moment $\frac{4}{3}\pi a^3\mathbf{I}$ were placed at the centre of the sphere.

8·62. The earth regarded as a uniformly magnetized sphere. Since the positive pole of a compass points to the magnetic north, the axis of the bipole equivalent to the earth's magnetism must point south. If a be the earth's radius and I its intensity of magnetization regarded as uniform, the equivalent bipole is of moment $M = \frac{4}{3}\pi a^3 I$ pointing to the south pole. Therefore in a place in the northern hemisphere of magnetic latitude l, the components of magnetic force are $\frac{2M}{a^3}\cos(\frac{1}{2}\pi + l)$ radially and $\frac{M}{a^3}\sin(\frac{1}{2}\pi + l)$ tangentially, i.e. a force $\frac{8}{3}\pi I\sin l$ vertically downwards, and a force $\frac{4}{3}\pi I\cos l$ horizontally towards the north. A magnet free to turn about its centre of gravity will therefore dip through an angle $\tan^{-1}(2\tan l)$ below the horizontal.

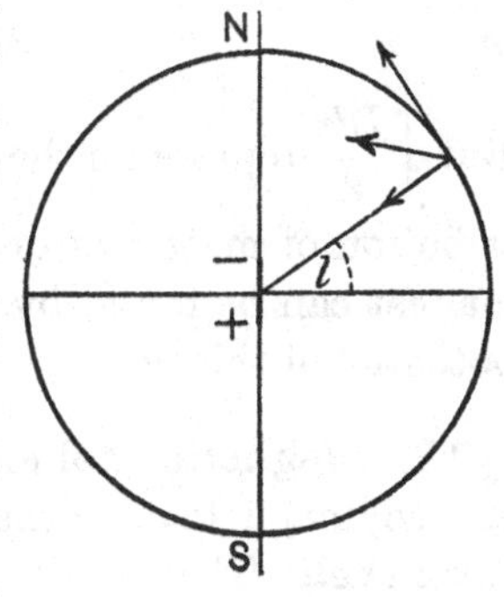

It must be remarked that the earth is very far from being a uniformly magnetized sphere, the earth's magnetic poles are not at opposite ends of a diameter and their positions are undergoing continual change, and there are disturbances of the earth's magnetism associated with the appearance of sun spots.

8·63. Uniformly magnetized solid. The following method may be adopted to find the potential at an external point due to a uniformly magnetized solid of any shape.

We may regard the magnet as composed of a large number of elementary magnets of moment $\mathbf{I}dv$, where $\mathbf{I}$ is the intensity of magnetization and dv an element of volume. Let x, y, z be the co-ordinates of a point Q of such an element, let P be an external point (ξ, η, ζ), and let $QP=r$ make an angle θ with the direction of $\mathbf{I}$. Take the axis of x in the direction of $\mathbf{I}$. Then the potential at P due to the elementary magnet is

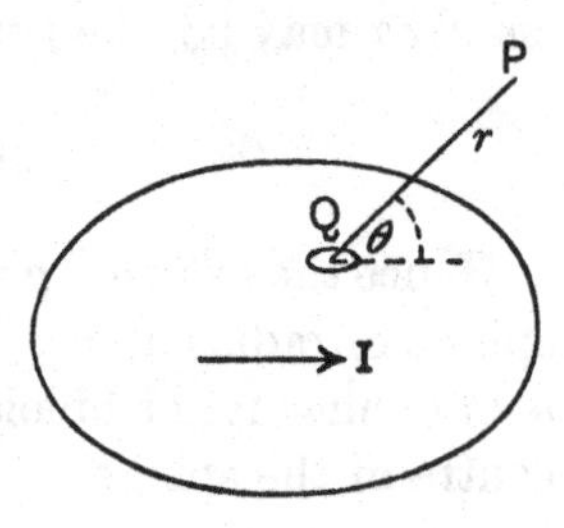

$I\,dv \cos\theta/r^2$ or $I\,dv \dfrac{\partial}{\partial x}\left(\dfrac{1}{r}\right)$. Hence the potential at P of the whole solid is given by

$$\phi = \int I \frac{\partial}{\partial x}\left(\frac{1}{r}\right) dv \quad \text{.....................(1)},$$

integrated through the volume.

But $$r^2 = (\xi - x)^2 + (\eta - y)^2 + (\zeta - z)^2,$$

so that $$\frac{\partial}{\partial x}\frac{1}{r} = -\frac{\partial}{\partial \xi}\frac{1}{r}.$$

Therefore $$\phi = -\int I \frac{\partial}{\partial \xi}\left(\frac{1}{r}\right) dv$$

or $$\phi = -\frac{\partial}{\partial \xi}\int \frac{I\,dv}{r} \quad \text{.....................(2)}.$$

But $\displaystyle\int \frac{I\,dv}{r}$ represents the potential at P due to a uniform distribution of matter of density I, so that the potential of the magnet can be found by applying the operator $-\partial/\partial\xi$ to the potential of this uniform distribution of matter.

8·7. Magnetic shells. An iron sheet of any form, plane or curved, magnetized normally at every point is called a **magnetic shell.** The **strength of the shell** at any point is defined to be the intensity of magnetization multiplied by the thickness of the shell at the point.

If the strength is constant at all points, the shell is called a **uniform** or **simple shell.**

Such a shell acts like two parallel layers of positive and negative magnetism at a short distance apart, for it can be conceived to be an agglomeration of small magnets all directed along normals to the sheet, with their positive poles on one surface and their negative poles on the other.

8·71. Potential due to a magnetic shell of uniform strength τ. Let I be the intensity of magnetization at a point Q on the positive face of the shell, t the thickness and dS a small area surrounding Q. The moment of the small magnet of which dS is the cross-section is, by **8·6**, $ItdS=\tau dS$, and its axis is normal to the surface.

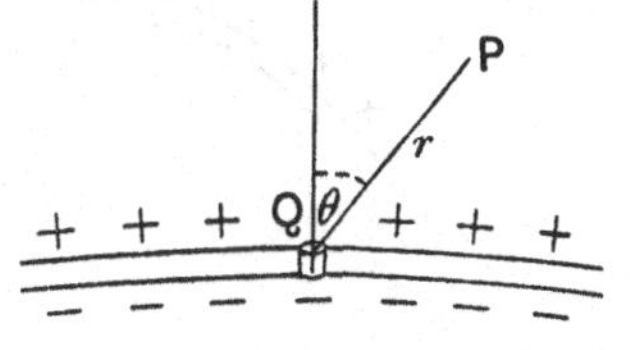

Let P be a point at a distance r from Q, and θ the angle which QP makes with the normal at Q drawn from the negative to the positive side of the surface. Then the potential at P due to this small magnet is $\frac{\tau dS \cos\theta}{r^2}$ or $\tau d\omega$, where $d\omega$ is the solid angle which dS subtends at P. *Therefore the potential at P due to the whole shell is $\tau\omega$, where ω is the solid angle which the boundary of the shell subtends at P*; and this formula is adapted to the case in which lines drawn from P to the shell meet it first on the positive side, and the sign must be changed if such lines meet the shell first on the negative side.

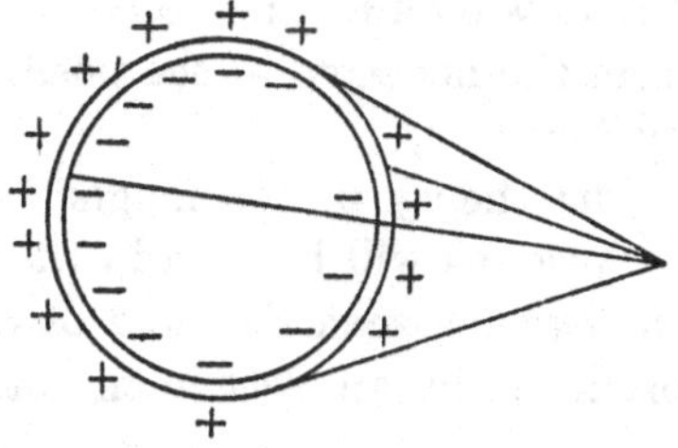

We notice that the expression for the potential of the shell does not depend on its form provided that its boundary is fixed. It is therefore the same for all shells having the same boundary.

If the shell is closed, e.g. a spherical shell, the potential at all external points is zero, and at all internal points it is $-4\pi\tau$.

Let P be a point close to the surface on the positive side of the shell, and ω the solid angle which the boundary subtends at P, so that $\tau\omega$ is the potential at P. Then at a point P' on the negative side and as near as possible to P the solid angle subtended by the boundary differs by little from $4\pi-\omega$, and in accordance with the rule of signs established above the potential at P' is

$$-\tau(4\pi-\omega)=\tau\omega-4\pi\tau.$$

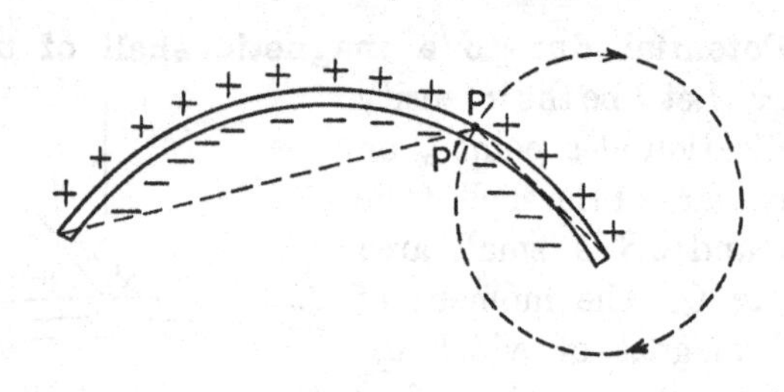

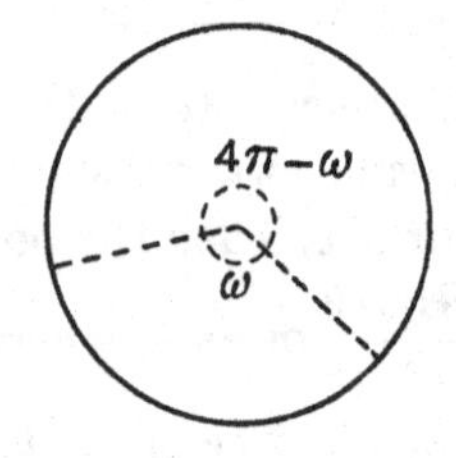

Therefore in passing from the positive to the negative side of the shell along a curve which goes round the edge the potential *decreases* by $4\pi\tau$, and conversely if we pass through the shell from the negative to the positive side the potential increases by $4\pi\tau$.

The theory of the magnetic shell is of great importance because, as will be seen in the next chapter, there is a close analogy between the field of a magnetic shell and the field produced by an electric circuit.

8·72. Example. *A uniform magnetic shell of strength τ is bounded by a circle of radius a. Find the magnetic force at a point on the axis of the shell at a distance z from the centre of the circle, and also find the mechanical force exerted on a small magnet placed along the axis.*

The potential of the shell is given by

$$\phi = \tau\omega,$$

where ω is the solid angle subtended by the shell.

At a point P on the axis at a distance z from the centre O of the boundary, on the positive side of the shell, this becomes

$$\phi = 2\pi\tau(1 - \cos OPA),$$

where OA is any radius of the circle;

or $$\phi = 2\pi\tau\left\{1 - \frac{z}{\sqrt{(a^2+z^2)}}\right\} \quad ....(1).$$

Hence the magnetic force at P is

$$H = -\partial\phi/\partial z = \frac{2\pi\tau a^2}{(a^2+z^2)^{\frac{3}{2}}} \quad ....(2),$$

directed along the axis.

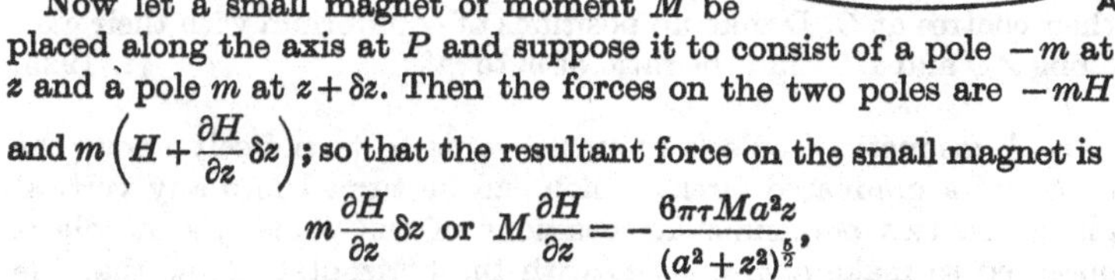

Now let a small magnet of moment M be placed along the axis at P and suppose it to consist of a pole $-m$ at z and a pole m at $z+\delta z$. Then the forces on the two poles are $-mH$ and $m\left(H + \frac{\partial H}{\partial z}\delta z\right)$; so that the resultant force on the small magnet is

$$m\frac{\partial H}{\partial z}\delta z \text{ or } M\frac{\partial H}{\partial z} = -\frac{6\pi\tau M a^2 z}{(a^2+z^2)^{\frac{5}{2}}},$$

the negative sign implying an attraction.

EXAMPLES

1. Two magnetic molecules lie in one plane, their axes making angles θ and θ' with the line joining the molecules. Shew that the couple on the second molecule due to the action of the first will vanish if $\tan\theta + 2\tan\theta' = 0$. [I. 1908]

2. The centres of two small horizontal magnets M, M' are at a distance r apart and their axes make angles θ, θ' with r. Reducing the forces which the magnet M exerts on M' to a force through the centre of M' and a couple, find the magnitude of the couple.

If the two magnets are fastened to a board which floats on water, explain why, although the couples exerted on each other are unequal, the board is not set in rotation. [I. 1907]

3. A small compass needle can swing about a vertical axis, and a bar magnet of length l and moment M is placed horizontally east and west with its centre vertically below the needle at a distance c. Shew that if l/c is small, the deviation of the needle from the magnetic meridian is given approximately by

$$\tan\theta = \frac{M}{Hc^3}\left(1 - \frac{3}{8}\frac{l^2}{c^2}\right),$$

where H is the horizontal component of the earth's magnetic field. [M. T. 1915]

4. Two small magnets of moments M_1, M_2 are placed with their centres at the corners B, C of an equilateral triangle ABC and their axes fixed along BA, CA. A third small magnet is free to turn about its centre which is at A. Determine the angle θ which the magnet at A makes with the bisector of the angle BAC in equilibrium.

Which of the two positions is stable? [I. 1933]

5. Magnets of moments μ, μ', μ' are pivoted at the vertices A, B, C respectively of a triangle in which the angles B and C are equal. They are in equilibrium when they lie along the internal angle bisectors with their positive poles inside the triangle. Shew that

$$\mu'/\mu = 8 \operatorname{cosec} \tfrac{1}{2}B \,.\, \cos^3 B\,(2 - 3\cos B). \qquad \text{[M. T. 1930]}$$

6. Two small magnets of moment m are placed at the corners A, B of a square $ABCD$ with their axes in the direction AB and BA respectively. If two other small magnets of moment m' free to turn about their centres at C, D take up positions of equilibrium with their axes along AC and BD, find the ratio of m to m'. [I. 1928]

7. A magnetic needle is mounted so as to turn freely about the centre of a graduated circle, which can be turned into any vertical plane. In two perpendicular positions of the plane, the needle is observed to make angles δ_1, δ_2 with the horizontal. Shew that the inclination of the earth's field to the horizontal is δ, where

$$\cot^2 \delta = \cot^2 \delta_1 + \cot^2 \delta_2,$$

and obtain a formula giving the position of the magnetic meridian with respect to the two positions of the circle. [M. T. 1916]

8. Three small magnets of equal moment are placed at equal distances from the origin along rectangular axes pointing away from the origin. Two are fixed and one is free to turn about its centre. Shew that in equilibrium the inclination of the latter to the line joining it to the origin is $\tan^{-1}(\sqrt{2}/6)$. [I. 1922]

9. Two doublets, each of moment M, are mounted with their centres at fixed points at a distance r apart. They can rotate in the same plane and are constrained by a frictionless mechanism to be always parallel and in the same sense. Find the positions of equilibrium, discuss their stability, and shew that the work which must be done to turn the system from a position of stable to one of unstable equilibrium is $3M^2/r^3$. [M. T. 1932]

10. The axis of a small magnet makes an angle ϕ with the normal to a plane. Prove that the line from the magnet to the point in the plane where the number of lines of force crossing it per unit area is a maximum makes an angle θ with the axis of the magnet, where

$$2\tan\theta = 3\tan 2(\phi - \theta). \qquad \text{[I. 1905]}$$

11. Prove that if there are two magnetic molecules of moments M and M' with their centres fixed at A and B, where $AB=r$, and one of the molecules swings freely, while the other is acted on by a given couple, so that when the system is in equilibrium this molecule makes an angle θ with AB, then the moment of the couple is

$$\tfrac{3}{2}MM' \sin 2\theta / r^3 (3\cos^2\theta + 1)^{\frac{1}{2}}. \qquad \text{[M. T. 1901]}$$

12. A small magnet of moment M is suspended with its centre at a point equidistant from the poles of a fixed bar magnet of length $2a$ and pole-strength m. The distance of the small magnet from the centre of the bar is b. If the small magnet be turned through two right angles from its equilibrium position, shew that the work performed is

$$4Mma/(a^2+b^2)^{\frac{3}{2}}. \qquad \text{[M. T. 1923]}$$

13. Prove that the potential energy of a small magnet of moment μ pivoted at a distance r from another small fixed magnet of moment μ' and in the direction of the axis of μ' is

$$-\frac{2\mu\mu'}{r^3}\cos\theta,$$

when the axis of μ makes an angle θ with the axis of μ'.

A uniform field H now acts in addition from the magnet μ towards the fixed magnet. Prove that the previously stable position of equilibrium becomes unstable when

$$H > \frac{2\mu'}{r^3}. \qquad \text{[M. T. 1929]}$$

14. M, M' are the moments of two small magnets, ϵ the angle between their axes and θ, θ' the angles their axes make with the line r joining their centres. Prove that, if $\theta = \frac{1}{2}\pi$, there is a couple on the magnet M' whose moment is $MM' \sin\epsilon / r^3$ and whose axis is parallel to the line of intersection of the planes perpendicular to the axes of the magnets.

[I. 1913]

15. Two small magnets of moment m are fixed with their axes in the same straight line pointing in opposite directions. A small magnet of moment m' has its centre fixed at a point in the plane bisecting at right angles the join of the other two magnets. Prove that the period of the small oscillations of the magnet m' in this plane is

$$2\pi (Ir^3/3mm' \sin\alpha)^{\frac{1}{2}},$$

where I is its moment of inertia, r its distance from either of the other two magnets and α is the angle subtended at m' by the line joining the other two magnets. [I. 1915]

16. The centres of two equal small magnets are fixed at a distance d apart. One magnet is fixed at an angle $\tan^{-1} 2\sqrt{3}$ to the line of centres and the other is free to move in the plane through its own centre and the first magnet. Find the positions of equilibrium of the second magnet and distinguish between the stable and unstable ones. [M. T. 1935]

17. A small magnet of moment μ is held fixed at the origin of coordinates, with its axis in the direction (l, m, n); another small magnet of moment μ' has its centre fixed at the point $(x, 0, z)$, and is free to turn so that its axis moves in a plane parallel to the plane $z=0$. Find the position of stable equilibrium of μ', and shew that the period of its free oscillations about this position is

$$2\pi I^{\frac{1}{2}} \mu^{-\frac{1}{2}} \mu'^{-\frac{1}{2}} (x^2+z^2)^{\frac{5}{4}} [\{l(x^2+z^2)-3x(lx+nz)\}^2+m^2(x^2+z^2)^2]^{-\frac{1}{4}},$$

where I is the moment of inertia of μ'. [M. T. 1913]

18. Two magnetic needles of equal moment M are pivoted at A, B and can turn in a horizontal plane, AB being horizontal. There is a uniform field H parallel to AB. Shew that if in equilibrium the needles make angles θ, ϕ with AB, where θ, ϕ are not zero, then either

$$\theta=\phi,\ 3\cos\theta=-\kappa, \quad \text{or} \quad \theta=-\phi,\ \cos\theta=-\kappa,$$

where $\kappa=\dfrac{a^3H}{M}$ and $a=AB$. [I. 1924]

19. Two small magnets of moments M and $6M$ are so mounted that they can turn about their centres, in the same horizontal plane, the line of centres being perpendicular to the earth's horizontal field H. Prove that the position of equilibrium in which their axes are in the direction of H is stable if $9M < c^3H$, where c is the distance between the centres. [I. 1925]

20. If two magnetic doublets are of equal strength m and are free to turn about their centres, which are fixed, and if they are in a field of uniform force H parallel to their line of centres, prove that the position in which the direction of each doublet is opposite to the direction of H is a position of stable equilibrium if $H < m/r^3$, where r is the distance between the centres. [I. 1914]

21. A compass-needle is in equilibrium in the magnetic meridian. Find the deflection (θ) produced by bringing up a small steel magnet to a known distance from the needle, in the direction of magnetic east and west.

In the above circumstances, shew that the period of small oscillations of the needle is diminished in the ratio

$$\sqrt{\cos\theta} : 1.$$

[I. 1908]

22. Obtain the equation $T=2\pi\sqrt{(I/mH)}$ for the time of small oscillation of a magnet of moment m swung about its centre of inertia in a horizontal plane in a uniform field of horizontal intensity H, the moment of inertia of the magnet being I.

Assuming the earth to be a sphere uniformly magnetized parallel to the axis of rotation with intensity M, shew that the time of small horizontal oscillation in latitude ϕ is $\sqrt{(3I\pi/mM\cos\phi)}$. [M. T. 1914]

23. A compass needle, which may be treated as a magnetic particle of moment μ, is free to turn about a vertical axis; and a long bar magnet is placed so that one end, which may be treated as a north-seeking pole of strength m, is in the horizontal plane passing through the needle, at a distance a from the centre of the needle, in a direction between north and west making an angle α with the magnetic meridian. Prove that, if the effect of the other pole of the bar magnet may be disregarded, the needle will rest in a direction between north and east making an angle β with the magnetic meridian so that

$$H \sin \beta = ma^{-2} \sin (\alpha + \beta),$$

where H denotes the horizontal component of the earth's magnetic force.

Prove also that, if the needle is slightly disturbed from this position, it will oscillate in the period

$$2\pi\sqrt{\{I \sin (\alpha + \beta)/\mu H \sin \alpha\}},$$

where I denotes the moment of inertia of the needle about the vertical axis. [M. T. 1912]

24. Prove that, if the earth be regarded as a sphere of radius a uniformly and rigidly magnetized to intensity I, the dip of the magnetic needle in latitude λ is $\tan^{-1}(2 \tan \lambda)$.

Find the intensity of the uniform field parallel to the earth's axis that would destroy the dip, and shew that when this field is applied the period of oscillation of a small magnet in any plane through the line of force is $\sqrt{\{\pi mk^2/IM \cos \lambda\}}$, where M is the moment of the magnet and mk^2 its moment of inertia about its centre. [I. 1932]

25. The portion of a sphere of radius a included between circles of latitude α, β forms a magnetic shell of uniform strength κ. Shew that the magnetic force at the centre of the sphere is

$$\frac{\pi\kappa}{a}(\cos 2\alpha - \cos 2\beta).$$ [Trinity Coll. 1898]

26. The periphery of a segment of a uniformly magnetized spherical shell of strength τ is a circle of radius a and centre A. O is the centre of the sphere and B is any point on AO or AO produced. A small magnet, free to swing in a horizontal plane, is placed at B. If AB is perpendicular to the magnetic meridian and θ is the angle between AB and the axis of the magnet when in equilibrium, prove that

$$2\pi a^2 \tau = (a^2 + b^2)^{\frac{3}{2}} H \cot \theta,$$

where b is the distance of B from A and H is the horizontal component of the earth's field. [M. T. 1933]

27. If the earth were a sphere and its magnetism due to two small straight bar magnets of the same strength, situated at the poles, with their axes in the same direction along the earth's axis, prove that the dip δ in latitude λ would be given by

$$8 \cot (\delta + \tfrac{1}{2}\lambda) = \cot \tfrac{1}{2}\lambda - 6 \tan \tfrac{1}{2}\lambda - 3 \tan^3 \tfrac{1}{2}\lambda.$$ [M. T. 1903]

28. Treating the earth as uniformly magnetized, and supposing the horizontal force to be 0·18 in magnetic latitude 52°, find the intensity of magnetization and the dip in this latitude. [St John's College, 1913]

29. A uniform plane magnetic shell is bounded by concentric circles of radii a, b. Shew that it exerts no force at a point distant $a^{\frac{2}{3}}b^{\frac{2}{3}}/\sqrt{(a^{\frac{2}{3}}+b^{\frac{2}{3}})}$ from its centre along its axis, and that the total work done in bringing a small magnet to this point from infinity is also zero. [I. 1929]

ANSWERS

2. See **8·34**.

4. $\tan^{-1}\{(M_1 \sim M_2)/\sqrt{3}\,(M_1+M_2)\}$. The stable position is the one in which the axis of the magnet is in the direction of the resultant field of the two small magnets.

6. $2:2\sqrt{2}-1$.

7. The meridian plane makes an angle $\tan^{-1}(\tan\delta_1 \cot\delta_2)$ with the first position of the circle.

9. Stable when in the line of centres, unstable when at right angles to it.

16. Stable when inclined at 60° to the line of centres on the opposite side to the first; unstable when reversed in direction.

17. The axis of μ' makes an angle
$$\tan^{-1} m\,(x^2+z^2)/\{l\,(x^2+z^2)-3x\,(lx+nz)\}$$
with the axis of x.

21. $\tan^{-1}\sqrt{\{2M/r^3H\}}$, where r is the distance of the magnet M from the compass needle and H is the earth's horizontal field.

28. 0·0698; 68° 39′.

Chapter IX

ELECTROMAGNETISM

9·1. It is found that an electric current produces a magnetic field in which the lines of force are closed curves surrounding the current. If the current is steady, so is the magnetic field.

9·11. Magnetic field due to a current in a long straight wire. If a cardboard disc is suspended with its plane horizontal, so that it is free to turn about a vertical wire as an axis, and a current passes along the wire, it is observed that when a small magnet is placed on the disc, with a suitable counterpoise to keep the disc from tilting, no motion ensues.

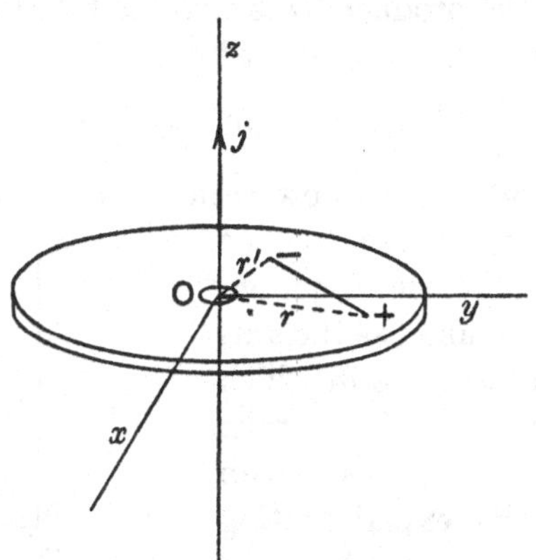

It follows that the forces exerted on the poles of the magnet by the magnetic field due to the current have equal and opposite moments about the axis.

It can further be shewn by sprinkling iron filings on the disc that the lines of force are circles round the current, for the filings become magnetized by induction (see **10·3**) and set themselves along the lines of force.

Let the poles of the magnet be of strength $-m$ and m at distances r', r from the wire, and let H', H denote the magnetic force at the points where the poles are situated. We assume that these forces are at right angles to the current and to the radii r', r, regarding this as a fact demonstrated by experiments with small magnets and with iron filings. Then, by taking moments about the axis, we have

$$-mH'r' + mHr = 0,$$

so that

$$Hr = H'r';$$

or the magnetic force varies inversely as the distance from the wire. The value of the constant Hr must depend on the strength of the current. This result provides another basis for the measurement of current, viz. that the current is proportional to Hr, where the symbols have the meanings assigned to them above. This introduces a new system of units, wherein currents are directly related to the magnetic fields they produce, and therefore called **electromagnetic units**. In this system of units the product Hr is equal to twice the current in the wire, or, denoting the current by j,

$$H = 2j/r \quad\text{.........................(1).}$$

9·12. Potential. Let us now examine whether the field due to a long straight current can be derived from a potential function.

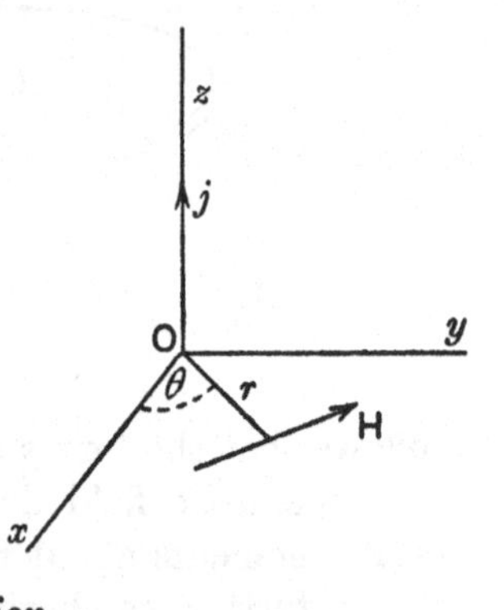

Let the wire be taken as the axis of z, positive in the sense in which the current is flowing. If we take a right-handed set of axes, it can then be shewn by experimenting with a small magnet that the positive sense of the magnetic force H is from x towards y, or *the sense of the field bears to the direction of the current a right-handed screw relation.*

Then the components of H parallel to the axes are

$$H_x = -2jy/r^2,\ H_y = 2jx/r^2,\ H_z = 0,$$

so that

$$H_x dx + H_y dy + H_z dz = \frac{2j\,(x\,dy - y\,dx)}{x^2 + y^2} = 2jd\left(\tan^{-1}\frac{y}{x}\right);$$

i.e. a perfect differential.

Putting $$2j\tan^{-1}y/x=-\phi,$$
we have $$H_x dx+H_y dy+H_z dz=-d\phi,$$
so that $$H_x,\, H_y,\, H_z=-\frac{\partial\phi}{\partial x},\ -\frac{\partial\phi}{\partial y},\ -\frac{\partial\phi}{\partial z};$$
and therefore $$\mathbf{H}=-\operatorname{grad}\phi,$$
where $$\phi=-2j\tan^{-1}y/x+C \qquad (1),$$
and C is a constant; and we note that the inverse tangent is a many-valued function.

9·121. It will be remembered that, when the potential function was first discussed (**2·4**), we postulated the existence in an electrostatic field of a *single-valued* potential function, and we have seen no reason to question the existence of such a function in a magnetostatic field, i.e. a field due to permanent magnets, but at the outset here of our investigation of the magnetic field produced by an electric current we are faced with the fact that *the field under consideration has no single-valued potential function.*

Further, an electrostatic field appears to be conservative, in that no work is done by the field on a unit charge which describes a closed path, but that is not so in the case we are now considering. Thus, let a unit magnetic pole travel round a circle of radius r with its centre on the wire and in a plane perpendicular to it. The magnetic force is constant and equal to $2j/r$ and tangential to the circle at every point of the path, so that the work done $=\dfrac{2j}{r}\times 2\pi r=4\pi j$. We shall see that if we attempt to measure potential in terms of work we cannot assign to it a single value at a particular point, because the work done in bringing a unit pole up to that point in the field will depend on the path by which it travels, $4\pi j$ being added to the amount every time the path makes a circuit round the current.

Putting θ for $\tan^{-1}y/x$, the formula **9·12** (1) can clearly be written
$$\phi=-2j\theta+4n\pi j \qquad (1).$$

Again, in any magnetic field the work done on a unit magnetic pole moving along any path is given by

$$\int(H_x dx + H_y dy + H_z dz),$$

and, for the field due to a long straight current, this is equal to

$$2j\int d\tan^{-1}\frac{y}{x} \text{ or } 2j\int d\theta.$$

For a closed path, which embraces the current once, θ increases by 2π and therefore the work done on a unit magnetic pole is $4\pi j$; but when a pole travels round a closed path which does not embrace the current the total increase in θ is zero, and therefore no work is done.

The theorem here established for a current in a straight wire is a special case of one of the fundamental theorems of electromagnetism; viz. *In a magnetic field due to steady currents, the work done on a unit pole as it travels round a closed path is* 4π *times the sum of the currents embraced measured in the right-handed screw sense.* This theorem is often called **Ampère's Circuital Relation.**

9·13. Let a current j flow in a straight wire whose cross-section is a circle of radius a. If we assume Ampère's circuital relation we can find the magnetic field inside or outside the wire. In a plane perpendicular to the wire draw a circle of radius r with its centre on the axis of the wire. Then, if $r < a$, this circle embraces current to the amount jr^2/a^2,* so that, if H denotes the force tangential to the circle,

$$2\pi r H = 4\pi j r^2/a^2 \text{ or } H = 2jr/a^2 \quad \text{.........(1);}$$

but, if $r > a$, then

$$2\pi r H = 4\pi j \text{ or } H = 2j/r, \text{ as before } \quad \text{.........(2).}$$

We notice that the force due to a straight current cannot become infinite, because to carry a finite current a wire must be of finite cross-section and the formula for H changes from (2) to (1) as we pass from the outside to the inside of the wire.

* We assume that the current is uniformly distributed over the cross-section of the wire. It can be proved that for a steady current this is necessarily the case.

9·2. Comparison with magnetic shell. Consider an infinite plane magnetic shell of strength j with a straight edge along the axis of z and occupying the half of the plane $x=0$ for which y is negative. Let the direction of magnetization of the shell be the positive direction of the axis Ox.

The potential at a point P on the positive side of the shell is j times the solid angle subtended by the shell at P, and this is the solid angle between two planes through P, one of which passes through Oz while the other is parallel to the plane of the

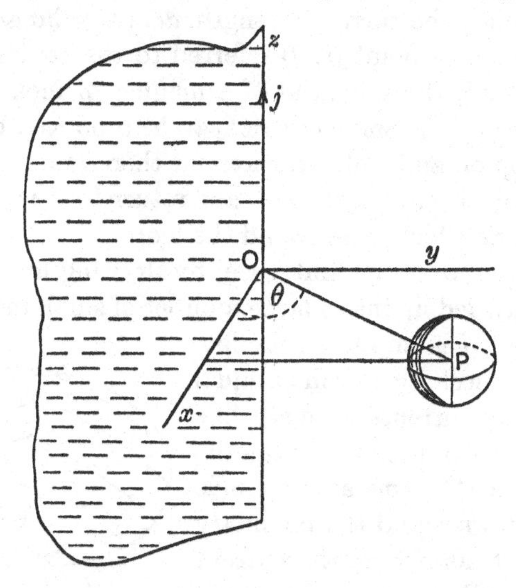

shell (in order to reach its boundary at infinity). If θ is the angle xOP, the solid angle is $2(\frac{1}{2}\pi-\theta)$, so that the potential of the shell is given by

$$\begin{aligned}\phi &= j(\pi-2\theta)\\ &= -2j\theta+\text{const.} \quad \text{.....................(1).}\end{aligned}$$

By comparison with **9·12** we see that the current produces the same magnetic field as would be produced by a magnetic shell of the same strength having the current for boundary.

When we consider the field of a current in a long straight wire we must bear in mind that all currents are closed circuits so that we assume that the rest of the current (apart from the straight part under consideration) is at a great distance, and this corresponds to the part of the plane shell at a great distance.

9·21. It can be shewn experimentally that the magnetic field of a small plane circuit at distances which are large compared with the dimensions of the circuit is the same as that of a small magnet whose axis is normal to the plane of the circuit, and whose moment is equal to the area of the circuit multiplied by the strength of the current. For if dS be the area of the circuit, j the current strength, $d\omega$ the solid angle which dS subtends at a point (r, θ) referred to the centre and axis of the magnet, the potential of a magnet of moment jdS is $jdS \cos\theta/r^2$ or $jd\omega$. Small circuits can be produced by making a small loop on an insulated wire and then doubling the wire back upon itself, so that the flow and return in the wire destroy one another's effects save round the loop.

Then for a circuit of finite size, by drawing lines across it, it can be divided up into a large number of small meshes, and, if a current j is supposed to flow round each mesh, there will be equal and opposite currents in every line common to two meshes so that the total effect is the same as if a single current j flows round the boundary circuit. But, as above, the current in each small mesh produces the same field as a small magnet, and the aggregate of these small magnets is a shell of strength j whose boundary coincides with the given circuit.

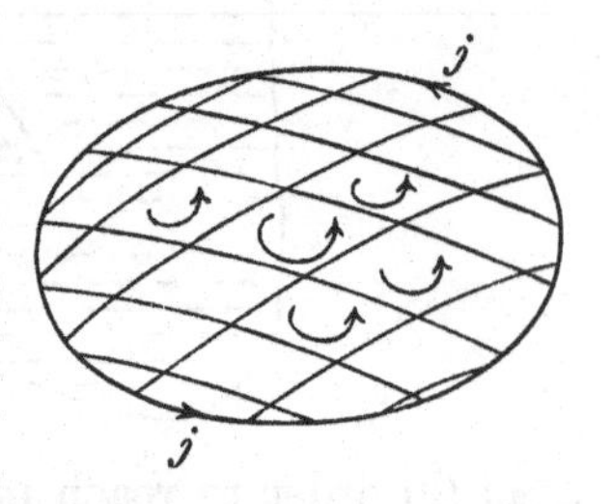

We have now arrived at the point at which we can enunciate, as the experimental basis of our theory of the magnetic fields produced by electric currents, the correspondence which we have established in **9·2** in a particular case; viz. *that a current flowing in a closed circuit produces the same magnetic field as a uniform magnetic shell having the circuit for boundary and of strength equal to that of the current, and the direction of the current bears to the direction of magnetization of the shell a right-handed screw relation.*

The reader will notice that we have left for later consideration the question of the field inside a magnet, and the equivalence of the fields due to a current and a shell must for the present be taken to refer to space outside the shell, but inasmuch as the field due to a uniform magnetic shell only depends upon the position of its boundary—in this case the current circuit—we are not imposing a serious limitation.

9·3. Magnetic field due to parallel straight currents. The field due to two or more parallel currents is found by superposition.

(i) **Flow and return current in parallel wires.** Let two straight parallel wires carrying currents j and $-j$ cut the plane of the paper at right angles at A and B. Then if radii from A, B to a point P make angles θ_1, θ_2 with the line BA, the potential is given by

$$\phi = -2j\theta_1 + 2j\theta_2 + \text{const.} \quad (1).$$

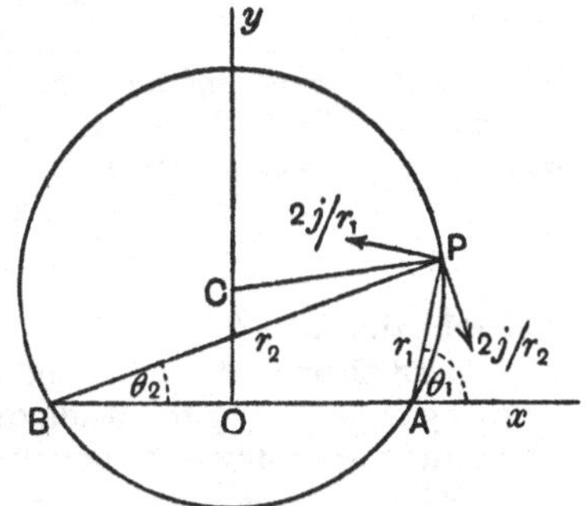

Therefore the equipotential curves are

$$\theta_1 - \theta_2 = \text{const.},$$

or $$APB = \text{const.};$$

i.e. the family of coaxial circles passing through A and B. And since the lines of force cut the equipotential curves at right angles, they must be the orthogonal family of coaxial circles which have A, B as limiting points.

Hence if C is the centre of the circle APB, PC is a tangent to the line of force at P; and the actual force at P is

$$H = \frac{2j}{r_1} \sin APC - \frac{2j}{r_2} \sin BPC$$

$$= \frac{2j}{r_1} \cos \theta_2 - \frac{2j}{r_2} \cos \theta_1$$

$$= \frac{2j \, . \, AB}{r_1 r_2},$$

where r_1, r_2 denote the distances AP, BP.

(ii) **Equal and parallel currents in the same sense.** With the same notation, the potential is given by

$$\phi = -2j(\theta_1 + \theta_2) + \text{const.} \quad \text{..............(2)},$$

so that the equipotential curves are $\theta_1 + \theta_2 = \text{const.}$; and as in **2·532** it can be shewn that these are rectangular hyperbolas passing through A, B, with Cassini's ovals $r_1 r_2 = \text{const.}$ for lines of force.

9·31. Examples. (i) *The significant part of an electric circuit consists of a current C flowing in the sense Oz in a wire along the line $x=0$, $y=a$, and in the sense zO in a wire along the line $x=0$, $y=-a$, the co-ordinate axes being right-handed. Shew that the magnetic potential at a point (x, y, z) due to the circuit is*

$$A + 2C \tan^{-1} \frac{2ax}{x^2 + y^2 - a^2},$$

where A is a constant, and that, if there is in addition a uniform field of strength H, parallel to Oz, a magnetic needle at $(k, 0, 0)$ will rest in equilibrium making an angle θ with Ox, where

$$\tan \theta = \frac{H(a^2 + k^2)}{4aC}. \qquad \text{[M. T. 1930]}$$

Let P be the point (x, y, z). Let a plane through P at right angles to the wires cut them in M and N and let PL be perpendicular to NM.

Then by superposing the fields due to the two currents and using **9·12**, we may write down the potential as

$$\phi = A - 2C(\widehat{MPL}) + 2C(\widehat{NPL}), \text{ where } A \text{ is a constant;}$$

or

$$\phi = A - 2C \tan^{-1} \frac{y-a}{x} + 2C \tan^{-1} \frac{y+a}{x}$$

$$= A + 2C \tan^{-1} \frac{2ax}{x^2 + y^2 - a^2} \quad \text{......................(1).}$$

Alternatively, we may regard the wires continued to infinity as forming a closed circuit. The magnetization of the equivalent shell will be in the positive direction of Ox, in accordance with the right-handed screw rule, so that P is on the positive side of the shell. Then since the solid angle which the shell or the circuit subtends at P is twice the angle NPM, the potential is

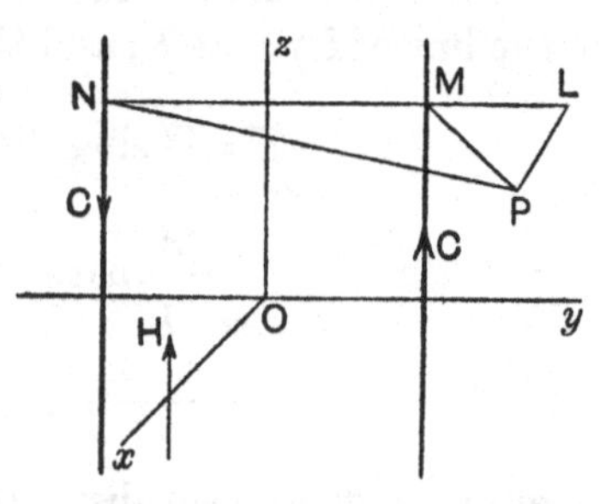

$$\phi = A + 2C(\widehat{NPM}),$$

giving the same result as before.

At the point $(k, 0, 0)$, the field due to the currents is found, from (1), to be

$$H_x = -\frac{\partial\phi}{\partial x} = \frac{4aC}{a^2+k^2}, \quad H_y = -\frac{\partial\phi}{\partial y} = 0, \quad H_z = -\frac{\partial\phi}{\partial z} = 0;$$

and on this is superposed a uniform field of strength H parallel to Oz, so that the resultant at this point makes with Ox an angle

$$\tan^{-1} H/H_x = \tan^{-1}\{H(a^2+k^2)/4aC\}$$

and this is the direction in which a magnetic needle at the point will rest in equilibrium.

(ii) *A current j flows in a long straight wire. Find the force and couple exerted on a small magnet and its period of oscillation about its equilibrium position.*

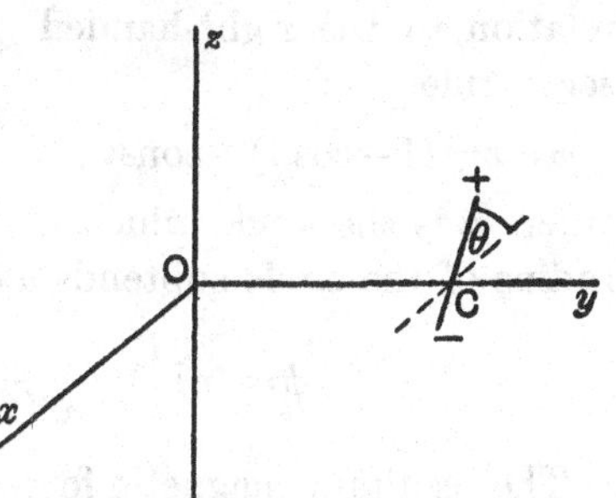

Let C be the centre of the magnet, M its moment and K its moment of inertia about C.

Take the axis Oz along the wire, the axis Oy through C and let $OC = y$.

Using right-handed axes, the field at C is

$$H_x = -2j/y, \ H_y = 0, \ H_z = 0.$$

Hence, if the axis of the magnet makes an angle θ with the line of force, its potential energy in the given field is

$$W = -MH\cos\theta \quad (8\cdot3)$$

$$= -\frac{2Mj\cos\theta}{y};$$

and the force tending to increase y is

$$-\frac{\partial W}{\partial y} = -\frac{2Mj\cos\theta}{y^2};$$

which means that a magnet whose axis makes an acute angle with the direction of the field at its centre is attracted towards the wire.

Again the couple on the magnet is [**MH**] by **8·12**, or in this case $(2Mj\sin\theta)/y$; and if it is free to turn about its centre its equation of motion is

$$K\ddot{\theta} + (2Mj\sin\theta)/y = 0,$$

and for small oscillations this gives a period

$$2\pi\sqrt{\{Ky/2Mj\}}.$$

As explained in **8·24** the magnet has two independent principal oscillations but they both have the same period.

9·4. Magnetic field due to a circular current. Let there be a steady current j in a circular wire of radius a. The potential at any point P is given by

$$\phi = j\omega + \text{const.},$$

where ω is the solid angle which the circuit subtends at P. This takes a simple form when P is on the axis of the circle, but for points off the axis more analysis is required than we have at our disposal.

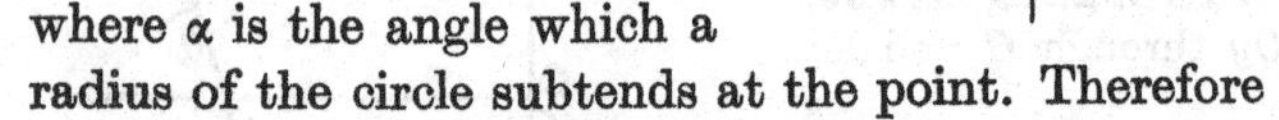

Let Oz be the axis of the circle and O its centre. Then at a point $(0, 0, z)$ on the positive side of the circuit in relation to the right-handed screw rule

$$\phi = 2\pi j\,(1 - \cos\alpha) + \text{const.},$$

where α is the angle which a radius of the circle subtends at the point. Therefore

$$\phi = 2\pi j \left\{1 - \frac{z}{\sqrt{(a^2+z^2)}}\right\} + \text{const.} \quad \text{.........(1).}$$

The resultant magnetic force at the point is directed along Oz and is given by

$$H = -\frac{\partial\phi}{\partial z} = \frac{2\pi j a^2}{(a^2+z^2)^{\frac{3}{2}}} \quad \text{..................(2),}$$

and this is independent of the sign of z, so that at all points of the axis the direction of the force bears to that of the current the right-handed screw relation.

At the centre of the circle the force is therefore

$$2\pi j/a \quad \text{...........................(3)}$$

acting at right angles to the plane of the circle in the right-handed screw sense in relation to the current. This result is of great importance in connection with galvanometers.

9·41. We may deduce from **9·4** (2) the mechanical force on a small magnet of moment M placed at a point P on the axis of the current with its axis along that of the current.

Suppose that the magnet consists of a pole $-m$ at $(0, 0, z)$ and a pole m at $(0, 0, z + \delta z)$, so that $M = m\,\delta z$. The force on the pole $-m$ is $-mH$, and the force on the pole m is $m\left(H + \frac{\partial H}{\partial z}\delta z\right)$.

Therefore the resultant force on the small magnet is $m\,\delta z\,\partial H/\partial z$ or $M\,\partial H/\partial z$; and taking the value of H given in **9·4** (2), this gives $-6\pi ja^2zM/(a^2+z^2)^{\frac{5}{2}}$, the minus sign indicating that a small magnet, whose axis is directed along the axis in the right-handed screw sense, will be attracted towards the centre of the circle.

The lines of force are closed curves linked with the circuit and lying in planes through the axis, since the field is clearly symmetrical about the axis. It will be seen in **10·12** that the cross-section multiplied by the force is constant along a tube of force, and since the tubes of force will be of narrower cross-section in the plane of the circle and inside it than anywhere else, so the field is strongest inside the circle.

9·42. Example. *A pair of circular coils of radii a, b and of m and n turns respectively carry the same current i in the same sense and are placed parallel on a common axis at a distance* $\frac{1}{2}(a+b)$ *apart. Find the magnetic force they produce at a point on the axis distant* $\frac{1}{2}a$, $\frac{1}{2}b$ *from the two coils and shew that on the axis near this point the field is so uniform that its first two differential coefficients vanish at the point if* $m/a^2=n/b^2$ *and its third in addition if* $a=b$, $m=n$.

[M. T. 1928]

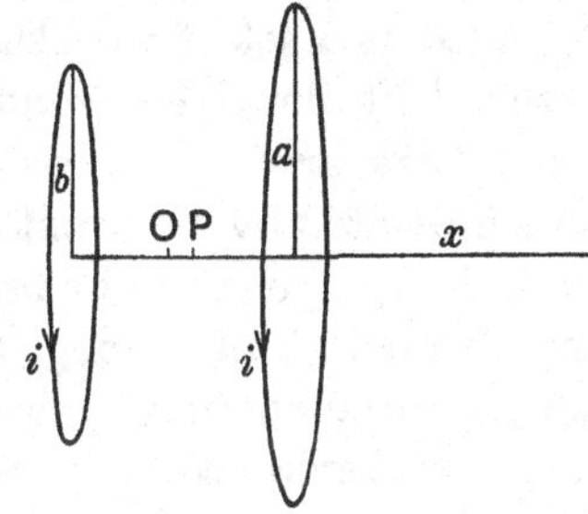

Let the point on the axis distant $\frac{1}{2}a$, $\frac{1}{2}b$ from the two coils be taken as the origin O, and let P be a point at a distance x from O on the axis regarded as the axis of x.

The coils produce the same effect as single turns of wire in which currents mi, ni are flowing. Then as in **9·4** the force at P is

$$H=\frac{2\pi mia^2}{\{a^2+(\frac{1}{2}a-x)^2\}^{\frac{3}{2}}}+\frac{2\pi nib^2}{\{b^2+(\frac{1}{2}b+x)^2\}^{\frac{3}{2}}}$$

$$=\frac{4\pi mi}{\sqrt{5}a}\left(1-\frac{4}{5}\frac{x}{a}+\frac{4}{5}\frac{x^2}{a^2}\right)^{-\frac{3}{2}}+\frac{4\pi ni}{\sqrt{5}b}\left(1+\frac{4}{5}\frac{x}{b}+\frac{4}{5}\frac{x^2}{b^2}\right)^{-\frac{3}{2}}.$$

At O, where $x=0$, $$H=\frac{4\pi i}{\sqrt{5}}\left(\frac{m}{a}+\frac{n}{b}\right);$$

and, at P,

$$H=\frac{4\pi i}{\sqrt{5}}\left\{\frac{m}{a}\left(1+\frac{6}{5}\frac{x}{a}-\frac{32}{25}\frac{x^3}{a^3}-\dots\right)+\frac{n}{b}\left(1-\frac{6}{5}\frac{x}{b}+\frac{32}{25}\frac{x^3}{b^3}-\dots\right).\right.$$

Hence, by taking the derivatives of H, we find that, when $x=0$, if

$m/a^2 = n/b^2$, both $\partial H/\partial x$ and $\partial^2 H/\partial x^2$ vanish; and further that $\partial^3 H/\partial x^3$ vanishes if $m/a^4 = n/b^4$, which is satisfied when $m/a^2 = n/b^2$ if $m = n$ and $a = b$.

9·43. Solenoids. A solenoid is a closely wound spiral coil generally of flat pitch. Its cross-section may be of any shape and it may form a cylinder or a ring.

A current j in a single plane circuit is equivalent to a plane magnetic shell of strength j whose boundary coincides with the circuit, and this is equivalent to two sheets of positive and negative magnetic matter of surface density σ and $-\sigma$, say, at a small distance d apart such that $\sigma d = j$.

Consider a pile of equal plane circuits in which equal currents j are flowing in the same sense. The pile of equivalent shells will be such that the positive and negative sheets neutralize one another's effects save at the top and bottom where there are a positive and a negative sheet.

Now consider a closely wound spiral coil of flat pitch and cylindrical form in which a current is flowing. This may be regarded as a set of parallel currents together with a drift along the cylinder, and if the coil is long compared to its breadth this drift could be represented by a current along the axis and could be compensated or balanced by leading the current back along the axis. This is **Ampère's solenoid** as shewn in the figure, consisting of a spiral coil of wire connected with a wire along its axis, through which the current enters and leaves the coil. The drift of current up the coil is compensated by the flow down the axis, so that so far as concerns its external field the solenoid is the equivalent of a pile of magnetic shells of the same strength, i.e. of a uniformly magnetized cylinder.

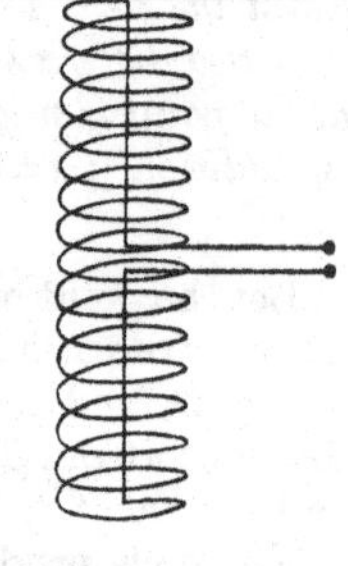

If the ends are so far distant that their effect is negligible, then the field outside the solenoid is negligible. To find the field *inside* the solenoid it is evident that there cannot be a component directed from or toward the axis; nor can there be a component at right angles to the radius in a plane perpendicular to the axis, for if there were such a component then work would be done on a unit pole describing a circle round the axis, though such a circle would embrace no current. Hence the force at every point inside the solenoid must be parallel to the axis. Let $ABCD$ be a rectangle with short sides AB, DC parallel to the axis inside the solenoid. Since the field has no component in the directions AD, BC, the only work done on a unit pole describing the rectangle arises from the sides AB and CD, and if H, H' denote the strength of the field along AB and DC, since no current is embraced, we have

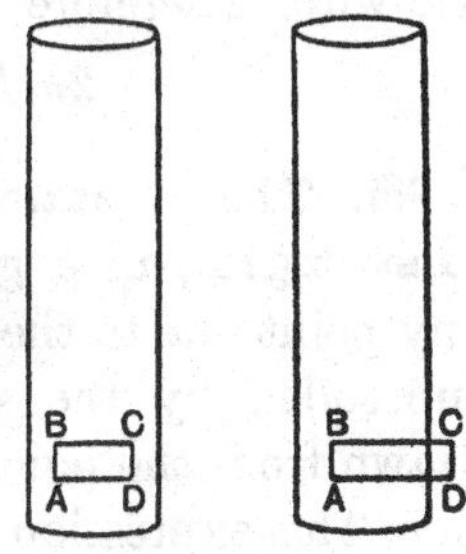

$$H.AB-H'.CD=0, \text{ or } H=H';$$

so that the field inside the solenoid is uniform. Next let the side AB of the rectangle be inside and the side CD outside the solenoid. Then H' is zero, but the rectangle embraces an amount of current $AB.nj$ if n is the number of turns of wire per unit length, so that

$$H.AB=4\pi AB.nj$$

or

$$H=4\pi nj$$

gives the strength of the field inside the solenoid and its direction is parallel to the axis.

9·44. A ring-shaped solenoid. In this case the magnetic shells equivalent to the separate turns of wire are bounded by planes passing through the axis of the ring and as the coil returns into itself there is no free polarity and no external field. To find the internal field, let H be its value at a point P at a distance r from the axis; then considering that an axial plane through P divides the solenoid into pairs of turns of wire symmetrically situated with regard to this plane and that any such pair would give a resultant force at right angles to the axial plane, it follows that H is at right angles to the axial plane. Then if a unit pole travels round a circle of

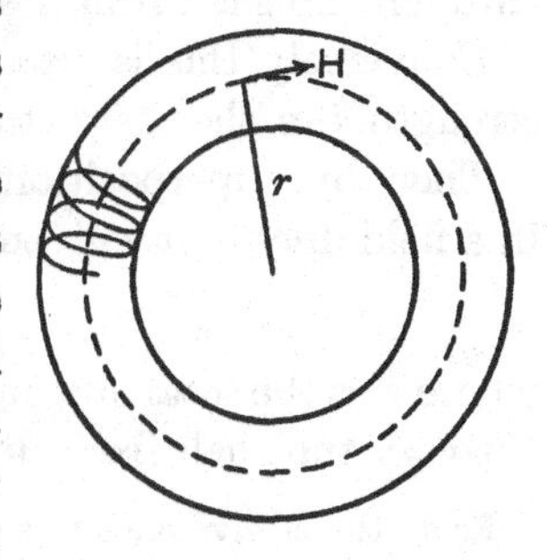

radius r with its centre on the axis and its plane perpendicular to the axis, the total current embraced is Nj, where N is the total number of turns of wire in the coil and j the current in the wire. Therefore

$$2\pi rH = 4\pi Nj \quad \text{or} \quad H = 2j/r.$$

9·5. The potential energy of a uniform magnetic shell of strength j in a given magnetic field. The potential at any point due to the shell is $j\omega$, where ω is the solid angle subtended by the shell at the point, provided that lines drawn from the point to the shell meet it first on its positive side. This expression $j\omega$ is therefore also the potential energy of a unit magnetic pole placed at the assigned point in the field of the shell, because it is the work which an operator would have to perform in order to place the unit pole in position.

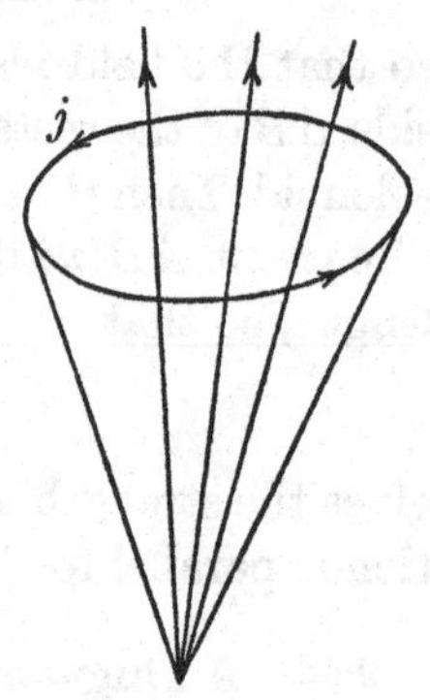

Now the total flux of force which proceeds from a unit pole is 4π and the amount of it which falls within a cone of solid angle ω, with its vertex at the pole, is ω; therefore the potential energy $j\omega$ of the unit pole is j times the flux of force from the unit pole which passes through the shell from its positive to its negative side or $-j$ times the flux of force from the unit pole which passes through the shell from its negative to its positive side.

Conversely this is also the potential energy of the shell of strength j in the magnetic field produced by the unit pole.

Then by superposition the potential energy of the shell in a field due to any given distribution of magnetic poles is

$$-jN,$$

where N is the total flux of force of the given field which passes through the shell from its negative to its positive side.

From the equivalence of a current circuit and a magnetic shell, we infer that the same expression would represent the potential energy of a closed circuit carrying a current j in a given magnetic field, N denoting the flux of force of the given field which threads the circuit in the right-handed screw sense. But if we use this expression in order to deduce the mechanical force exerted by the field on the circuit, we must be careful to note that the current must be supposed to remain constant during a virtual displacement, for it is only on this hypothesis that $-jN$ represents the potential energy of the equivalent shell.

9·51. We may also obtain the result of **9·5** by considering the uniform magnetic shell to be a number of small magnets with their axes normal to the surface. Consider a small element of the shell of area δS. If I is the intensity of magnetization, the element of the shell is equivalent to a small magnet whose poles are of strength $-I\,\delta S$ and $I\,\delta S$ at a distance t apart, where t is the thickness of the shell; i.e. a magnet of moment $It\delta S$ or $j\,\delta S$, since $It=j$.

By **8·3** the potential energy of this small magnet is $-jH_n\delta S$, where H_n is the component of the force of the given field in the direction of the axis of the magnet, i.e. normal to the shell from its negative to its positive side. Therefore the potential energy of the whole shell

$$= -j\int H_n dS = -jN,$$

where N is the total flux of force of the given field which traverses the shell from its negative to its positive side, or in the right-handed screw sense in relation to the equivalent current.

9·6. To find the mechanical force on a circuit carrying a current in a given magnetic field. Since the potential energy of such a circuit is given by

$$W = -jN \quad \text{.........................(1)},$$

therefore a displacement which increases N decreases the potential energy.

Let the circuit be displaced through a small distance δx in any direction, an element PP' of length δs tracing out a small parallelogram $PQQ'P'$. If s, s' denote the two positions of the circuit and S, S' surfaces of which they are the boundaries, then S, S' and the aggregate S'' of the small parallelograms such as $PQQ'P'$ bound a closed region of space and, as will be seen in **10·11**, the total flux of magnetic induction (or, in air, magnetic force) from

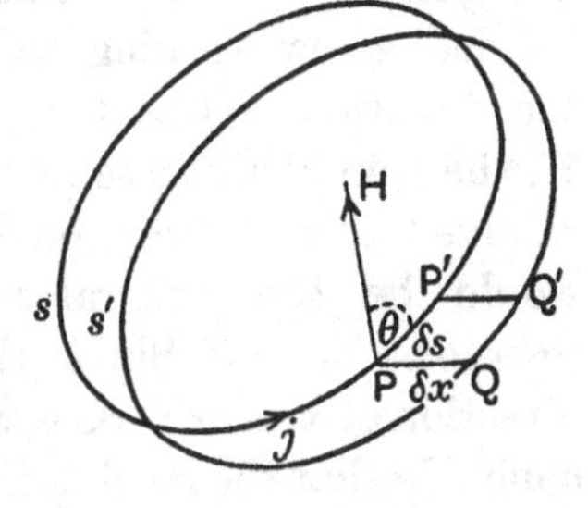

any such region is zero. Hence, if N, N', N'' denote the flux of force of the given field through S, S' S'' taken in the same sense, we have

$$N - N' + N'' = 0 \quad \text{or} \quad \delta N = N' - N = N''.$$

Hence if X denotes the resultant force on the circuit in the direction δx,

$$X\delta x = -\delta W = j\delta N = jN''.$$

But the contribution to N'' arising from the element of area $PQQ'P'$

$= \text{area } PQQ'P' \times$ component of $\mathbf{H}$ perpendicular to this area
$=$ volume of parallelepiped of edges δx, δs, $\mathbf{H}$
$= \delta s . H \sin\theta . \delta x \cos\epsilon$,

where θ is the angle between δs and $\mathbf{H}$, ϵ is the angle between δx and the perpendicular to δs and $\mathbf{H}$, and the positive direction of this perpendicular is along the axis of a right-handed screw twisting from the direction of δs to that of $\mathbf{H}$. Therefore

$$X\delta x = j\,\delta x \int H \sin\theta \cos\epsilon\, ds$$

integrated round the circuit, or

$$X = j\int H \sin\theta \cos\epsilon\, ds.$$

But if each element δs of the circuit were acted upon by a force

$$j\delta s H \sin\theta \quad \text{.......................(2)}$$

at right angles to δs and $\mathbf{H}$ in the sense of the axis of a right-handed screw turning from the direction of j to that of $\mathbf{H}$, the sum of all these forces resolved in the direction δx would be the integral expression for X. Since the direction of δx is arbitrary, we conclude that the mechanical effect of the field on the circuit is the same as if every element δs was subject to a force $j\,\delta s\, H \sin\theta$ in the direction and sense just indicated.

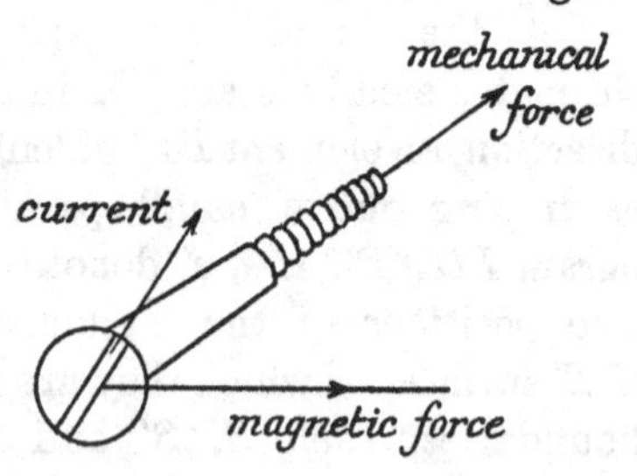

The formula (2) with its accompanying rule of signs is known as **Ampère's Law** for the force on an element of a circuit.

It will be noticed that Ampère's Law does not give a representation of the resultant reaction of a magnetic field on a circuit, but leaves the resultant to be calculated by compounding the forces on the different elements. There are however cases in which the resultant can be calculated directly from the expression for the potential energy, as will be seen in the next article.

9·61. Couple acting on a plane circuit in a uniform magnetic field. A plane circuit of area A carries a current j in a uniform magnetic field of intensity H. The potential energy of the circuit is

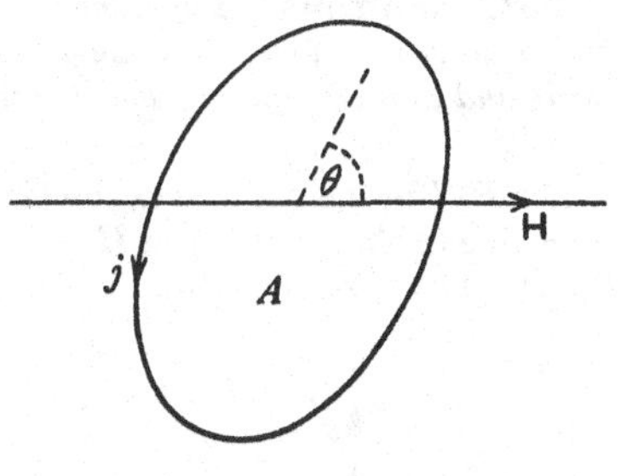

$$W = -jN$$
$$= -jAH \sin\theta \quad \text{...(1)},$$

where θ is the angle which the plane of the circuit makes with the direction of the field H.

Any displacement which keeps the plane of the circuit parallel to itself does not alter N, so that there can be no force tending to produce such a displacement. But there is in general a couple tending to increase θ measured by

$$-\frac{dW}{d\theta} = jAH\cos\theta \quad \text{..................(2)}.$$

If the circuit is free to rotate, it will set itself in equilibrium so that this couple vanishes; i.e. so that $\theta = \frac{1}{2}\pi$, or the plane of the circuit is perpendicular to the direction of H and the circuit then embraces as large a flux of force of the external field as possible.

9·62. Force between parallel straight currents. Let two parallel wires carry currents j, j' flowing in the same direction and be at a distance a apart.

Let PP' be an element δs of the wire carrying the current j.

The magnetic force H at P in the field produced by the current j' is given by $H=2j'/a$, and it is at right angles to the plane of the parallel wires. Hence in accordance with Ampère's Law (**9·6**) the element PP' is acted upon by a force $j\,\delta s\,H$ or $2jj'\delta s/a$ directed towards the other wire. Therefore *when the currents are in the same sense there is a mutual attraction of amount $2jj'/a$ per unit length between the wires.*

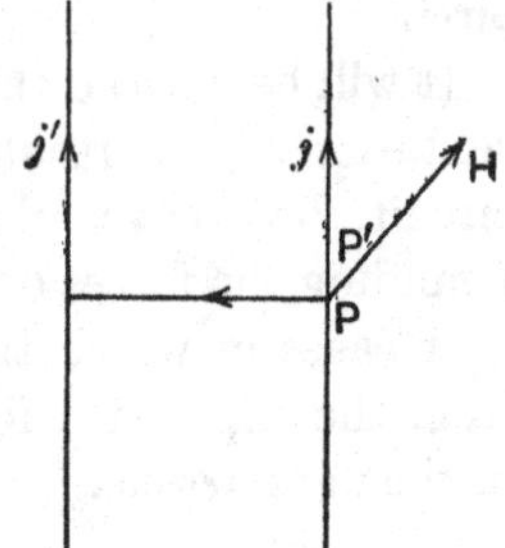

If one of the currents is reversed, it is easy to see that the force is a repulsion.

9·63. Example. *A current j flows round a circular wire of radius a, and a current j' flows in a long straight wire in the same plane as the circle and at a distance c from its centre. Determine the force between the circuits.*

Let PP' be a small arc of the circle of length δs. At P the magnetic force due to the current j' is $H=2j'/r$, where r is the distance of P from the straight wire. The direction of H is at right angles to the plane of

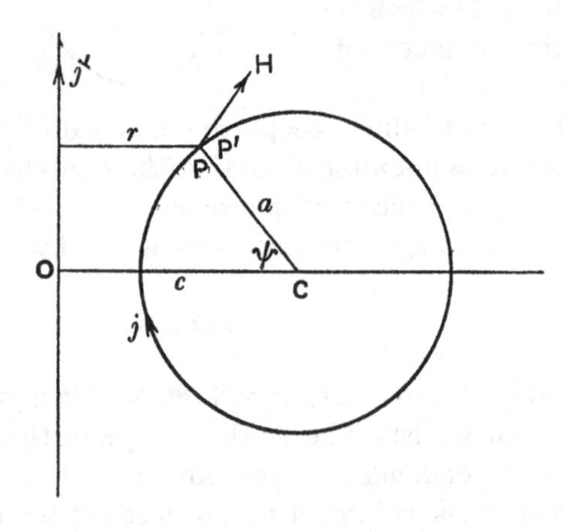

the paper so that it is at right angles to the element $j\,\delta s$ of current, and the mechanical force on the element PP' of the wire is therefore $j\,\delta s\,H$, or $2jj'\delta s/r$, and in accordance with Ampère's Law it is directed along the radius outwards, provided that j is in the sense indicated in the figure.

With ψ denoting the angle between the radius CP and the perpendicular CO to the straight wire, we have $r=c-a\cos\psi$ and $\delta s=a\,\delta\psi$

Also the resultant force on the circle can easily be shewn to act in the line CO and its magnitude is

$$2jj'\int_0^{2\pi}\frac{a\cos\psi\,d\psi}{c-a\cos\psi}=2jj'\int_0^{2\pi}\left(\frac{c}{c-a\cos\psi}-1\right)d\psi=4\pi jj'\left\{\frac{c}{\sqrt{(c^2-a^2)}}-1\right\}.$$

This force is an attraction when the currents are directed as in the figure, i.e. so that the direction of the flux of magnetic force through the circle is related to the direction of the current round it in the right-handed screw sense; and we notice that, as in **9·61**, the closed circuit tends to move so as to increase the flux through it of magnetic force of the field in which it is located.

9·7. Magnetic field produced by a given circuit. Instead of deducing the magnetic force from the potential function $j\omega$, we may often make use of Ampère's Law as given in **9·6**. Taking the formula $j\delta s.H\sin\theta$ for the mechanical force on the element δs of the circuit, suppose that the field H is due to a single unit pole. Then the reaction of each element of the circuit on the pole will be equal and opposite to the reaction of the pole on the element. Consider for example the magnetic force at a point on the axis of a circular current (**9·4**).

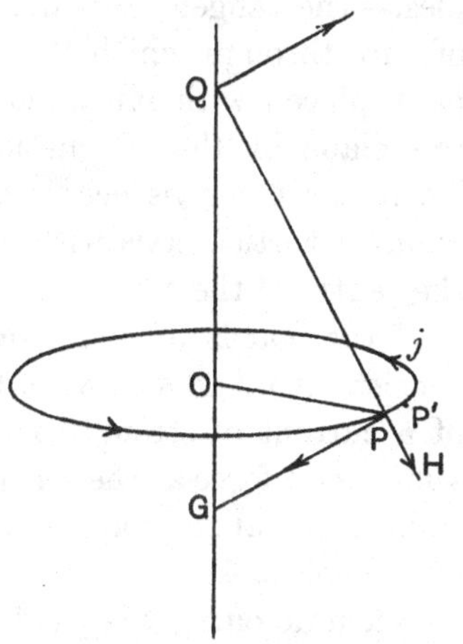

Let O be the centre of the circle, a its radius and Q a point on the axis at a distance z from O. A unit pole at Q would produce at P a force $H=1/QP^2$ along QP. Therefore the mechanical force on an element PP' of the circle carrying a current j is $j.PP'/QP^2$, at right angles to both QP and PP', i.e. along PG, in the figure. Hence the element $j.PP'$ of the circuit would exert on the pole at Q a force of the same magnitude in the opposite direction. To get the resultant force on the pole we must resolve along the axis, i.e. multiply by $\sin OQP$ or OP/QP; so that the element $j.PP'$ makes a contribution $j.PP'.a/QP^3$ to the resultant force on the pole, and by summing for the whole circle we see that the force is $2\pi ja^2/(a^2+z^2)^{\frac{3}{2}}$ as in **9·4**.

It must be observed however that there is something

unsatisfactory about this line of argument although it leads to the correct result. We assume that the reaction between a magnetic pole and an element of a current circuit consists of equal and opposite forces acting *at right angles* to the line joining them—rather a curious form of action at a distance; and we cannot demonstrate that the resultant force due to the whole circuit is necessarily resolvable into components assigned to its elements in this way. It is as well to remember that an element of current, except in so far as it is a stream of electrons, has no independent existence, nor has a solitary magnetic pole. Both are convenient mathematical fictions.

9·8. The tangent galvanometer. Galvanometers are instruments for measuring the strength of currents. The simplest—the tangent galvanometer—consists of a circular coil of wire through which the current can pass, placed with its plane vertical in the plane of the magnetic meridian. There is a compass needle free to turn about a vertical axis with its centre at the centre of the coil.

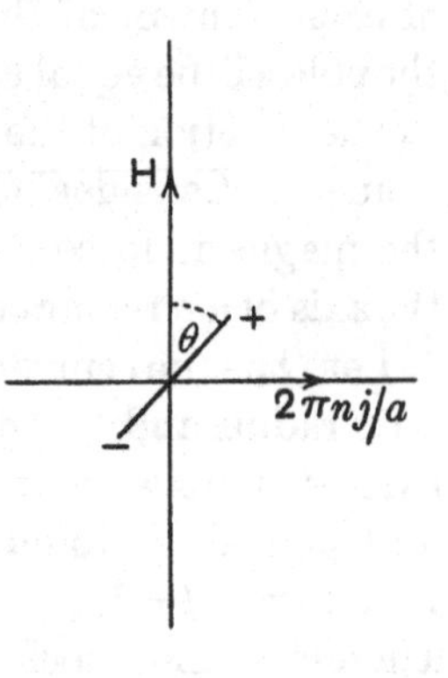

If the coil is of mean radius a and consists of n turns of wire, the passage of a current of strength j will produce a magnetic force at the centre of magnitude $2\pi nj/a$ at right angles to the plane of the coil (9·4).

When no current is passing the needle rests in the plane of the coil, but when a steady current j is passing the needle is deflected and takes up an equilibrium position in which the couple due to the earth's magnetic field is balanced by the couple due to the magnetic field produced by the current. Thus if H denotes the horizontal component of the earth's magnetic field and θ the angle of deflection, sometimes called the 'throw of the needle', we have

$$\tan\theta = 2\pi nj/aH \quad \text{.....................(1).}$$

Thus j can be determined by measurement of θ when the other constants are known. If we write $j = G \tan\theta$, G may be called the constant of the instrument.

9·81. The sine galvanometer. If the coil of the instrument described in **9·8** can turn about a vertical axis through its centre, and if, when a current passes, the plane of the coil is rotated until it overtakes the needle, then since the direction of the needle in equilibrium is that of the resultant of the two fields H and $2\pi nj/a$, we must have

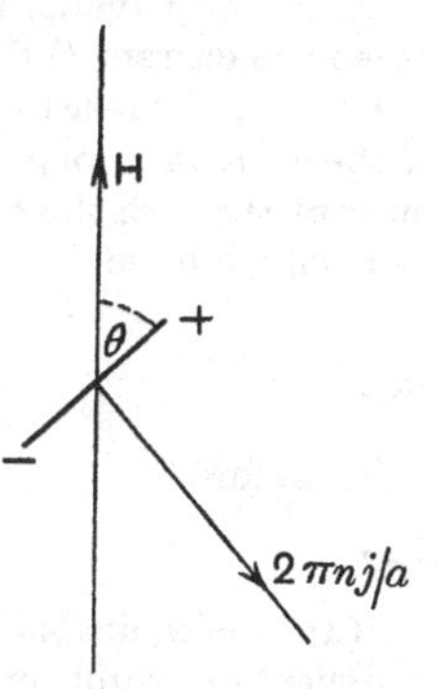

$$H \sin\theta = 2\pi nj/a,$$

or
$$\sin\theta = 2\pi nj/aH;$$

and an instrument of this kind is called *a sine galvanometer.*

9·82. Examples. (i) *The poles of a battery* (E.M.F. 2·9 *volts and internal resistance* 4 *ohms*) *are joined to those of a tangent galvanometer whose coil has* 20 *turns of wire and is of mean radius* 10 *cm. Shew that the deflection of the galvanometer is approximately* 45°. *The horizontal intensity of the earth's magnetic force is* 0·18 *dyne and the resistance of the galvanometer is* 16 *ohms.* [M. T. 1898]

The current j is determined by Ohm's law; thus

$$j(4+16) = 2{\cdot}9$$

gives
$$j = 0{\cdot}145 \text{ amp.}$$
$$= 0{\cdot}0145 \text{ absolute E.M.F. electromagnetic unit of current } (\mathbf{7{\cdot}33}).$$

Then
$$\tan\theta = \frac{2\pi nj}{aH} = \frac{2\pi \times 20 \times 0{\cdot}0145}{10 \times 0{\cdot}18}$$
$$= 1{\cdot}01,$$

so that θ is approximately 45°.

(ii) *In a tangent galvanometer, the sensibility is measured by the ratio of the increment of deflection to the increment of current, estimated per unit current. Shew that the galvanometer will be most sensitive when the deflection is* $\frac{1}{4}\pi$, *and that in measuring the current given by a generator whose electromotive force is* E, *and internal resistance* R, *the galvanometer will be most sensitive if there be placed across the terminals a shunt of resistance* $HRr/\{E - H(R+r)\}$, *where* r *is the resistance of the galvanometer, and* H *the constant of the instrument.*

What is the meaning of the result if the denominator vanishes or is negative? [M. T. 1899]

Let $j = H \tan\theta$ give the current in terms of the deflection. The increment of current per unit current is $\delta j/j$, so that the sensibility is

$$\frac{j\,\delta\theta}{\delta j} = \frac{H\tan\theta}{H\sec^2\theta} = \sin\theta\cos\theta = \tfrac{1}{2}\sin 2\theta,$$

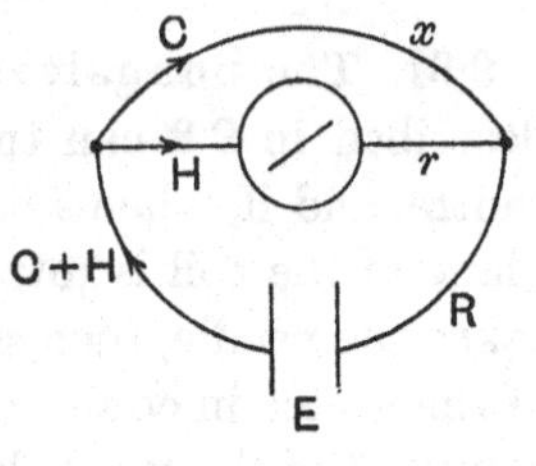

and this is greatest when $\theta = \frac{1}{4}\pi$.

Hence the galvanometer is most sensitive when the current through it is H.

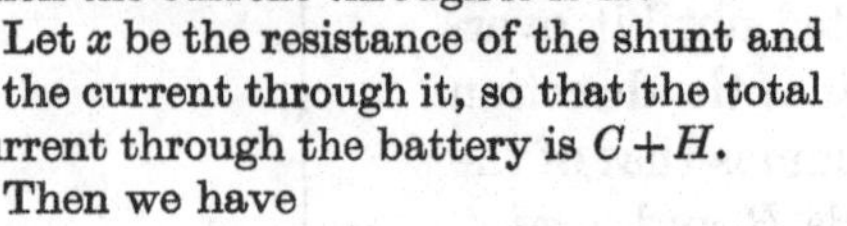

Let x be the resistance of the shunt and C the current through it, so that the total current through the battery is $C+H$.

Then we have

$$E = R(C+H) + rH \quad \text{.....................(1)}$$

and

$$xC = rH \quad \text{.....................(2).}$$

Therefore

$$E = H\left(\frac{Rr}{x} + R + r\right)$$

and

$$x = HRr/\{E - H(R+r)\}.$$

If the denominator vanishes or is negative we see that E is only just sufficient or insufficient to drive a current H through the galvanometer without a shunt.

9·83. The moving coil galvanometer. The disadvantages of the tangent galvanometer are that it requires an accurate determination of the horizontal component of the earth's magnetic field in an assigned position in the building in which it is to be used and that the instrument may be much affected by other local magnetic fields such as those due to neighbouring electric currents.

In the moving coil galvanometer the coil carrying the current moves in a fixed strong magnetic field. A coil, rectangular in shape, containing many turns of fine wire is suspended between the poles of a horse-shoe magnet. The current enters the coil through a fine wire by which the coil is suspended; and it is the elastic properties of this suspending wire, whose upper end is fixed, which resist the rotation of the coil. The current leaves the coil through a fine wire attached to its lower end.

When a current passes through the coil it tends to set itself at right angles to the magnetic field. For as was shewn in **9·5** and **9·61**, if A be its mean area and H the magnetic force supposed uniform, the potential energy when a current j passes through the coil is

$$W = -njAH\sin\theta \quad\text{.................(1)},$$

where θ is the angle between H and the plane of the coil and n is the number of turns of wire.

It follows that there is a couple tending to increase θ given by

$$-dW/d\theta = njAH\cos\theta \quad\text{..............(2)}.$$

This couple is resisted by the torsion of the suspending wire. The torsion in a straight wire one end of which is fixed while the other is twisted about the axis of the wire produces a couple proportional to the angle turned through.

Suppose that the coil is suspended in such a way that there is no torsion in the wire when the plane of the coil is approximately parallel to the magnetic field, and that we measure θ from the position in which the torsion is zero. Then when the coil is turned through an angle θ there will be a torsion couple proportional to θ, say $\lambda\theta$, opposing its rotation; and if the turning is due to the passage of a current j, the coil will take up a position of equilibrium in which the couple given by (2) is balanced by the torsion couple, i.e.

$$njAH\cos\theta = \lambda\theta,$$

or

$$j = \lambda\theta/nAH\cos\theta \quad\text{.....................(3)}.$$

For small deflections $\cos\theta$ may be taken as unity, and then j varies as θ, or

$$j = k\theta \quad\text{..........................(4)},$$

where k is the constant of the instrument.

A good instrument will measure a millionth of an ampère and a very sensitive one will measure a billionth or 10^{-12} amp.

9·84. The ballistic galvanometer. This is an instrument for measuring the total quantity of electricity which passes in a transient current; for example, in the discharge of a condenser.

If we use a moving-coil galvanometer for this purpose and let I denote the moment of inertia of the coil about its axis, its equation of motion during the passage of a current j is

$$I\ddot{\theta} = njAH\cos\theta - \lambda\theta \quad\text{.................(1)},$$

using the notation of **9·83**.

For a transient current, the passage of the current is completed before the coil has time to move, so that we may put $\cos\theta = 1$, in the preceding equation, and write

$$I\ddot{\theta} = Gj - \lambda\theta \quad \text{.......................(2)},$$

where $G = nAH$.

Integrating with regard to t through the duration of the current, we have

$$I\dot{\theta} = G\int j\,dt - \lambda\int \theta\,dt,$$

were $\int j\,dt =$ the total charge which passes $= e$, say; and $\int \theta\,dt$ is negligible, so that

$$I\dot{\theta} = Ge \quad \text{..........................(3)}$$

gives the initial velocity with which the coil begins to oscillate after the passage of the current.

There is now no longer any current passing and the coil oscillates under the influence of the torsion couple alone, with an initial velocity given by (3).

The equation of motion is now

$$I\ddot{\theta} = -\lambda\theta \quad \text{.......................(4)},$$

whence we get $\qquad I\dot{\theta}^2 = C - \lambda\theta^2$,

and $\dot{\theta} = Ge/I$ when $\theta = 0$, so that $C = G^2e^2/I$, and

$$I\dot{\theta}^2 = \frac{G^2e^2}{I} - \lambda\theta^2 \quad \text{....................(5)}.$$

Hence, if α be the amplitude of the oscillation, i.e. the maximum value of θ, $\quad G^2e^2/I = \lambda\alpha^2$;

or

$$e = \sqrt{(I\lambda)}\,\frac{\alpha}{G} \quad \text{......................(6)}$$

gives the quantity of electricity which passed through the coil.

Again, from (4), the period T of oscillation of the coil is given by $T = 2\pi\sqrt{(I/\lambda)}$, so that (6) may also be written

$$e = \frac{T}{2\pi}\cdot\frac{\lambda}{G}\alpha,$$

or

$$e = \frac{T}{2\pi}k\alpha \quad \text{......................(7)},$$

where $k = \lambda/G$ is the constant of the instrument.

9·85. To use a tangent galvanometer as a ballistic galvanometer. Let M be the moment of the magnet, I its moment of inertia, H the horizontal component of the earth's magnetic field, and A the magnetic force at the centre of the coil when a unit current passes.

Let a transient current j pass through the coil. This produces a field jA perpendicular to the plane of the coil and while the magnet is still in this plane its equation of motion is

$$I\ddot{\theta}=jAM \quad \text{.............................(1).}$$

By integrating with regard to t through the duration of the current, we get

$$I\dot{\theta}_0=AM\int j\,dt=AMe \quad \text{....................(2),}$$

where $\dot{\theta}_0$ is the angular velocity with which the needle begins to oscillate and e is the total charge which passed through the coil.

In the subsequent motion there being no longer any current the needle oscillates in the earth's magnetic field and its equation of motion is

$$I\ddot{\theta}=-HM\sin\theta \quad \text{.......................(3),}$$

which gives on integration

$$I\dot{\theta}^2=I\dot{\theta}_0^{\,2}-2HM(1-\cos\theta) \quad \text{................(4),}$$

since $\dot{\theta}=\dot{\theta}_0$ when $\theta=0$.

Hence if α be the amplitude, or maximum value of θ,

$$I\dot{\theta}_0^{\,2}=4HM\sin^2\tfrac{1}{2}\alpha,$$

or, from (2),

$$A^2M^2e^2/I=4HM\sin^2\tfrac{1}{2}\alpha;$$

i.e.

$$e=\frac{2}{A}\sqrt{\left(\frac{HI}{M}\right)}\sin\tfrac{1}{2}\alpha \quad \text{.......................(5).}$$

But, from (3), the time T of small oscillations is given by

$$T=2\pi\sqrt{(I/HM)},$$

so that (5) can also be written

$$e=\frac{TH}{\pi A}\sin\tfrac{1}{2}\alpha \quad \text{..........................(6).}$$

EXAMPLES

1. Shew that the magnetic potential at P due to a current i in a closed circuit is equal to $i\Omega_P$, where Ω_P is the solid angle subtended at P by the circuit. Deduce the value of the magnetic potential and magnetic force near a long straight wire, the rest of the circuit being very distant. [M. T. 1929]

2. Two equal magnetic poles repel one another with a force of 80 dynes when a decimetre apart. A current through a coil of wire of radius 3 cm. consisting of 4 turns of wire exerts a force of 5 dynes on one of the poles placed at the centre. Find the strength of the current in ampères. [I. 1906]

3. Two infinitely long conducting wires A and B are parallel to the line ZOZ', and intersect the plane XOY (perpendicular to ZOZ') in $x=a$, $y=0$, $z=0$ and $x=-a$, $y=0$, $z=0$ respectively. Find the force acting on a unit north magnetic pole situated at (i) $(0, 0, 0)$ and (ii) $(3a, 0, 0)$, when A carries a current I, and B a current $2I$ in the opposite direction. Draw a diagram to illustrate the distribution of the lines of magnetic force in the plane XOY around the wires when they carry equal currents (a) in the same direction, (b) in opposite directions. [M. T. 1926]

4. Currents of strength i flow in opposite directions in two long parallel wires at distance $2a$ apart. A small magnet, of magnetic moment M and moment of inertia I, is mounted with its centre equidistant from the two wires and distant b from the plane containing them. If the magnet is free to swing in a plane perpendicular to the wires, shew that its period of oscillation is $\pi\{I(a^2+b^2)/(Mia)\}^{\frac{1}{2}}$.
[M. T. 1935]

5. Four long vertical wires intersect a horizontal plane in A, B, C and D which are the vertices of a rectangle such that $AB=2AD=2a$. The currents in the wires through A, B, C and D are $-i$, $-j$, $+j$ and $+i$ respectively. Prove that the period of oscillation of a small magnet of moment μ and moment of inertia I about its position of equilibrium when placed at the mid-point of AB and free to move in the plane $ABCD$ is

$$2\pi\left\{\frac{a^2I^2}{2\mu^2(i^2+j^2)}\right\}^{\frac{1}{4}}.$$ [M. T. 1933]

6. Three long parallel straight wires carrying currents of the same strength j meet a plane at right angles in the corners of an equilateral triangle ABC and O is the centre of the triangle. Shew that the magnetic force at any point P in the plane is

$$6j \cdot OP^2/AP \cdot BP \cdot CP.$$

Also shew that the wires are all acted upon by forces tending to move them inwards, the force per unit length on each being $2j^2/OA$.
[M. T. 1924]

7. A current I flows in a circular wire of radius a. A circular coil of radius $2a$ coplanar and concentric with the wire is added. It has eight turns, and the same current I circulates in it in the sense opposite to that in the wire. Shew that the magnetic field at the centre is numerically three times as strong as it was before, and that the field at points on the axis in the neighbourhood of the centre is nearly uniform. [M. T. 1932]

8. If the force normal to the plane of a circular coil of n turns and radius a which carries a current i is measured on the axis of the coil at a small distance h from the centre of the coil, shew that it is smaller than at the centre by the fraction $\frac{3}{2}h^2/a^2$. [M. T. 1927]

9. A coil of wire is wound closely on a cylinder of n turns per unit length. The diameter of the cylinder is equal to its length. Shew that the intensity at the centre of the cylinder is

$$2\pi n i \sqrt{2}.$$ [M. T. 1934]

10. Two circular coils A and B are set in planes at right angles to one another; the centres of the coils are coincident and their common diameter is vertical, and the coil A is set with its plane in the magnetic meridian. Under the influence of a current in A a small horizontal magnet at the centre is in equilibrium at an angle of 60° with the meridian. A current is then passed through B such that the deflection of the magnet is reduced to 45°. If H is the horizontal component of the earth's magnetic field and the intensities of the fields at the centre produced by these currents separately are respectively H_1 and H_2, find the values of the ratios $H_1 : H_2 : H$. [M. T. 1911]

11. Two equal circular coils A and B are mounted on a horizontal board with their planes vertical and perpendicular to one another, their centres being in the same horizontal plane and equidistant from the line of intersection of their planes. A compass needle is pivoted at the intersection of the axes of the coils. When the board is placed so that the plane of A is in the magnetic meridian and currents C_A, C_B are passed through A, B respectively, the deflection of the compass needle is θ_1. The board is now rotated through a right angle in the direction from north to west, and the deflection is found to be θ_2. (Deflections of the needle are counted positive when its north pole moves toward the west.) Shew that

$$\frac{C_A}{C_B} = \left| \frac{1 + \cot \theta_2}{1 - \cot \theta_1} \right|.$$

In what circumstances does this fail as a method of comparison of the two currents? [M. T. 1923]

12. A battery of electromotive force E and internal resistance A, when connected to a tangent galvanometer of resistance R, gives a deflection α. A second battery of the same E.M.F. but of internal resistance B, when connected to the galvanometer gives a deflection β. When the two batteries are connected in series to the galvanometer, the deflection is γ. Shew that

$$A = R \frac{2 \cot \gamma - \cot \beta}{\cot \alpha + \cot \beta - 2 \cot \gamma},$$

and that the current through the galvanometer when the batteries are connected to it in series is

$$\frac{E}{R} \cdot \frac{\cot \alpha + \cot \beta - 2 \cot \gamma}{\cot \gamma}.$$ [M. T. 1915]

13. A given current sent through a tangent galvanometer deflects the magnet through an angle θ. The plane of the coil is slowly rotated

round the vertical axis through the centre of the magnet. Prove that, if $\theta > \frac{1}{4}\pi$, the magnet will describe complete revolutions, but if $\theta < \frac{1}{4}\pi$, the magnet will oscillate through an angle $\sin^{-1}(\tan\theta)$ on each side of the meridian. [M. T. 1906]

14. A tangent galvanometer is to be constructed with a coil consisting of five turns of thick copper wire, so that the tangent of the angle of deflection may be equal to the number of ampères flowing in the coil.

Shew that the radius of the coil must be about 17·45 cm., assuming the earth's horizontal magnetic force to be 0·18 dyne. [I. 1895]

15. Shew that, if a slight error is made in reading the angle of deflection of a tangent galvanometer, the percentage error in the deduced value of the current is a minimum if the angle of deflection is $\frac{1}{4}\pi$. [M. T. 1907]

16. The circumference of a sine galvanometer is 1 metre, the earth's horizontal magnetic force is 0·18 C.G.S. unit. Shew that the greatest current which can be measured by the galvanometer is 4·56 ampères approximately. [M. T. 1896]

17. If there be an error α in the determination of the magnetic meridian, find the true strength of a current which is i as ascertained by means of a sine galvanometer. [M. T. 1903]

18. Find the electromotive force of a Daniell cell, in C.G.S. units, from the following data. Five Daniell cells were connected in series to a tangent galvanometer, the coil of which had 10 turns of 11 cm. radius. The deflection produced was 45°. The total resistance of the circuit was $16{\cdot}9 \times 10^9$ C.G.S. units, and the horizontal component of the earth's magnetic field at the point was 0·180 C.G.S. unit. [M. T. 1917]

19. A battery of 8 cells, placed 2 in series in 4 sets in parallel, is connected with a galvanometer of 100 ohms resistance, which is shunted with a resistance of 100/9 ohms. Each cell has E.M.F. 2 volts and internal resistance 12 ohms. Find the deflection of the galvanometer, if a current 0·0005 ampère gives a deflection of one division.
[M. T. 1916]

20. Two galvanometers are wound respectively with 2 turns of wire of resistance 0·1 ohm per turn and 20 turns of wire of resistance 1 ohm per turn. The coils can be regarded as all having the same radius and their centres at the centre of the suspended magnet. Which galvanometer will shew the greater deflection when connected to a cell whose internal resistance is 1 ohm? [M. T. 1908]

21. The coil of a tangent galvanometer consists of 2 turns of wire of radius 20 cm. If the horizontal component of the earth's magnetic

field is equal to 0·17 dyne per unit pole, find the deflection due to a current of one-tenth of an electromagnetic unit.

Assuming that the magnetic field due to the earth is that due to a small magnet placed at its centre, shew that the horizontal component in magnetic latitude λ varies as $\cos\lambda$. Hence find how the deflection of the galvanometer varies with latitude. [M. T. 1924]

22. If the coil of a tangent galvanometer consists of n turns of wire of radius a and cross-section A and specific resistance κ, and a current is sent through it from a cell of internal resistance R, and electromotive force E, find the deflection of the galvanometer.

Shew that if the radius, specific resistance and the total cross-section nA are given, there will be definite values of A and n for which the deflection is a maximum. [St John's Coll. 1913]

23. The plane of the coil of a tangent galvanometer made a small angle with the magnetic meridian. A current was passed through the galvanometer and the deflection of the needle from the magnetic meridian was θ. A second reading was made with an equal current but with half as many turns of wire as were used before, and this time the deflection was θ'.

Shew that the current passing through the galvanometer was

$$\frac{H}{G}\cdot\frac{1}{\cot\theta'-\cot\theta}\text{ approximately,}$$

where G is the constant of the galvanometer when the full number of turns are used, and H is the horizontal component of the earth's magnetic force. [M. T. 1934]

24. A tangent galvanometer has two parallel coaxial coils of radius a cm., each wound with N turns of fine wire. Their common axis points E. and W. They are distant a cm. apart and are connected in series. The magnet is suspended on the axis of the coils, midway between them. Its length is small compared with a. Shew that if H is the horizontal component of the earth's magnetic field, in dynes per unit pole, and if θ is the deflection of the magnet, the current flowing is $\dfrac{1\cdot 1aH}{N}\tan\theta$ ampères. [M. T. 1925]

25. The terminals of a battery (E.M.F. E), a tangent galvanometer and a voltameter are connected in parallel arc, the resistance of each arc being B, G, and U respectively. The deflection of the galvanometer is observed to be θ. After a time the battery arc is thrown out of the arrangement. Prove that the deflection of the galvanometer drops to ϕ, where

$$\tan\phi=\tan\theta-\frac{(E-\kappa G\tan\theta)\,U}{\kappa B(G+U)},$$

and κ is the reduction factor of the galvanometer. [I. 1902]

26. The coil of a moving-coil galvanometer is a square of side 1 cm. with 200 turns. The coil can rotate about a vertical axis and the

restoring couple is 10^{-3} gm.-cm. per degree. The field in which the coil swings is 2×10^3 C.G.S. electromagnetic units. Find approximately the angular deflection due to a current of 10^{-4} ampères. [M. T. 1935]

27. If M is the moment of the magnet of a ballistic galvanometer, K its moment of inertia about its axis of rotation, H the intensity of the earth's magnetic field and A the magnetic intensity at the centre of the coil due to unit current, shew that neglecting frictional forces the quantity of electricity passing through the coil which causes the magnet to swing through a small angle θ is Q, where

$$Q = \frac{\theta}{A}\sqrt{\frac{HK}{M}}.$$

Find the relation connecting Q with θ, if there is a retarding frictional couple $\mu\dot{\theta}$ on the magnet. [M. T. 1931]

28. Find (i) the couple exerted on a small circuit of m turns of area A carrying a current i placed at the centre of a large circular coil of radius a and n turns carrying a current j, the planes of the coils being at right angles;

(ii) the attraction between two circuits of areas A and B carrying currents i and j placed parallel to one another at a large distance apart and on a line perpendicular to them both;

(iii) the attraction between two long parallel straight wires carrying a flow and return current i. [M. T. 1934]

29. A thin bar magnet of moment M and length $2l$ lies along the axis of a circular wire of radius a in which a current i is flowing, the plane of the circle bisecting the bar. Prove that the reaction of the magnet on the circuit causes a longitudinal stress in the wire of magnitude $Mia/(a^2+l^2)^{\frac{3}{2}}$. Determine in what cases the stress is a tension or a thrust. [I. 1927]

ANSWERS

2. 0·00066 amp. **3.** (i) $6I/a$, (ii) 0. **10.** $\sqrt{3} : \sqrt{3}-1 : 1$.

17. $i(1+\alpha \cot\theta)$, where θ is the reading of the instrument.

18. 1·06 volts. **19.** 50 divisions. **20.** The former.

21. 20° 16′. **22.** $\tan^{-1}\{2\pi nAE/Ha(RA+2\pi\kappa na)\}$.

26. 4°.

27. $Q = \frac{\theta}{A} e^{\frac{\mu t}{2K}} \sqrt{\left(\frac{HK}{M}\right)}$, where $\frac{\cos\sqrt{(4HMK-\mu^2)}\,t}{2K} = \frac{\mu}{2\sqrt{(HKM)}}$.

28. (i) $2\pi mnAij/a$. (ii) $6ABij/d^4$, where d is the distance between the circuits. (iii) $-2i^2/d$, where d is the distance between the wires.

29. Thrust or tension according as the sense of the current in relation to the axis of the magnet is that of a right-handed or a left-handed screw.

Chapter X

MAGNETIC INDUCTION AND INDUCED MAGNETISM

10·1. Magnetic induction. In Chapter VIII we limited our considerations to the magnetic fields external to the magnets which produce them; and we saw in **8·12** that the magnetic force **H** was really the mechanical force acting upon a unit pole at the point under consideration and we have assumed that in air **H** is the negative gradient of a potential function. In order to define in this way the magnetic force at a point inside a magnet, it is necessary to imagine there to be at the point under consideration a small cavity inside the magnet, in which a unit pole might be placed.

Let the cavity be cylindrical with its axis in the direction of the intensity of magnetization **I**. Let the length be $2l$, and let the cross-section be such that its linear dimensions are small compared to l. Let α denote the area of the cross-section. The removal of the cylindrical portion of the magnet will leave free polarity on the ends of the cylinder, equivalent to poles of strengths $I\alpha$ and $-I\alpha$ at the ends. The unit pole at the centre of the cavity may be regarded as in air and subject to a force **H** together with the force $2I\alpha/l^2$ in the direction of **I** due to the polarity of the ends of the cylinder. But by hypothesis α/l^2 is negligible, so that inside such a 'needle-shaped' cavity the force per unit pole is independent of **I** and is simply the vector **H**. Hence we may define *the magnetic force inside a magnet as the force per unit pole in a needle-shaped cavity whose axis is in the direction of the magnetization* **I**.

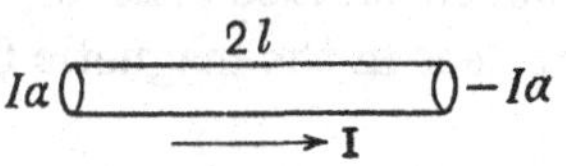

If however we take the length of the cylindrical cavity to be short compared to the linear dimensions of its cross-section, the free polarity on its ends produces a field of force on the axis of the cavity comparable with that produced by two infinite plane sheets of surface density I and $-I$, viz. $4\pi I$. Hence the total force per unit pole at the centre of this cavity is the sum of two vectors $\mathbf{H}$ and $4\pi\mathbf{I}$. This vector is called the **magnetic induction** and denoted by $\mathbf{B}$. Thus

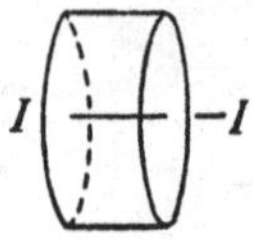

I $-I$

$$\mathbf{B}=\mathbf{H}+4\pi\mathbf{I} \quad\text{..........................(1)};$$

and the **magnetic induction** is therefore *the resultant force per unit pole in a disc-shaped cavity whose axis is in the direction of the magnetization.*

We note that $\mathbf{B}$ and $\mathbf{H}$ are not necessarily in the same direction, but that they will be so whenever $\mathbf{I}$ coincides in direction with $\mathbf{H}$.

We notice that outside the magnets, $\mathbf{B}=\mathbf{H}$ since $\mathbf{I}=0$; i.e. in air magnetic induction is the same as magnetic force.

This is Clerk-Maxwell's method of defining magnetic induction. The vector $\mathbf{B}$ can also be introduced by an analytical process and defined by the relations which it has to satisfy.

10·11. The total flux of magnetic induction out of any closed surface is zero. We have to prove that for any closed surface in the magnetic field

$$\int B_n dS=0,$$

where B_n denotes the outward normal component of the magnetic induction $\mathbf{B}$.

Since $\mathbf{B}=\mathbf{H}+4\pi\mathbf{I}$, therefore

$$\int B_n dS=\int H_n dS+4\pi\int I_n dS \quad\text{...........(1)}.$$

Since the magnetic force $\mathbf{H}$ is related to magnetic polarity in exactly the same way as the electric intensity $\mathbf{E}$ is related to electric charge, it follows that, by analogy from Gauss's Theorem,

$$\int H_n dS=4\pi M \quad\text{....................(2)},$$

where M is the algebraical sum of the strengths of the magnetic poles inside S. But the sum of the strengths of the poles of every elementary magnet is zero. Therefore the only elementary magnets which contribute to the total M are those which are intersected by the surface S in such a way that one pole lies inside and one outside the surface.

Consider a small magnet whose section by the surface is an element of area dS. The axis of the magnet is in the direction of the intensity of magnetization **I**; let this, for the sake of precision, make an acute angle ϵ with the outward normal to dS. The negative pole is then inside S and the positive pole outside. If l be the small length of the magnet, its volume is $l dS \cos \epsilon$ and its moment is therefore $IldS \cos \epsilon$, so that the strength of its negative pole is $-I \cos \epsilon dS$ or $-I_n dS$. Hence the sum of the poles inside S is $-\int I_n dS$, so that, by substituting this expression for M in (2), we get

$$\int H_n dS = -4\pi \int I_n dS,$$

and therefore from (1)

$$\int B_n dS = 0 \quad \text{......................(3).}$$

Since this relation is true for integration over every closed surface that can be drawn in the field, it follows from the definition of divergence (**1·52**) that

$$\operatorname{div} \mathbf{B} = 0 \quad \text{.........................(4)}$$

at every point of the field; therefore magnetic induction is a solenoidal vector throughout the whole field (see **5·21**).

10·12. Tubes of induction. Lines and tubes of induction may be defined like lines and tubes of force. Applying the theorem of **10·11** to a tube of induction bounded by two cross-sections as in **2·31** (ii), we see that the induction multiplied by the cross-section is constant along the tube. But whereas in an electrostatic field tubes of force terminate on the surfaces of charged conductors, in a magnetic field a tube of induction cannot have an end, for if such tubes could terminate there could be closed surfaces for which the theorem of **10·11** would not be true. Consequently all such tubes are re-entrant, all lines of induction forming closed curves. In air of course lines of induction are the same as lines of force, since in air $\mathbf{B} = \mathbf{H}$.

In the case of a bar magnet whose positive and negative ends are P, Q, by experimenting with iron filings, it is easy to shew that the lines of force in air proceed from P to Q; but in air these are also lines of induction. It follows that tubes of induction pass from the end P in air round to the end Q and then enter the magnet and pass along it back to P, so that inside the magnet the induction **B** is directed from Q towards P, though the magnetic force **H** inside the magnet is clearly directed from the positive end P towards the negative end Q. Also, since **H** has a definite value at all points of the field, it follows that the tubes of induction fill the whole field, and whilst all of them are crowded into the magnet, passing along it from Q to P, they widen out considerably in air.

10·13. The amount of induction across any surface in a given magnetic field depends only on the boundary curve. For if S_1, S_2 denote any two surfaces having a common boundary curve s, then since S_1, S_2 together form the boundary of a closed region, by **10·11**

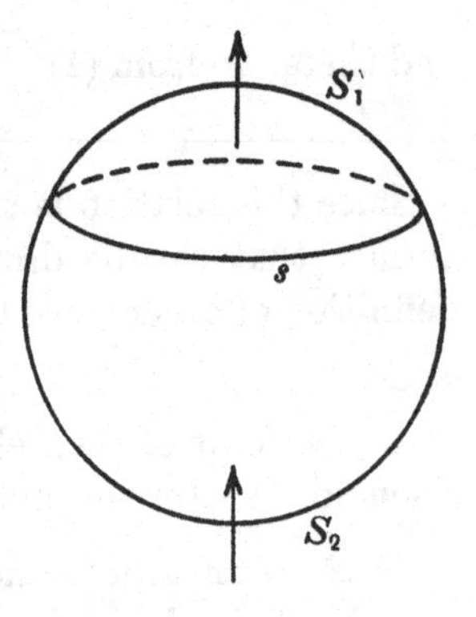

$$\int B_n dS = 0$$

when the integral is taken over S_1 and S_2 and B_n is the outward normal component over the whole surface. If therefore we reverse the sign over S_2 so that induction is measured in the same sense over both S_1 and S_2 in relation to the curve s, we have

$$\int_{S_1} B_n dS = \int_{S_2} B_n dS$$

so that the amount of induction is the same for all surfaces having the same boundary curve.

10·2. Continuity of normal induction. Let S be the surface of separation of two media in a magnetic field, such as the boundary surface of a magnet in air or in contact with a block of iron. Let $\mathbf{B}_1$, $\mathbf{B}_2$ denote the magnetic induction at points close to one another and on opposite sides of the surface. By applying the theorem of **10·11** to a cylinder with ends parallel to the surface and so short that the induction through its sides is negligible, we get

$$(B_n)_1 = (B_n)_2 \text{ ....}(1),$$

or the normal component of induction is continuous.

This result might also be inferred from the properties of tubes of induction. For if the induction on opposite sides of the surface makes angles θ_1, θ_2 with the normal, and a tube of induction in the first medium cuts the surface in an element of area dS, the cross-section is $\cos\theta_1 dS$ in the first medium, and it is continued as a tube of cross-section $\cos\theta_2 dS$ in the second medium; but the induction multiplied by the cross-section is constant, so that

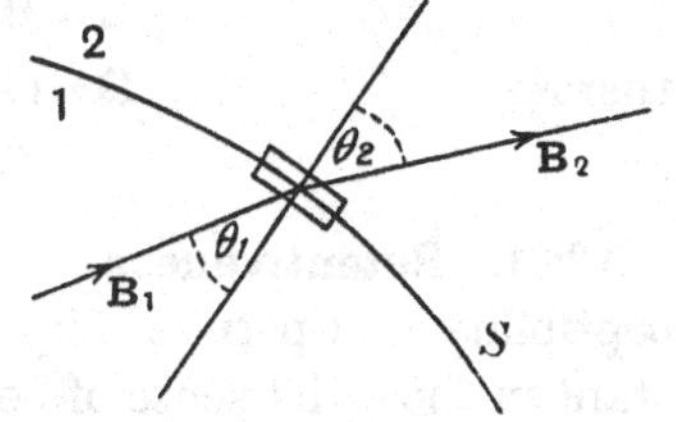

$$B_1 \cos\theta_1 dS = B_2 \cos\theta_2 dS$$

or

$$(B_n)_1 = (B_n)_2.$$

10·3. Induced magnetism. If a piece of soft iron is brought near to the positive pole of a permanent magnet it is found to acquire magnetic properties temporarily, exhibiting negative polarity in its parts which are nearest to the positive pole of the permanent magnet and positive polarity in its remoter parts, so that if free to move it is attracted towards the permanent magnet. The soft iron is then said to be **magnetized by induction** and the magnetism which it acquires is '**induced magnetism**'. Its tendency is clearly to move from the weak part of the field to the strong. Nickel and cobalt possess the same property in this respect as iron, and such bodies are called **Paramagnetic**. There is another class of bodies called **Diamagnetic** which have the opposite property of tending to move from the strong part of the field to the weak. Bismuth is the principal member of this class but its susceptibility is almost infinitesimal as compared with that of iron.

In the elementary theory of induced magnetism it is

assumed that the intensity of magnetization induced in a piece of metal by a given magnetic field is proportional to the magnetic force of the field; i.e. that

$$\mathbf{I} = \kappa \mathbf{H} \quad \text{.........................(1)},$$

where the constant κ is called the **susceptibility** of the metal.

If we put $\mu = 1 + 4\pi\kappa$, then μ is called the **permeability** of the metal, and in the case in which the magnetism of a substance is entirely induced magnetism, since

$$\mathbf{B} = \mathbf{H} + 4\pi\mathbf{I},$$

therefore

$$\mathbf{B} = (1 + 4\pi\kappa)\,\mathbf{H},$$

or

$$\mathbf{B} = \mu\mathbf{H} \quad \text{.........................(2)}.$$

10·31. Retentiveness. The 'constants' κ and μ, the susceptibility and permeability of a piece of metal, are not constant in the strict sense of being independent of the magnetic force. If they were independent of **H**, then it might be expected that the withdrawal of a field of magnetic force would involve the instant disappearance of induced magnetism from any piece of metal which had been exposed to the action of the field; since if $\mathbf{I} = \kappa\mathbf{H}$, then $\mathbf{I} = 0$ when $\mathbf{H} = 0$. But this is found not to be the case. The induced magnetization does not wholly disappear. What remains is called **residual magnetism**, and such metals as retain residual magnetism when the external magnetic field is withdrawn are said to possess **retentiveness**.

Residual magnetism produces a magnetic field the tendency of which is in general to reduce the residual magnetism. For example, an iron bar placed along the lines of force in a magnetic field becomes magnetized longitudinally, the direction of magnetization from the negative to the positive end being that of the field. But when the bar is removed from the field the residual magnetism will produce a field in the iron from the positive towards the negative end and this field will tend to reduce the residual magnetism. But the longer the bar the less effective will be the demagnetization since the central parts will be at a distance from the ends.

10·4. Equations for the potential. In magneto-static as in electrostatic fields we assume that the magnetic vector **H**

is the negative gradient of a potential function ϕ, i.e. we assume that

$$\mathbf{H} = -\operatorname{grad}\phi \quad \text{.....................(1)},$$

or that
$$H_x, H_y, H_z = -\frac{\partial\phi}{\partial x}, -\frac{\partial\phi}{\partial y}, -\frac{\partial\phi}{\partial z}.$$

But in any case in which the magnetism is entirely induced magnetism, we have

$$\mathbf{B} = \mu\mathbf{H} \quad \text{..........................(2)}$$

and, from **10·11**,
$$\operatorname{div}\mathbf{B} = 0 \quad \text{.........................(3)},$$

i.e.
$$\frac{\partial}{\partial x}(\mu H_x) + \frac{\partial}{\partial y}(\mu H_y) + \frac{\partial}{\partial z}(\mu H_z) = 0$$

or
$$\frac{\partial}{\partial x}\left(\mu\frac{\partial\phi}{\partial x}\right) + \frac{\partial}{\partial y}\left(\mu\frac{\partial\phi}{\partial y}\right) + \frac{\partial}{\partial z}\left(\mu\frac{\partial\phi}{\partial z}\right) = 0 \quad \text{.........(4)}.$$

By comparing with **5·3** we see that **B**, **H** and μ play the same part in a magnetic field which contains pieces of soft iron, as do **D**, **E** and K in an electric field containing dielectrics, and the equations for the potential are similar in form.

At an interface between two media of permeabilities μ_1 and μ_2, since the normal induction is continuous, we have

$$(B_n)_1 = (B_n)_2$$

or
$$\mu_1 (H_n)_1 = \mu_2 (H_n)_2,$$

or
$$\mu_1 \frac{\partial\phi_1}{\partial n} = \mu_2 \frac{\partial\phi_2}{\partial n} \quad \text{.....................(5)},$$

where ϕ_1, ϕ_2 denote the potentials on opposite sides of the surface, and the differentiations are along the normal in the same sense on both sides. If either medium is air, the corresponding μ is unity.

10·41. When permanent magnetism is present we may denote its intensity at any point of the field by $\mathbf{I}_0$, so that the total magnetization is given by

$$\mathbf{I} = \kappa\mathbf{H} + \mathbf{I}_0 \quad \text{........................(1)},$$

where $\kappa\mathbf{H}$ represents the induced magnetization. It is not to be supposed that $\kappa\mathbf{H}$ and $\mathbf{I}_0$ both necessarily have non-zero values at the same point; **H** has a value at all points of the field, $\kappa \neq 0$ in such substances as iron and $\mathbf{I}_0 \neq 0$ in permanent magnets.

Again the induction is given by

$$\mathbf{B} = \mathbf{H} + 4\pi\mathbf{I}$$
$$= \mathbf{H} + 4\pi(\kappa\mathbf{H} + \mathbf{I}_0)$$
$$= \mu\mathbf{H} + 4\pi\mathbf{I}_0 \quad \text{.........................(2).}$$

But $\qquad \operatorname{div} \mathbf{B} = 0 \quad$ (**10·11**),

therefore $\qquad \operatorname{div}(\mu\mathbf{H}) = -4\pi \operatorname{div} \mathbf{I}_0,$

or, as in **10·4**,

$$\frac{\partial}{\partial x}\left(\mu\frac{\partial\phi}{\partial x}\right) + \frac{\partial}{\partial y}\left(\mu\frac{\partial\phi}{\partial y}\right) + \frac{\partial}{\partial z}\left(\mu\frac{\partial\phi}{\partial z}\right) = 4\pi \operatorname{div} \mathbf{I}_0 \quad \text{...(3).}$$

This is the *characteristic equation for the potential* in a field which contains permanent magnets; and **10·4** (4) is the special case in which $\mathbf{I}_0 = 0$.

At the boundary of a permanent magnet, let the suffix 1 denote the magnet and the suffix 2 the adjoining medium. Then, as in **10·4**, we have

$$(B_n)_1 = (B_n)_2,$$

where $\mathbf{B} = \mu\mathbf{H} + 4\pi\mathbf{I}_0$, and $\mathbf{I}_0$ only exists in medium 1, so that

$$(\mu H_n)_1 + 4\pi I_{0n} = (\mu H_n)_2.$$

In terms of potential this gives

$$\mu_1\frac{\partial\phi_1}{\partial n} - \mu_2\frac{\partial\phi_2}{\partial n} = 4\pi I_{0n} \quad \text{.................(4),}$$

where the normal differentiation is outwards from the magnet.

In every case the potential is continuous at the interface, i.e. $\phi_1 = \phi_2$ over the interface, so that the tangential component of magnetic force is continuous. It is easy to shew, as in **5·26**, that since normal induction and tangential forces are continuous, at the interface between two media at which there is no permanent magnetism the lines of force undergo a refraction in accordance with a definite law.

10·5. Examples. (i) *An infinite mass of soft iron has a plane face. To find the magnetization induced by a magnetic particle outside the iron.*

(α) *When the axis of the magnet is normal to the plane face.* Let m be the moment of the magnet and A the position of its centre. Let B be the point such that AB is bisected at right angles at O by the plane face. Let r_1, r_2 denote distances of any point P from A, B and θ_1, θ_2 the angles made by AP, BP with BA.

There will be different potential functions ϕ_1, ϕ_2 in air and in iron, and the conditions to be satisfied at the boundary are that

$$\phi_1 = \phi_2 \qquad \text{.........(1)}$$

and that

$$\frac{\partial \phi_1}{\partial n} = \mu \frac{\partial \phi_2}{\partial n} \qquad \text{.........(2),}$$

where μ is the permeability of the iron, and ∂n is normal to the plane face.

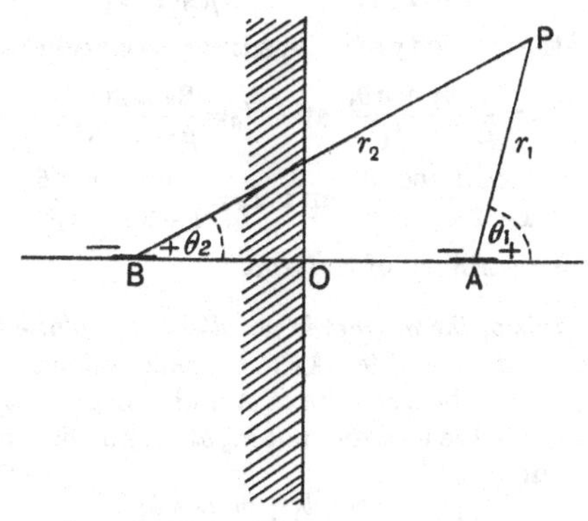

Let us assume that the field in air is due to the magnet m at A and an image m' at B with their axes in the same direction BA, and that the field in iron is due to a magnet m'' at A with its axis in the same direction.

Also take an axis Ox along OA and axes Oy, Oz in the plane face, let $OA = a$ and let P be the point (x, y, z).

Then

$$\phi_1 = \frac{m \cos \theta_1}{r_1^2} + \frac{m' \cos \theta_2}{r_2^2}$$

and

$$\phi_2 = \frac{m'' \cos \theta_1}{r_1^2};$$

or

$$\phi_1 = \frac{m(x-a)}{\{(x-a)^2 + y^2 + z^2\}^{\frac{3}{2}}} + \frac{m'(x+a)}{\{(x+a)^2 + y^2 + z^2\}^{\frac{3}{2}}} \qquad \text{.........(3)}$$

and

$$\phi_2 = \frac{m''(x-a)}{\{(x-a)^2 + y^2 + z^2\}^{\frac{3}{2}}} \qquad \text{.........(4).}$$

Substituting from (3), (4) in (1) and (2) and putting $x = 0$, we find that

$$m - m' = m''$$

and

$$m + m' = \mu m'',$$

so that

$$m' = \frac{\mu - 1}{\mu + 1} m \text{ and } m'' = \frac{2m}{\mu + 1} \qquad \text{.........(5)}$$

(compare 6·7).

Since all the necessary conditions are satisfied, this represents the solution of the problem.

Hence the field inside the iron is such as would be produced by a magnetic particle of moment $\frac{2m}{\mu+1}$ in air; i.e. its potential is

$$\phi_2 = \frac{2m}{\mu+1}\frac{\cos\theta_1}{r_1^2},$$

and the magnetic force has radial and transverse components

$$H_r = \frac{4m}{\mu+1}\frac{\cos\theta_1}{r_1^3}, \quad H_\theta = \frac{2m}{\mu+1}\frac{\sin\theta_1}{r_1^3}.$$

It follows that the intensity of induced magnetization has components

$$I_r = \frac{4\kappa m}{\mu+1}\frac{\cos\theta_1}{r_1^3} \text{ and } I_\theta = \frac{2\kappa m}{\mu+1}\frac{\sin\theta_1}{r_1^3}$$

or

$$I_r = \frac{2\kappa m}{1+2\pi\kappa}\frac{\cos\theta_1}{r_1^3} \text{ and } I_\theta = \frac{\kappa m}{1+2\pi\kappa}\frac{\sin\theta_1}{r_1^3},$$

where κ is the susceptibility of the iron.

(β) *When the axis of the magnet is parallel to the plane face.* Take Oz parallel to the axis of m, and let AP, BP make angles ψ_1, ψ_2 with Oz.

Let the field ϕ_1' in air be due to m at A and a magnet m_1 at B, and let the field ϕ_2' in iron be due to a magnet m_2 at A, all the axes pointing in the same direction.

Then

$$\phi_1' = \frac{m\cos\psi_1}{r_1^2} + \frac{m_1\cos\psi_2}{r_2^2}$$

and

$$\phi_2' = \frac{m_2\cos\psi_1}{r_1^2}$$

or

$$\phi_1' = \frac{mz}{\{(x-a)^2+y^2+z^2\}^{\frac{3}{2}}} + \frac{m_1 z}{\{(x+a)^2+y^2+z^2\}^{\frac{3}{2}}} \quad \text{.........(6)}$$

and

$$\phi_2' = \frac{m_2 z}{\{(x-a)^2+y^2+z^2\}^{\frac{3}{2}}} \quad \text{.....................(7);}$$

and as before the boundary conditions are that when $x=0$

$$\phi_1' = \phi_2' \quad \text{..............................(8)}$$

and

$$\frac{\partial\phi_1'}{\partial x} = \mu\frac{\partial\phi_2'}{\partial x} \quad \text{..........................(9).}$$

Substituting (6) and (7) in (8) and (9), we get

$$m + m_1 = m_2$$

and

$$m - m_1 = \mu m_2,$$

so that

$$m_1 = -\frac{\mu-1}{\mu+1}m, \quad m_2 = \frac{2m}{\mu+1} \quad \text{..................(10);}$$

whence the field in the iron and the components of induced magnetization can be written down as before.

(ii) *To find the magnetization induced in an iron sphere placed in a uniform field of force.* This problem is the exact parallel of that of a dielectric sphere in a uniform field of electric force (**6·71**).

If H_0 denotes the uniform field of force, and we take an origin O at the centre of the sphere and the axis Ox in the direction of the field, the potential of the field in the absence of the sphere is

$$\phi = -H_0 x = -H_0 r \cos\theta \quad \text{.....................(1)}.$$

Then arguing as in **6·71** when the sphere is present, we assume for the potential in air

$$\phi_1 = -H_0 r \cos\theta + \frac{M\cos\theta}{r^2} \quad \text{.....................(2)}$$

and in iron

$$\phi_2 = Nr\cos\theta \quad \text{.........................(3)}.$$

If a denotes the radius of the sphere, then ϕ_1, ϕ_2 have to satisfy the conditions

$$\phi_1 = \phi_2 \text{ and } \frac{\partial\phi_1}{\partial r} = \mu\frac{\partial\phi_2}{\partial r} \quad \text{.....................(4)}$$

when $r=a$, where μ is the permeability of the iron. These give

$$M = \frac{\mu-1}{\mu+2}H_0 a^3 \text{ and } N = -\frac{3H_0}{\mu+2} \quad \text{..............(5)}.$$

The potentials are therefore

$$\phi_1 = -H_0 r\cos\theta + \frac{\mu-1}{\mu+2}\frac{H_0 a^3\cos\theta}{r^2} \quad \text{..............(6)}$$

and

$$\phi_2 = -\frac{3H_0}{\mu+2} r\cos\theta \quad \text{........................(7)}.$$

From (6) we see that the disturbance of the given field produced by the sphere is such as would be caused by a magnetic particle of moment $\frac{\mu-1}{\mu+2}H_0 a^3$ placed at its centre.

From (7) it follows that since $\phi_2 = -\frac{3H_0 x}{\mu+2}$ the magnetic force **H** or $-\partial\phi_2/\partial x$ is constant and equal to $3H_0/(\mu+2)$ inside the iron. The intensity of induced magnetization is given by

$$I = \kappa H = \frac{3\kappa H_0}{\mu+2} \quad \text{........................(8)}.$$

Since $\mu = 1+4\pi\kappa$, therefore

$$I = \frac{3\kappa}{3+4\pi\kappa}H_0 = \frac{3}{4\pi+3/\kappa}H_0;$$

and as the susceptibility κ increases, I tends to a limit $3H_0/4\pi$. It follows that no matter how great the susceptibility of the sphere the intensity of magnetization in a given field cannot exceed a fixed amount.

The induction inside the sphere is given by

$$B = \mu H = \frac{3\mu}{\mu+2}H_0$$

and, as μ or κ increases, the induction tends to $3H_0$ and cannot exceed this amount.

In air at a great distance from the sphere the uniform field is undisturbed and $B=H_0$. Suppose that dS_i and dS_0 denote the cross-sections of the same tube of induction inside the sphere and at a great distance from it, then since the product of the induction and cross-section is constant

$$\frac{3\mu}{\mu+2}H_0 dS_i = H_0 dS_0,$$

so that

$$dS_i = \frac{\mu+2}{3\mu} dS_0.$$

For large values of μ it follows that $dS_i = \frac{1}{3}dS_0$ so that the magnetic induction is gathered up and concentrated in the sphere, in the manner indicated in the figure of 6·71.

In this connection we note that for paramagnetic substances such as iron, cobalt and nickel the susceptibility κ is positive so that $\mu > 1$, and that for some samples of iron μ may be as great as 1000. But for diamagnetic bodies κ is negative and $\mu < 1$. For all such bodies κ is very small, its largest value is for bismuth, for which $\kappa = 2{\cdot}5 \times 10^{-6}$. The direction of magnetization in a diamagnetic body is opposite to that of the inducing field, and a diamagnetic sphere would take up fewer tubes of induction than would occupy the same space in an undisturbed field, its general effect being to push the tubes apart instead of gathering them together.

EXAMPLES

1. Find the force with which a small magnet of moment m is attracted towards a large block of iron of permeability μ having a plane face at a distance a from the magnet, when the axis of the magnet is (i) perpendicular, (ii) parallel to the plane face.

2. Shew that, if the small magnet in Ex. 1 be free to turn about its centre, either position in which its axis is normal to the plane face is one of stable equilibrium, and that the period of a small oscillation is $2\pi\sqrt{\left\{\frac{\mu+1}{\mu-1}\frac{8a^3K}{m^2}\right\}}$, where K is the moment of inertia of the magnet.

3. Shew that, when the small magnet of Ex. 1 makes an angle α with the normal to the plane face, the mutual potential energy of the magnet and the block of iron is

$$-\frac{\mu-1}{\mu+1}\frac{m^2}{16a^3}(1+\cos^2\alpha).$$

4. A small magnet is placed at the centre O of a spherical cavity in a large mass of soft iron. Prove that at any point P in the iron the direction of the induced magnetism lies in a plane through the axis of the magnet and makes with OP an angle $\tan^{-1}(\frac{1}{2}\tan\theta)$, where θ is the angle between OP and the axis of the magnet.

5. A small spherical cavity is cut in a permanent magnet. Shew from the expression for the potential that the force within it is $H+\frac{4}{3}\pi I$, where H is the magnetic force and I the intensity of magnetization. [I. 1923]

6. Shew that inside a sphere uniformly magnetized to intensity I there is a constant demagnetizing force $\frac{4}{3}\pi I$. [I. 1922]

7. A long cylinder of circular section, of radius a and of given magnetic permeability μ, is placed across a given uniform magnetic field. Find the ratio in which the intensity of the field inside the cylinder is reduced. [M. T. 1911]

8. Shew that the force at the centre of a long narrow circular cylindrical cavity, whose axis is perpendicular to the direction of I, tends, as the cavity is indefinitely diminished, to half the vector sum of the magnetic force and the magnetic induction. [I. 1909]

9. A sphere of permeability μ is placed in a uniform magnetic field. A circle has its centre on the line of force which passes through the centre of the sphere and its plane perpendicular to this line and lies outside the sphere. Shew that the presence of the sphere increases the induction through the circle in the ratio

$$1+\frac{2(\mu-1)}{\mu+2}\sin^3\alpha : 1,$$

where α is the greatest angle subtended by a radius of the sphere at a point on the circle.

10. Prove that if P is the centre of a circular cylindrical cavity whose diameter and length are equal, with generators parallel to the intensity of magnetization I in a magnet, the limit of the magnetic force at P as the cavity tends to vanishing is

$$H+4\pi\left(1-\frac{1}{\sqrt{2}}\right)I.$$

11. A specimen of soft iron is placed in a uniform magnetic field, in the form of (i) a sphere, (ii) a long rod with its axis parallel to the external field, (iii) a long rod of circular cross-section with its axis at right angles to the field. Compare the intensities of magnetization in the three cases, estimating the values in cases (ii) and (iii) near the middle of the rod. If the magnetic susceptibility be 20, shew that the intensities are approximately as

$$3:252:2.$$ [M. T. 1915]

ANSWERS

1. (i) $\frac{3}{8}\frac{\mu-1}{\mu+1}\frac{m^2}{a^4}$; (ii) $\frac{3}{16}\frac{\mu-1}{\mu+1}\frac{m^2}{a^4}$.

7. $2:\mu+1$.

11. $\frac{3}{\mu+2}:1:\frac{2}{\mu+1}$.

Chapter XI

ELECTROMAGNETIC INDUCTION

11·1. Faraday's experiments. About the year 1832 Michael Faraday was making experiments at the Royal Institution which led to the discovery of the law of electromagnetic induction, which made possible all the modern applications of electric power. He made two coils of insulated wire, each about two hundred feet in length, winding them on the same wooden block, attaching the ends of one coil to the terminals of a battery and the ends of the other to a galvanometer. Calling the circuit which contains the battery the 'primary' circuit and the other the 'secondary', he observed that whenever the primary circuit was completed the galvanometer indicated the passage of a transient current in the secondary circuit in the opposite direction to that in the primary; also that whenever the primary circuit was broken a transient current passed through the secondary in the same direction as the current in the primary. It was further observed that so long as the current in the primary was steady there was no current through the secondary. The currents in the secondary circuit are called **induced currents.**

Other experiments were performed with primary and secondary circuits separated from one another and situated in parallel planes. It was found that, when there was a steady current in the primary, relative motion of the circuits resulted in an induced current in the secondary; and that the two currents were in the same sense when the motion was causing increased separation of the circuits, and in opposite senses when the circuits were approaching one another.

Other experiments shewed that the relative motion of a circuit and a magnet resulted, in general, in the induction of a current in the circuit; that is to say, if a closed loop of wire is in motion in a magnetic field there is, in general, an electric current induced in the wire.

11·2. Law of electromagnetic induction. Faraday attributed the phenomena described in **11·1** to what he called the *electrotonic state of the medium.* The precise formulation of the experimental law which accounts for the phenomena is due to the mathematician F. E. Neumann.* It is as follows:

Whenever the number of tubes of magnetic induction which pass through a circuit is varying, there is an electromotive force produced in the circuit equal to the rate of decrease of the number of tubes of magnetic induction; and the positive direction of this electromotive force and of the tubes of induction have a right-handed screw relation.

As this law is based upon Faraday's experiments, it is commonly called **Faraday's circuital relation.**

The reader will have no difficulty in following the way in which the law accounts for the results of the experiments. In the first case when the primary current is steady the magnetic field caused by it is also steady and there is no secondary current; but when the primary circuit is first completed the primary current has to increase from zero to its steady value so that its magnetic field is also growing, i.e. the flux of induction through the secondary circuit is *increasing* and in consequence there is an E.M.F. in the secondary in the left-handed screw sense in relation to the direction of the induction, and this produces a current in the secondary in the opposite sense to that in the primary. The break of the primary circuit involves a dying-away of the magnetic field through the secondary and therefore an E.M.F. in the secondary in the right-handed screw sense producing a current in the same sense as the original current in the primary.

In relation to the second experiment the separation of the parallel circuits decreases the amount of induction due to the primary which passes through the secondary and the approach of the circuits increases it, and hence the currents are induced as stated.

* F. E. Neumann (1798–1895), German mineralogist, physicist and mathematician. *Berlin Akad.* 1845 and 1847.

11·21. Lenz's Law. An induced current of course produces a magnetic field of its own and this affects the total amount of magnetic induction threading the circuit. From what has already been said it will be observed that things happen as though an effort were being made to preserve the *status quo* as regards the amount of induction which threads the circuit. For example, in Faraday's first experiment, before the primary circuit is closed there is no current and no induction through the secondary; but directly the circuit is closed the magnetic field begins to grow and so does the induction through the secondary; and a current is induced in the secondary in such a sense that its magnetic field gives induction through the circuit in the opposite sense to that produced by the primary, as though making an effort to keep the total induction through the secondary zero as long as possible.

This is in accord with a law formulated by Lenz,* viz. *If a current is induced in a circuit B by the relative motion of circuits A and B, the direction of the induced current will be such that by its electromagnetic action on A it tends to oppose the relative motion.* Thus if the circuits are equal rings in parallel planes and B is made to approach A, i.e. to increase the amount of induction due to A which passes through B, the current induced in B will be in the opposite sense to that in A so that they would repel one another, i.e. the relative motion would be opposed.

We remark that, if Lenz's law were not true, stability of equilibrium of any arrangement of circuits would be impossible. For relative displacement would start a current which would produce a force tending to increase the displacement and if unchecked the motion would increase indefinitely.

11·3. Self-induction and mutual induction. The law of induction of currents applies to all circuits whether they contain batteries or condensers or are simply closed loops of wire in a varying magnetic field. Thus in the case of a single circuit containing a battery, directly the circuit is closed the growth of the current means an increasing amount of magnetic

* Poggendorff, *Annalen der Physik und Chemie*, xxxi, p. 483 (1834).

induction threading the circuit, and, by the law of induction, there must be an induced E.M.F. tending to produce a current in the opposite sense to that produced by the battery. The effect will be more marked if the wire is coiled round a bar of soft iron, for this collects and concentrates the magnetic induction within the coil. It constitutes in fact an electromagnet.

It is necessary to have a symbol to denote the amount of induction which passes through a circuit when a unit current flows round it and there are no other currents in the field. This amount of induction is called **the coefficient of self-induction** or the **inductance** of the circuit and is denoted by L; so that if a current j is passing through the circuit Lj is the amount of induction passing through the circuit in the right-handed screw sense in relation to the current. The constant L depends of course on the form of the circuit and varies if the shape of the circuit is changed, and is in general larger for coils containing iron cores because they concentrate the induction.

In the same way if a unit current round a circuit A causes an amount of induction M through a circuit B, then M is called the **coefficient of mutual induction**, or the **mutual inductance** of the two circuits. It can be proved that a unit current round B produces the same amount M of induction through A. Consequently if L' denotes the inductance of B and currents j, j' flow round A and B, the total amounts of induction through A and B respectively are

$$Lj + Mj' \text{ and } Mj + L'j'.$$

11·31. The phenomenon of self-induction is well illustrated by the following experiment. An electric lamp F of large resistance is in parallel with a coil G of small resistance surrounding an iron bar—i.e. an electromagnet. The terminals P, Q are connected to a battery. Owing to the difference of the resistances most of the current passes along PGQ and there is not sufficient along PFQ to make the lamp incandescent. But when the battery circuit is broken the lamp instantly flashes

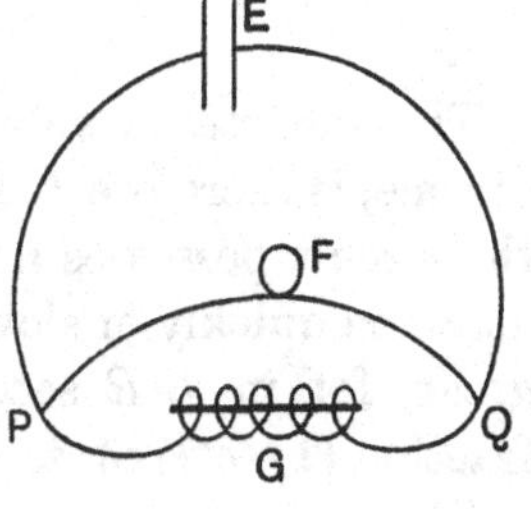

up, because the tendency described in **11·21** to keep constant the flux of induction through the circuit $PFQG$ results in the passage of a transient current round this circuit much larger than the steady current which passed along PFQ before the battery circuit was broken.

11·4. Single circuit with self-induction. When a single circuit of resistance R and inductance L contains a battery of electromotive force E, the current j in the circuit is determined by the combination of Ohm's Law and Faraday's circuital relation. The magnetic induction through the circuit due to a current j is Lj, so that in addition to the electromotive force of the battery there is an induced electromotive force $-\frac{d}{dt}(Lj)$, or $-L\frac{dj}{dt}$ and therefore by Ohm's Law

$$Rj = E - L\frac{dj}{dt} \quad \text{.....................(1).}$$

Thus we have a differential equation for the current

$$L\frac{dj}{dt} + Rj = E \quad \text{.....................(2).}$$

When E is constant, the solution of this equation is

$$j = \frac{E}{R} + Ce^{-Rt/L} \quad \text{.....................(3),}$$

where C is a constant of integration.

From (2) we see that Ldj/dt is finite when j is zero. We infer that a finite current is not started instantaneously but by a gradual growth, though it may be that a finite value is attained in a very short time. In (3) therefore we put $j = 0$ when $t = 0$, giving $C = -E/R$, so that

$$j = \frac{E}{R}(1 - e^{-Rt/L}) \quad \text{.....................(4).}$$

The final steady value of the current, after a sufficient lapse of time, is therefore E/R, the value given by Ohm's Law when the circuit possesses no self-induction. Whether this value is attained quickly or slowly depends on whether L/R is small or great, for in L/R seconds the current attains the definite fraction $(1 - e^{-1})$ of its final value.

When the circuit is first closed the current is at first very small and the battery is doing work in creating the magnetic field of the circuit, but after the permanent state is reached dj/dt is negligible and the work done by the battery is entirely spent in heating the circuit.

The inductance L would be increased by inserting some iron in the circuit and this would retard the attainment of the permanent state.

It appears therefore that in a single circuit with a battery of constant E.M.F. the self-induction of the circuit is only effective during the interval of time that is occupied in setting up the permanent current as given by Ohm's Law.

11·41. Flow of electricity caused by the sudden creation of a uniform magnetic field. Consider a plane circuit of wire of area A and suppose that a uniform magnetic field is brought into existence and that B denotes the induction normal to the plane of the circuit. The amount of induction of this field which passes through the circuit is AB and the amount due to self-induction when a current j is passing is Lj, so that the total amount of induction through the circuit is $AB+Lj$. Therefore by Ohm's Law and the law of induction

$$Rj = -\frac{d}{dt}(AB+Lj) \quad \text{.......................(1)},$$

where R denotes the resistance of the circuit. If we integrate this equation with regard to t from $t=0$ to $t=\infty$, we have

$$R\int_0^\infty j\,dt = -\Big[AB+Lj\Big]_0^\infty.$$

But $\int_0^\infty j\,dt = Q$, the total quantity of electricity which crosses any section of the wire while the magnetic field is being created. The initial value of B is zero and its final value is the strength of the induction of the ultimate uniform field, which we may still denote by B; the initial value of j is zero and its final value when the permanent state of the field is attained is zero also. Hence it follows that

$$RQ = -AB \quad \text{..........................(2)}.$$

11·5. A single circuit with a periodic electromotive force. We can easily see that, if a wire ring were to rotate steadily in a uniform magnetic field, the amount of magnetic induction passing through the ring would be continually changing, so that, in accordance with the law of induction, there would be an electromotive force induced in the wire, which would be periodic in the sense that its direction along the wire would alternate with every half revolution and its magnitude would vary between zero and a definite maximum.

Let us therefore investigate the current which would be produced in the circuit of **11·4** if, instead of a constant electromotive force E, the circuit is subject to a periodic electromotive force $E\cos pt$.

Instead of **11·4** (2) we now have the equation

$$L\frac{dj}{dt}+Rj=E\cos pt \quad \text{..................(1).}$$

This equation has an integrating factor $e^{Rt/L}$, and the solution is

$$je^{Rt/L}=C+\frac{E}{L}\int e^{Rt/L}\cos pt\,dt$$

or

$$j=Ce^{-Rt/L}+\frac{E\,(R\cos pt+Lp\sin pt)}{R^2+L^2p^2} \quad \text{........(2),}$$

where C is a constant of integration. After the lapse of sufficient time the term $e^{-Rt/L}$ will disappear and the current will contain periodic terms only, viz.

$$j=\frac{E\,(R\cos pt+Lp\sin pt)}{R^2+L^2p^2} \quad \text{..............(3)}$$

or

$$j=\frac{E\cos(pt-\alpha)}{\sqrt{(R^2+L^2p^2)}} \quad \text{....................(4),}$$

where

$$\tan\alpha=Lp/R.$$

The current has the same frequency (number of alternations per second) $p/2\pi$ as the electromotive force, but its phase $pt-\alpha$ lags behind that of the electromotive force by an amount $\tan^{-1}(Lp/R)$, and this will be approximately $\frac{1}{2}\pi$ when Lp is large compared to R, i.e. if the alternations of E.M.F. are of high frequency. In this case we have, approximately,

$$j=\frac{E\sin pt}{Lp} \quad \text{......................(5),}$$

shewing that the current is independent of the resistance, in this case, and depends only on the inductance and the frequency.

In the general case the maximum value of the E.M.F. is E and the maximum value of the current is $E/\sqrt{(R^2+L^2p^2)}$, and the ratio of these, viz. $\sqrt{(R^2+L^2p^2)}$, is called the **impedance** of the circuit.

11·51. Plane circuit rotating uniformly in a uniform magnetic field of force *H*. Let A be the area of the circuit and at time t let θ be the angle between the plane of the circuit and the direction of the magnetic field H. The amount of induction threading the circuit is then $AH \sin\theta$. By the law of induction of currents there is therefore an E.M.F. in the circuit of amount $-AH\cos\theta.\dot{\theta}$; and if we put $\theta = \omega t$, where ω is the angular velocity, the induced E.M.F. is $-AH\omega\cos\omega t$.

Hence, as in **11·5**, the equation for the current j is

$$L\frac{dj}{dt} + Rj = -AH\omega\cos\omega t \quad\text{.....................(1)},$$

where L, R denote the inductance and resistance of the circuit. As in **11·5** the solution is

$$j = -\frac{AH\omega\cos(\omega t - \alpha)}{\sqrt{(R^2+L^2\omega^2)}} \quad\text{.....................(2)},$$

where

$$\tan\alpha = L\omega/R.$$

As in **9·5**, the potential energy of the circuit in the given field H is

$$W = -jAH\sin\theta,$$

so that the field exerts on the circuit a couple tending to *increase* θ of moment

$$-\frac{dW}{d\theta} = jAH\cos\theta \quad\text{.........................(3)}.$$

Substituting the value of j from (2) and taking note of the minus sign, we see that there is a couple *opposing* the rotation of moment

$$\frac{A^2H^2\omega\cos\omega t\cos(\omega t-\alpha)}{\sqrt{(R^2+L^2\omega^2)}}.$$

This may be written

$$\frac{1}{2}\frac{A^2H^2\omega\{\cos(2\omega t-\alpha)+\cos\alpha\}}{\sqrt{(R^2+L^2\omega^2)}},$$

and since the mean value of $\cos(2\omega t-\alpha)$ is zero, the mean value of the moment of the couple is

$$\frac{1}{2}\frac{A^2H^2\omega\cos\alpha}{\sqrt{(R^2+L^2\omega^2)}} \quad\text{or}\quad \frac{1}{2}\frac{A^2H^2R\omega}{R^2+L^2\omega^2}.$$

When the rotation is so rapid that $L\omega$ is large compared to R, α is approximately $\frac{1}{2}\pi$ and the current is approximately

$$j = -\frac{AH\sin\omega t}{L},$$

so that

$$Lj + AH\sin\theta = 0,$$

or the total induction passing through the circuit is zero.

11·6. Circuit containing a condenser. Discharge of a condenser. Let a circuit contain a condenser of capacity C and a battery of electromotive force E; let R denote the total resistance and L the inductance of the circuit. At time t let Q and $-Q$ denote the charges on the plates of the condenser and ϕ the difference of their potentials. Also let j denote the current in the circuit from the positive towards the negative plate of the condenser and suppose that the electromotive force is in the same sense. By the law of induction there is an induced E.M.F. equal to $-L\,dj/dt$, therefore by Ohm's Law, as in **7·3** (5),

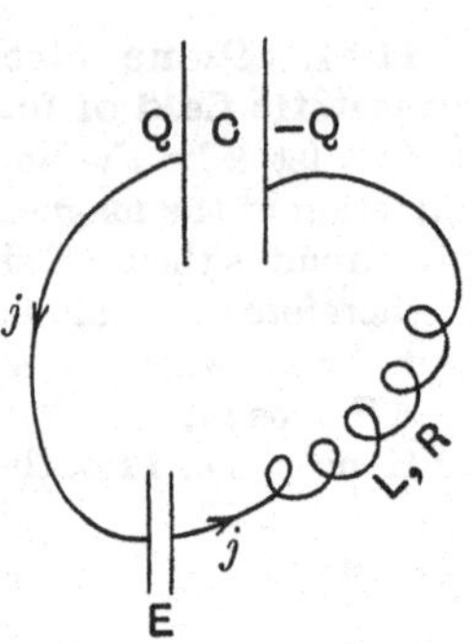

$$Rj = E + \phi - L\frac{dj}{dt},$$

or

$$L\frac{dj}{dt} + Rj - \phi = E \quad \text{..................(1).}$$

But $j = -dQ/dt$, and $Q = C\phi$, so that (1) is equivalent to

$$L\frac{d^2Q}{dt^2} + R\frac{dQ}{dt} + \frac{Q}{C} = -E \quad \text{..............(2).}$$

When E is constant, this equation admits of simple solution, for the left-hand side retains the same form if we write $Q = -CE + Q'$. But the only case of special interest is when $E = 0$, i.e. when there is no battery in the circuit and we are simply concerned with the discharge of a condenser through a wire of given resistance and inductance. The equation is then

$$L\frac{d^2Q}{dt^2} + R\frac{dQ}{dt} + \frac{Q}{C} = 0 \quad \text{................(3),}$$

and, if we substitute for Q in terms of either j or ϕ, we find that j and ϕ satisfy the same differential equation. The ordinary process of solution gives

$$Q = A_1 e^{-\lambda_1 t} + A_2 e^{-\lambda_2 t} \quad \text{................(4),}$$

where $-\lambda_1$, $-\lambda_2$ are the roots of

$$L\lambda^2 + R\lambda + \frac{1}{C} = 0 \quad \text{..................(5).}$$

When $R^2 > 4L/C$, the roots of this quadratic are real and negative, i.e. λ_1, λ_2 are positive numbers and the solution (4) shews that the charge dies away continuously.

For the determination of the constants A_1, A_2, we note that if Q_0 is the initial charge, since the initial current $-dQ/dt$ is zero, we have by putting $t=0$

$$Q_0 = A_1 + A_2$$

and
$$0 = -\lambda_1 A_1 - \lambda_2 A_2,$$

so that $A_1 = \lambda_2 Q_0/(\lambda_2 - \lambda_1)$ and $A_2 = -\lambda_1 Q_0/(\lambda_2 - \lambda_1)$,

and
$$Q = \frac{Q_0}{\lambda_2 - \lambda_1}(\lambda_2 e^{-\lambda_1 t} - \lambda_1 e^{-\lambda_2 t}) \quad \text{..............(6).}$$

The current j is then given by

$$j = -\frac{dQ}{dt} = \frac{Q_0 \lambda_1 \lambda_2}{\lambda_2 - \lambda_1}(e^{-\lambda_1 t} - e^{-\lambda_2 t}) \quad \text{...........(7).}$$

From this we find that j increases from zero to a maximum, which it attains at time t given by $\lambda_1 e^{-\lambda_1 t} = \lambda_2 e^{-\lambda_2 t}$, and then dies away continuously.

When $R^2 = 4L/C$, the quadratic (5) has equal roots $-R/2L$ and the solution of (4) is

$$Q = e^{-Rt/2L}(A_1 + A_2 t) \quad \text{....................(8),}$$

and the initial values of Q and dQ/dt give

$$Q_0 = A_1, \quad 0 = -\frac{R}{2L}A_1 + A_2,$$

so that
$$Q = e^{-Rt/2L} Q_0 (1 + Rt/2L) \quad \text{....................(9)}$$

and
$$j = -\frac{dQ}{dt} = e^{-Rt/2L} Q_0 \frac{R^2 t}{4L^2} \quad \text{.................(10);}$$

and j attains a maximum value after a time $t = 2L/R$ and then dies away continuously.

The more interesting case is when $R^2 < 4L/C$; the quadratic (5) then has imaginary roots and the solution of (3) has the form

$$Q = Ae^{-Rt/2L} \sin\left\{\sqrt{\left(\frac{1}{CL} - \frac{R^2}{4L^2}\right)} t + \alpha\right\} \quad \text{......(11),}$$

representing harmonic oscillations, Q changing sign regularly and vanishing at regular intervals of $\pi \Big/ \sqrt{\left(\frac{1}{CL} - \frac{R^2}{4L^2}\right)}$ as the charge is transferred from one plate of the condenser and back again, but all the time dying away because of the damping factor $e^{-Rt/2L}$.

As a special case, if the condenser be 'short-circuited', e.g. if its plates were allowed to come into contact, this is equivalent to making its capacity infinite, so that (3) becomes

$$L\frac{d^2Q}{dt^2}+R\frac{dQ}{dt}=0,$$

or

$$L\frac{dj}{dt}+Rj=0;$$

with a solution $$j=j_0e^{-Rt/L}$$

shewing that the current would decrease exponentially with the time.

It frequently happens that R^2 is small compared to $4L/C$. The damping is then small and the period of the oscillations is approximately $2\pi\sqrt{(CL)}$. This is known as *Thomson's Formula*, as the theory of the discharge was first given by Sir William Thomson (Lord Kelvin) in 1853. The actual oscillations were first detected by Feddersen in 1857.

11·61. It should be remarked that currents only flow in closed circuits and that in the case of the discharge of a condenser the circuit is to be regarded as completed by a displacement current, i.e. a time rate of change of the displacement in the dielectric medium between the plates. But this does not necessitate any changes in the foregoing results.

It should be noted that while the inductance of a wire bent into a single loop may be small, it may be greatly increased by being coiled in spiral form and it is still more increased if an iron bar is inserted in the coil.

11·62. Unit of inductance. The practical unit of inductance is the **henry**,* which is equal to 10^9 electromagnetic absolute C.G.S. units of inductance.

11·7. Examples. (i) *A Wheatstone's bridge is adjusted so that there is no permanent current through the galvanometer. One of the arms whose resistance is P contains a coil of inductance L. If the resistance of the galvanometer is negligible, prove that when the battery circuit is broken a quantity*

$$EQL/[(P+Q)\{(P+Q)B+(P+R)Q\}]$$

of electricity will flow through the galvanometer.

* Joseph Henry (1797–1878), American physicist.

[*P, Q, R, S are the resistances of the arms, R being opposite to Q; E is the* E.M.F. *of the battery which connects the junction of P and Q with the junction of R and S; and B is the resistance of the battery.*] [M.T. 1923]

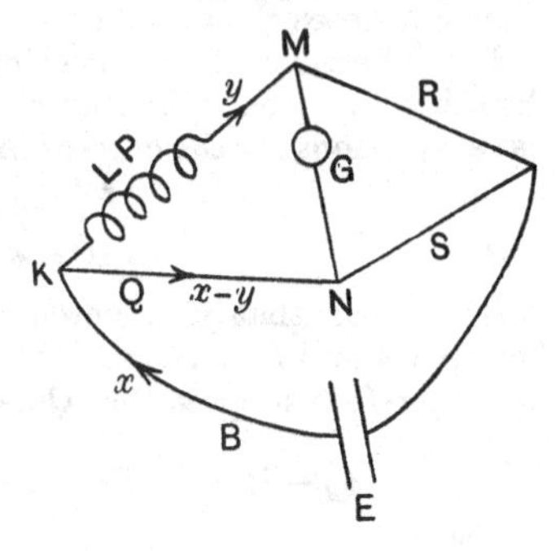

Since in the permanent state there is no flow through the galvanometer, therefore $PS = QR$ ....(1).

Let the steady currents before the battery circuit is broken be x, y and $x-y$ as indicated in the diagram. Then from the two circuits containing the battery we have

$$Bx + (P+R)\,y = E = Bx + (Q+S)\,(x-y) \quad \text{.........(2);}$$

so that $$(P+Q+R+S)\,y = (Q+S)\,x$$

or, substituting for S from (1),

$$(P+Q)\,y = Qx.$$

Hence from (2) $$y\{(P+Q)\,B + (P+R)\,Q\} = EQ \quad \text{...............(3).}$$

After the battery circuit is broken, the change in the amount of induction through the coil will cause a current z, say, in the arm KM and since MN is of negligible resistance the whole current will flow along MN and return along NK. Hence for the current z

$$(P+Q)\,z = -L\frac{dz}{dt},$$

and z has an initial value y.

The total flow through the galvanometer is then

$$\int_0^\infty z\,dt = -\frac{L}{P+Q}\int_0^\infty \frac{dz}{dt}\,dt$$

$$= -\frac{L}{P+Q}\int_y^0 dz = \frac{Ly}{P+Q},$$

and from (3) the required result follows.

(ii) *Between A and B there is a coil of wire of resistance R and inductance L, and a condenser of capacity C has its plates connected to B and to D by wires the sum of whose resistances is r, and a potential difference*

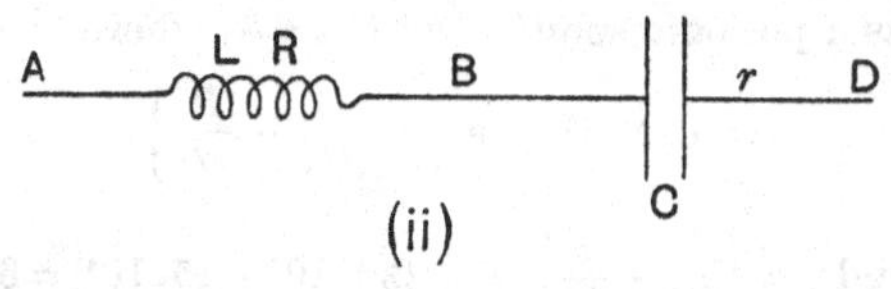

(ii)

$E\cos pt$ is maintained between A and D. If E_1 and E_2 are the amplitudes of the potential differences between A and B and between B and D respectively and if $L/C = Rr$, shew that $E_1^2 + E_2^2 = E^2$. I. 1911]

Let Q denote the charge on the plate of the condenser nearer to B, then $\dot{Q}$ denotes the current.

The difference of potential of the condenser plates is Q/C, and denoting the other potential differences by $\phi_A-\phi_B$ and $\phi_B-\phi_D$, we have equations for the current in AB and in BD

$$R\dot{Q}=\phi_A-\phi_B-L\ddot{Q}$$

and

$$r\dot{Q}=\phi_B-\phi_D-Q/C.$$

It is clear that if a potential difference $E\cos pt$ is maintained between A and D all the variables must have the same period $2\pi/p$; we may therefore suppose that $Q\propto e^{ipt}$ so that $\dot{Q}=ipQ$ and $\ddot{Q}=-p^2Q$ and

$$\phi_A-\phi_B=Q\,(iRp-Lp^2),\quad \phi_B-\phi_D=Q\left(\frac{1}{C}+irp\right).$$

Hence

$$\frac{\phi_A-\phi_B}{-Lp^2+iRp}=\frac{\phi_B-\phi_D}{\frac{1}{C}+irp}=\frac{\phi_A-\phi_D}{\left(-Lp^2+\frac{1}{C}\right)+i(R+r)p}\quad\ldots\ldots(1).$$

But a complex number $x+iy$ may be written $r(\cos\theta+i\sin\theta)$ or $re^{i\theta}$ and its amplitude is r or $\sqrt{(x^2+y^2)}$. Hence from (1)

$$\frac{E_1^2}{L^2p^4+R^2p^2}=\frac{E_2^2}{\frac{1}{C^2}+r^2p^2}=\frac{E^2}{\left(\frac{1}{C}-Lp^2\right)^2+p^2(R+r)^2};$$

and

$$E_1^2+E_2^2=E^2$$

if

$$L^2p^4+R^2p^2+\frac{1}{C^2}+r^2p^2=\left(\frac{1}{C}-Lp^2\right)^2+p^2(R+r)^2;$$

or if

$$Lp^2/C=p^2Rr,\quad \text{i.e. } L/C=Rr.$$

(iii) *A condenser is formed of a pair of parallel circular plates of* 1 *metre radius* 1 *centimetre apart. If it is charged up to* 100 *volts and discharged through a circuit of induction* $\frac{1}{10}$ *henry and resistance* 100 *ohms, find the period of the oscillations and the maximum current.*

Neglecting edge effects the capacity C of the condenser is equal to 'area/$4\pi\times$ thickness' $=10^4\,\pi/4\pi=10^4/4$ absolute electrostatic units $=10^4/4.9.10^{11}$ Farads; i.e. $C=1/36.10^7$ Farads. Also $L=\frac{1}{10}$ henry and $R=100$ ohms.

As in **11·6** (3) we can shew that the differential equation for the potential difference ϕ is

$$L\ddot{\phi}+R\dot{\phi}+\phi/C=0\quad\ldots\ldots\ldots\ldots\ldots\ldots\ldots(1)$$

and if this has a periodic solution with $\phi=\phi_0$ where $t=0$, it is of the form

$$\phi=\phi_0e^{-Rt/2L}\cos\sqrt{\left\{\frac{1}{CL}-\frac{R^2}{4L^2}\right\}}t\quad\ldots\ldots\ldots\ldots(2),$$

where

$$R/2L=500,\text{ and }\sqrt{\left\{\frac{1}{CL}-\frac{R^2}{4L^2}\right\}}=\sqrt{\{36.10^8-25.10^4\}}=6.10^4\text{ approx.},$$

so that

$$\phi=\phi_0e^{-500t}\cos(6.10^4t).$$

The period of the oscillations is therefore $2\pi/6.10^4$, or approximately

$\frac{1}{9550}$ second, and the damping factor drops to e^{-1} in $\frac{1}{500}$ second or about 19 oscillations.

Again the current is given by

$$j = -C\dot{\phi} = C\phi_0 e^{-500t}\{500\cos(6.10^4 t) + 6.10^4 \sin(6.10^4 t)\}$$

and the first term is negligible compared with the second, and, since $\phi_0 = 100$,

$$j = 6.10^6 Ce^{-500t}\sin(6.10^4 t)$$

with a maximum value approximately $6.10^6 C = \frac{1}{60}$ amp.

It can be shewn that electromagnetic effects are propagated with the velocity of light, i.e. 3.10^{10} cm. per sec. Hence the length of the waves in the above oscillations being 'velocity × period' is

$$\frac{2\pi}{6.10^4} \times 3.10^{10} = 10^6\pi \text{ cm.}$$

or about 31·5 kilometres.

In reference to the units in this example, the capacity of the condenser was actually calculated as so many centimetres but the capacity so found was essentially an electrostatic measure, and in electrostatic units the Farad is 9.10^{11} absolute units, and as the other measures in the data were all in practical units so the capacity was the only measure which needed conversion.

EXAMPLES

1. A coil of resistance R and self-induction L is joined to a battery of electromotive force E. Shew that the current after a time t is $(1-x)E/R$, and that if contact be then broken the current after a further time t is $(x-x^2)E/R$, where $L\log x + Rt = 0$. [I. 1898]

2. A coil is rotated with constant angular velocity ω about an axis in its plane in a uniform field of force perpendicular to the axis of rotation. Find the current at any time, and shew that it is greatest when the plane of the coil makes an angle $\tan^{-1}(L\omega/R)$ with the lines of magnetic force. [M. T. 1897]

3. A metal ring rotates uniformly round a horizontal diameter which is at right angles to the magnetic meridian. If the effects of self-induction be neglected, in what parts of the revolution is the induced current strongest and when does it vanish? [Trinity Coll. 1895]

4. A condenser of capacity C is discharged through a resistance R so high that inductance is inappreciable; prove that the charge falls according to the law $e^{-tC/R}$.

If the two plates of the condenser are each circular of radius 10 cm., and are a millimetre apart, and are discharged through a damp thread of resistance 10^9 ohms, shew that the charge will fall in the ratio of 1 to e in about a quarter of a second. Would the same mode of calculation apply if the resistance were only 10^6 ohms? [M. T. 1911]

5. A circuit of resistance R and of self-induction L has a condenser of capacity C inserted in it, and is acted on by an electromotive force $E\sin pt$. Prove that the current at any time is

$$\frac{E}{\sqrt{R^2+\left(\frac{1}{Cp}-Lp\right)^2}}\cos(pt-\alpha),$$

where $$\tan\alpha = R\bigg/\left(\frac{1}{Cp}-Lp\right).$$ [I. 1903]

6. A circular wire of radius a and resistance R is spun in a magnetic field of strength H with angular velocity ω about an axis in the plane of the wire and at right angles to the lines of force. Shew that the average rate of dissipation of energy is $\pi^2a^4\omega^2H^2/2R$ approximately.

Shew that if the speed of rotation becomes larger the rate of dissipation of energy would be less than that given by the above formula. [M. T. 1919]

7. The resistance and self-induction of a coil are R and L and its ends A and B are connected with the electrodes of a condenser of capacity C by wires of negligible resistance. There is a current $I\cos pt$ in a circuit connecting A and B, and the charge of the condenser is in the same phase as this current; shew that the charge at any time is $\frac{LI}{R}\cos pt$ and that $C(R^2+p^2L^2)=L$. Obtain also the current in the coil. [M. T. 1897]

8. Shew that, when a coil of inductance L and resistance R is attached to two terminals at which an electromotive force $E\cos pt$ is maintained, the average rate of consumption of energy is

$$\tfrac{1}{2}E^2R/(R^2+L^2p^2).$$

Shew also that, if $p=500$, $L=0{\cdot}2$ henry and $R=75$ ohms and energy is consumed at the rate of 24 watts, E must be 100 volts.

9. Four points A, B, C, D are connected up as follows: A, B are joined through a coil of self-induction L and resistance P; A, D through a resistance Q; B, C through a resistance R; C, D through a resistance S and through a condenser of capacity K, the resistance and the condenser being *in parallel*; B, D through a galvanometer; A, C through a source of current of period $2\pi/p$. Shew that, if no current passes through the galvanometer,

$$PS=QR \quad \text{and} \quad L=QRK.$$

(The resistances of the connecting wires may be neglected.) [M. T. 1908]

10. Two insulated conductors whose coefficients of potential are p_{11}, p_{12}, p_{22} have charges Q_0, Q_0' given to them, and are then connected by a coil of resistance R and self-induction L; shew that the

current will be oscillatory if $L(p_{11}-2p_{12}+p_{22})-\frac{1}{4}R^2$ is positive; and, denoting this expression by K^2, shew that, in this case, the current at time t will be

$$\frac{1}{K}\{Q_0(p_{11}-p_{12})-Q_0'(p_{22}-p_{12})\}e^{-\frac{Rt}{2L}}\sin\frac{Kt}{L},$$

and that the difference of phase of charge and current will be

$$\tan^{-1}\frac{2K}{R}.$$ [M. T. 1904]

11. The arms AB, AD of a Wheatstone's bridge are of inductances L_1, L_2 and of resistances R_1, R_2; the arms DC, BC contain condensers of capacities C_1, C_2 and are of resistances R_3, R_4. BD contains the galvanometer and AC an alternating source of E.M.F. Shew that the bridge will not be balanced for all frequencies unless

$L_1R_3=L_2R_4$, $C_1R_2=C_2R_1$ and either $R_1R_3=R_2R_4$ or $L_1=C_1R_2R_4$.

[I. 1926]

12. A thin bar magnet of moment M and length $2c$ lies along the axis of a circular ring of radius a and has a small longitudinal vibration $\beta\sin pt$ about its mean position in which the centre of the magnet coincides with the centre of the ring. Shew that there is an induced current in the ring of strength

$$\frac{3\pi Ma^2\beta^2p}{(c^2+a^2)^{\frac{5}{2}}}\frac{\sin(2pt-\alpha)}{(R^2+4L^2p^2)^{\frac{1}{2}}},$$

where R, L are the resistance and inductance of the ring and

$$\tan\alpha=2Lp/R.$$ [M. T. 1924]

ANSWERS

2. See **11·51**.

3. Strongest when the plane of the ring contains the lines of force and zero when at right angles.

7. $(R\cos pt+Lp\sin pt)\,I/R$.

CAMBRIDGE: PRINTED BY WALTER LEWIS, M.A., AT THE UNIVERSITY PRESS

Printed in the United States
By Bookmasters